COMMUNITY ECOLOGY OF A CORAL CAY

MONOGRAPHIAE BIOLOGICAE

VOLUME 43

Editor
J. ILLIES
Schlitz, F.R.G.

Dr W. Junk Publishers The Hague–Boston–London 1981

COMMUNITY ECOLOGY OF A CORAL CAY

A study of One-Tree Island, Great Barrier Reef, Australia

HAROLD HEATWOLE / TERENCE DONE
ELIZABETH CAMERON

Dr W. Junk Publishers The Hague–Boston–London 1981

Distributors:

for the United States and Canada

Kluwer Boston, Inc.
190 Old Derby Street
Hingham, MA 02043
USA

for all other countries

Kluwer Academic Publishers Group
Distribution Center
P.O. Box 322
3300 AH Dordrecht
The Netherlands

Library of Congress Cataloging in Publication Data
Main entry under title:
Community ecology of a coral cay.
(Monographiae biologicae; v. 43)
Bibliography: p.
Includes indexes.
1. Coral reef ecology—Australia—One Tree Island (Qld.) 2. One Tree Island (Qld.) 3. Great barrier Reef (Qld.) I. Heatwole, Harold. II. Done, Terence. III. Cameron, Elizabeth. IV. Series
QP1.P37 vol. 43 [QH197] 574s 81-8138
[574.5′26367′09943] AACR2

ISBN 90 6193 096 0 (this volume)
ISBN 90 6193 881 3 (series)

Cover design: Max Velthuijs

Printed in the Netherlands

One-Tree Island

An island slumbering –
Clasped in a shimmering circlet
Of waters turquoise and blue.
Guarding her jewel from the pounding surf
On her coral rim.
Herons, hunched on her shore
Gulls, huddled in a bunch
Turnstones picking ’mongst the rocks
And o’erhead hovering frigate birds.
The trees stand languid
The bushes still
Sol casts his warmth
From a canopy of blue
Slumber on – never wake
Isle of peace and beauty.

June Chilvers

Dedicated to:

Professor S. W. Witmer
Goshen College, Goshen, Indiana, U.S.A.

Professor F. H. Test
University of Michigan, Ann Arbor, Michigan, U.S.A.

Professor A. F. O'Farrell
University of New England, Armidale, N.S.W., Australia

Preface

It is not enough to know the general body of theory about island ecology. It is also important to know just how well it applies in its various aspects to particular geographic areas. Such knowledge can only be gained empirically and over a long period of time. Formulation of policies for ecological management of an archipelago based solely on theory, without regard to its specific application in the area in question, invites disaster.

The entire Great Barrier Reef area is now being assessed by the Great Barrier Reef Marine Park Authority of the Australian federal government with the aim of zoning it for a variety of recreational, scientific, touristic and economic uses. The Council of the Great Barrier Reef Committee, under consultancies to the Marine Park Authority, has summarized the previously available information on the Capricorn–Bunker region. The present book treats in more detail one of the terrestrial insular ecosystems in the area and presents data previously unavailable.

The book is written in four major sections. Parts I and II treat the history and development of the theory of insular biogeography and ecology, and by examining modern concepts, provides a background against which the later sections can be interpreted.

Parts III and IV present the results of a detailed analysis of the terrestrial ecology of One-Tree Island. The approach is that of a quantitative community study and the vegetation, above-ground invertebrates, soil arthropods, and the land and sea bird faunas are considered individually and as an interactive community. The effect of drought on the island is discussed and problems of dispersal and extinction examined.

Although the project was conceived of as an integrated study, and carried out that way, certain parts have been, or are being used in theses. The section on soil microarthropods is summarized from a thesis submitted for a Ph.D. degree in Zoology at the University of New England by the second author (T.D.). Part of the vegetation data will be incorporated as part of a Ph.D. thesis in Botany being prepared by the senior author (H.H.) for submission to the University of Queensland.

The book is aimed at the professional biologist who wishes to explore the topic of island ecology. However, the first two sections provide sufficient background, expressed in language relatively free from technical jargon, for educated non-professionals to gain insight into ecological properties of islands and to proceed to the more technical parts of the book.

Many people have contributed directly or indirectly to the success of the One-Tree project. A number of unpaid volunteers helped us either in the field or in handling specimens or processing data. They are Barbara Saylor Done, Judy Herrington, Rosemaree Dowling, David Vleck, Carol Masters Vleck, Bruce Firth, Grahame Webb, Leon Zann, Robert Newby, Robert Done, Roland Scollay, Gayle Urquhart, Alan Richards, Ross Davey, Audry, Eric and Miguel Heatwole, and Michael MacQuillen.

We are grateful to the Heron Island Board and to Peter Woodhead and A. J. Bruce, successive directors of the Heron Island Research Station, for permission to use the laboratories and accommodation at the station. Reginald and Mary McMahon and Ted and June Chilvers showed many kindnesses to us during our trips to Heron and One-Tree Islands, and they expedited our work in various ways. We wish to thank Peter Sale, director of the One-Tree Research Station for permission to use the facilities at One-Tree Island during our 'follow-up' visits. Steve Domm and Ben Cropp provided boat transportation to One-Tree when commercial or research vessels were unavailable. We are grateful for lifts by helicopter pilots, especially one rescue occasioned by seas unsuitable for small boats.

Identification of specimens was assisted by the following specialists: Plants, Robert W. Johnson and the late S. D. Blake; Fish, David Woodland; Gastropoda, Rodney Simpson and John Stanisik; Araneida, Valerie Davies, Peter Raven, Michael Gray and the late Fr. Chrysanthus; Pseudoscorpionida, W. B. Muchmore; Odonata, A. F. O'Farrell; Dermaptera, E. T. Giles; Orthoptera, K. H. L. Key; Psocoptera, C. N. Smithers; Neuroptera and Lepidoptera, the late C. W. Frazier; Coleoptera, E. B. Britton; Diptera, D. H. Colless and D. K. McAlpine; Braconidae, P. M. Marsh; Ants, Robert Taylor.

Competent technical services and advice were rendered by Harry Wadleigh and Russel Hobbs of the University of New England. The many drafts of the manuscript were typed by Viola Watt, Neva Walden, Audry Heatwole, Gladys Slade and Sharon Porter.

The study was funded by the National Science Foundation (U.S.A.), the Myer Foundation, and the Internal Research Funds of the University of New England.

Special thanks go to Jared Diamond who read the entire manuscript and made a number of important suggestions, and to Robert W. Johnson, Hugh Ford, David Lamb, Peter Sale, Harry Recher, Trevor Clifford and Frederic Wagner who read parts of the manuscript and offered useful comments and criticism. John Williams took time out from his busy schedule and made himself available for consultation on many occasions. Prof. N. Beadle advised on the assigning of plant species to life form categories. Finally we appreciate stimulating feedback from colleagues too numerous to mention individually.

8th August 1980

Harold Heatwole, Armidale, N.S.W.
Terence Done, Townsville, Qld.
Elizabeth Cameron, Sydney, N.S.W.

Contents

PART FOUR: SYNTHETIC CHARACTERISTICS

PART ONE

Introduction

1. Background and rationale of the study

For a long time biologists have been aware that biotic communities have attributes over and above the collective autecologies of their component species. The network of interactions within the community, and the alternate pathways through which flow of energy or materials can be routed in response to subtle environmental influences, provide for an extraordinary dynamism, intriguing because of its intricacy but at the same time perplexing by virtue of its complexity. In order to understand how communities work it is important to identify their properties, elucidate their organization in time and space, and assess the effect of interaction of component species.

Early community ecologists set out to define the nature of communities and were divided as to their concepts of the distinctiveness of communities. Biotic communities were variously likened to 'superorganisms' functioning as discrete identities with definite, rather constant, species composition, or at the opposite extreme, as continua of ever-changing species-groupings along a gradient lacking well defined boundaries. Attempts were made to name and describe specific communities in terms of their contained species and to arrange them along a successional sequence up to a climax type for each type of climate and/or soil. Recognition that communities from like climates had similar physiognomy even though the taxonomic entities comprising them differed greatly, gave rise to the concept of the biome. Different communities of the same biome type often had taxonomically unrelated species that played similar roles in the community. Identification of such ecological equivalents suggested that community structure and functional organization were not entirely random. The concept of the ecological niche was an extension of this concept as it implies that in a given community there are specific roles to be played, and that adaptive radiation will tend to provide species to fill vacant niches. The corollary that two species could not fill the same niche (play the same role in the community) led to a long, and still not completely resolved, controversy as to the role of interspecific competition in community structure. Attention also was directed to the role played by other types of interactions, such as predator-prey relations, in structuring communities. The organization of species into trophic interactions (food chains or food webs) was recognized as an important part of community structure. The number of species participating in a particular community appeared to influence community properties, such as stability. In addition there was a surge of interest in the measurement of species diversity, and the factors influencing it.

One of the more recent developments has been an increasingly synthetic and quantitative approach to community function. Communities together with the physical environments impinging upon them, are collectively treated as systems (ecosystems) and the flow of energy, carbon, nitrogen or some other commodity through the system is measured, much as an engineer might study a purely physical system.

The application of mathematical and computer techniques to the modelling of ecosystems and communities has led to rapid, major advances in the theory of community function. New insights into species interactions, community structure, energy flow, cycling of materials, and community stability have been gained.

The optimum sequence for a study of the ecology of a particular community is (1) to analyze all available relevant empirical data as a basis for (2) constructing a model of community structure and function, and (3) to test the model empirically by field studies. The resulting field data should then be used to generate an improved model for further testing, and so on, until a theoretical framework has been constructed that requires little alteration when subjected to empirical testing.

Empirical testing is extremely time-consuming and often expensive because specific weather, or other important conditions may recur only after long intervals. Consequently, theoretical advances in community ecology have outstripped their relevant empirical base and many models have not been validated for any particular community, much less for communities in general.[1] Not only is such a state intellectually unsatisfying but it is potentially dangerous. If management practices or conservation policies are based upon unvalidated models, the results may very well prove to be disastrous.

Communities on small islands have a number of unique properties (see Chapter 5) and it is unlikely that theories of community function arising from studies of less isolated situations could be applied unaltered to them. Consequently either a specific theory relating to island communities may have to be developed, or the more general theory modified. At present the empirical base for such a procedure is wholly lacking as there have been few long-term detailed studies of the community ecology of small islands.[2] A

[1] The International Biological Programme initiated relatively long-term ecosystem studies in a number of biomes. These studies collectively should provide the required empirical base for testing many models which are currently not assessable. Eventually, when the enormous quantities of data that have been collected are summarized and analyzed the gap between theory and reality perhaps can be narrowed.

[2] Signy Island, Antarctica, has been studied from the ecosystem standpoint as part of the International Biological Programme. However, the communities there have been studied

recent review of the ecology and biogeography of coral cays ended with the conclusion: 'Treatment of islands as ecosystems promises to be one of the most fruitful approaches to insular ecology. Clearly long-term detailed studies of individual islands are required' (Heatwole 1976). The voluminous literature on island ecology has been restricted largely either to short-term studies or to only a few aspects of community ecology. The kind of study needed is one which would have the following characteristics: (1) To be as complete a community study as possible, involving as many taxa and trophic levels as feasible, (2) to encompass the full range of environmental perturbations experienced by the community, including unusual weather events as well as normal seasonal changes, (3) to treat a relatively common island type so as to be characteristic of many islands rather than a unique situation, and (4) to deal with a relatively natural community so as to be free from the necessity of interpreting the effect of human disturbance. The need for such a study gave rise to the present investigation.

Not all of the above aims were achieved completely in this study. The selection of One-Tree Island as the site of the study was influenced by its relative lack of previous human disturbance, yet proximity to inhabited areas. Not all aspects could be studied; sampling of some taxa did not prove to be feasible given the time and effort which could be expended, and those same limitations prevented treatment of micro-organisms – an obviously important component of the community. One-Tree Island was not entirely typical of coral cays as it consists of coral rubble rather than sand. The duration of intensive study was limited by the number of quantitative samples that could be removed without the sampling itself affecting the community; we were fortunate, however, in that during the study a drought and a tropical storm both occurred.

This book reports the empirical results of the study of the community ecology of One-Tree Island from 1968 to 1978, and interprets the results in general terms. The data have formed the basis of a formal exercise in ecosystem modelling by a team of researchers at Griffith University, Brisbane, Australia, led by Dr. Roger Kitching. The model will be presented as a separate publication.

---

because they were Antarctic communities rather than because they were insular; because of the specialized conditions, the results from Signy Island may not have application to islands generally. Research on Surtsey Island (Fridriksson 1975) and on Aldabra (Royal Society of London, 1979) has covered many ecological topics and contributes toward an understanding of the communities on those islands. Much of the general biological literature on islands contains ecological data on particular topics of interest to community ecologists (see especially the Atoll Research Bulletin).

2. Classification of islands

Traditionally islands are classified into two major types, continental and oceanic. Continental islands are those which lie on a continental shelf and are geologically similar to the adjacent mainland. In most cases there have been previous land connections between them and the mainland during periods when the local land levels were higher and/or when there was eustatic sea level lowering. By contrast, oceanic islands are not on a continental shelf and are not geologically related to mainlands. They have arisen by volcanic action directly from the oceanic depths and have never had a land connection with a larger land mass.

Continental islands may have possessed the ancestors of a portion of their present flora and fauna at the time they were part of an extended mainland; oceanic islands of necessity have received their terrestrial biota via some form of over-sea dispersal.

There is, however, a third type of island that satisfies neither of these definitions. That type consists of the sand or coral cays that lie on continental shelves or on oceanic atolls. They differ from oceanic and continental islands by being composed of coral or sand rather than rock. In most cases they are relatively young; those occurring on continental shelves have usually formed so recently that they have never been directly connected to a mainland, that is, they developed after the part of the continental shelf on which they are located was submerged. Thus, the cays both of atolls and continental shelves resemble oceanic islands in having acquired their animals and plants across a water barrier even though many are quite close to land.

In this book we treat in some detail the community ecology of a coral cay on a continental shelf. However, the applicability of the findings from One-Tree to other cays, especially those on atolls, will require additional investigation. Continental rocky islands or oceanic islands are even more likely to depart from the conditions at One-Tree. Nonetheless, we believe the One-Tree study to be a valuable step in a much needed direction and hope it stimulates investigation of other island ecosystems.

PART TWO

The development of insular ecology and biogeography

3. Early theories

Two historical concepts of the origins of living creatures have a bearing on ideas about insular distributions. One was the belief by some of the Ionian, Eleatic and other early Greek natural philosophers that ultimately organisms were formed spontaneously be generative forces residing in such elemental matter as air, water, fire or earth (mud, slime)[1] (Nordenskiöld 1928). This view was also championed in ΠΕΡΙ ΖΩΙΩΝ ΓΕΝΕΣΕΩΣ (Generation of Animals) by Aristotle who believed that although some animals reproduced by copulation, others habitually 'came into being not as the result of the copulation of living animals, but out of putrescent soil and out of residues' (Peck's 1963 English translation). Upon Aristotle's authority the view of spontaneous generation of life enjoyed wide acceptance for many centuries and found some adherents as late as the 19th century (Nordenskiöld 1928). Clearly, insular distributional patterns are compatible with the acceptance of such a theory, as animals could presumably be generated from mud as easily on remote islands as on continents.

The second view, less easily reconciled with the presence of organisms on distant islands, was the Judaeo-Christian one that after an initial creation by God, sexually reproducing terrestrial animals were destroyed by the great flood except for those preserved in the ark. Upon recession of the flood waters, animals repopulated the earth by dispersal from the final resting site of the ark, Mt. Ararat. Proponents of this view shared with later zoogeographers the problem of the necessity of explaining how animals reached remote islands from a continental source.

In antiquity, the biota of remote islands on a world-wide scale was not appreciated, of course, and philosophers seldom considered the problems of overwater dispersal. A notable exception was St. Augustine in his *De Civitate Dei* (413–426; Chapter VII, Book XVI). He did not find the two views of animal origin incompatible and for his time, made a remarkably broad

[1] This idea is in a sense similar to the Judaeo-Christian one. In the Genesis account of the creation by divine command, the *waters* brought forth aquatic animals and birds (1:20) and the *earth* brought forth terrestrial animals (1:24).

assessment of the various possible mechanisms of dispersal to islands. He excluded as not problematical animals (such as frogs) which he believed to be generated spontaneously from slime, presumably for the reasons stated above, i.e. they could appear *de novo* on islands. For sexually reproducing animals he considered swimming as a possible means of reaching nearby islands, and either human transport of favoured hunting species, or transport by angels under the command of God as likely ways of reaching distant ones. He concluded his analysis of the problem by additionally postulating a type of localized creation from the earth by divine command. He stated (English translation, Tasker 1945) 'But if they were produced from the earth, as at first, because God said: "Let the earth bring forth the living soul": then is it most apparent that the diversity of beasts were preserved in the ark rather for a figure of the divers nations, than for a restoration, if the earth brought them forth in those isles to which they could not otherwise come'. It seems that St. Augustine believed a combination of natural causes and divine whim lay behind insular distributional patterns.

Little interest in the problem was expressed subsequently until a scientific knowledge of the insular biotas of the world became possible with the voyages of discovery and exploration begun by Europeans in the latter part of the 15th century. Then traditional interpretations began to be questioned. For example, the Jesuit missionary, Joseph Acosta in his *Historia Natural y Moral de las Indias* in 1590 attempted a refutation of St. Augustine's views on dispersal mechanisms (White 1895).

With the increasing number of species known throughout the world it became evident that the ark could not possibly have held a pair of each sexually reproducing species, a dilemma which in the 18th century prompted the Benedictine, Dom Calmut, to suggest that all species of a genus had originally formed one species. This would have reduced the number of species needing accommodation in the ark to the number of present day genera rather than the number of present day species (White 1895). In essence this view embodied an idea of evolution, at least at the specific level.

A further problem was the existence in remote areas of species that were very different from anything in the vicinity of the presumed point of dispersal, Mt. Ararat. Milius in 1667 in *De Origine Animalium et Migratione Popularum* resorted to postulating spontaneous generation for even the more complex animals. He suggests that 'the earth and the waters, and especially the heat of the sun and of the genial sky, together with that slimy and putrid quality which seems to be inherent in the soil, may furnish the origin for fishes, terrestrial animals and birds' (English translation in White, 1895). He concluded that plants and animals take their origin from such generative forces in the lands in which they are found, and provided biblical support by quoting Genesis 1:24, 'Let the earth bring forth the living creatures after its kind.'

It appears that at least some of the scholastically-oriented scholars and theologians began to re-think traditional views and to attempt to reconcile them with the new information on natural history. However, the greatest impetus for change came from the naturalists themselves. In the 18th and 19th centuries such gifted biologists as Joseph Banks and Charles Darwin officially were attached to expeditions and a more organized study of the world's natural history was attempted. Patterns of nature emerged which caused scientists to seek explanations beyond those inherent in the traditional beliefs of the time.

Although objective scientific investigation had begun to develop during the period of geographical exploration, and some theologians were exploring alternative views, philosophical thought continued to be dominated by dogma. Not only was there militant objection to mechanistic explanations but the very prevalence of ecclesiastical thinking stifled enquiry. Nevertheless, as advances in knowledge accumulated and evidence became overwhelming, bold new theories were proposed and defended, not only against attacks from fellow scientists (who rightfully demanded rigorous support for new interpretations) but against the onslaughts of a clergy and a public unwilling to be convinced.

Such new, mechanistic ideas developed simultaneously in several disciplines, especially geology and biology, and were gradually accepted by more and more scholars. Although the middle and latter part of the 19th century witnessed continued heated debate, the intellectual mood altered and paved the way for a greater freedom of thought and for immense strides in development of scientific theory.

Probably the most important development in biology during this period of intellectual flux was the propounding of the theory of evolution. The central idea developed independently in the minds of Charles Darwin and Alfred Wallace and in 1858 they jointly submitted essays outlining their theory, through Charles Lyell and Joseph Hooker, to the Linnean Society (Darwin and Wallace 1859; reprinted in Bates and Humphrey 1956). Island biology played an important part in the thinking of both men.

Charles Darwin was able to examine the fauna, flora and geology of many parts of the globe as part of his duties as naturalist on the round-the-world voyage of H.M.S. Beagle, December 1831 to October 1836. On that trip he visited islands and archipelagos in the major oceans of the world and was impressed particularly by the Galapagos Islands. The insights he received from these and other islands contributed greatly to the formation of the theory of evolution. Darwin's ideas were conveyed privately to various scholars but it was more than two decades after his voyage before he could be persuaded by them to make public his views.

Several examples of insular conditions were used to illustrate points in the original, brief essay of 1858, and considerable, detailed attention was given to island biology in the more complete exposition published the following year (Darwin 1859). Although the latter work was most famous for its expanded statement of the theory of evolution via natural selection, it also laid the foundation of modern insular ecology and biogeography, and developed several theses still tenable today. The implication of the theory of evolution for insular biogeography was that the resident species were not created *in situ* but were organic descendents from progenitors occurring on mainlands. In some cases the organisms had changed because of isolation and the different environments encountered on the islands. Those on oceanic islands he considered to have reached their present location by dispersal across sea barriers, and those on continental islands he believed to have arrived over land-connections when the islands were still connected to mainlands, and/or after separation by dispersal across the sea barrier. He was interested in the ability of organisms to disperse across sea barriers and conducted extensive experiments on the buoyancy and survival of plant seeds when exposed to sea water. He calculated that the dried seeds of about 10% of a flora could still germinate after floating across nine hundred miles of sea. He also discussed at some length the possibility of transport via flotsam, icebergs, or birds (either in their digestive tracts or externally on their bodies); he noted examples of long-distance flight to islands by locusts and extracted viable seeds from locusts' fecal pellets.

Important contributions by Darwin to the development of island biogeography and ecology are summarized below and exemplified by quotations from his 'Origin of Species' (6th ed. 1872). Some of these are his original ideas, some he gleaned from other sources; the synthesis, however, is his.

1. *Island biotas are depauperate (the island holds fewer taxa than an equivalent area of mainland).*
   'The species of all kinds which inhabit oceanic islands are few in number compared with those on equal continental areas.'
   'He who admits the doctine of the creation of each separate species, will have to admit that a sufficient number of the best adapted plants and animals were not created for oceanic islands: for man has unintentionally stocked them far more fully and perfectly than did nature.'
2. *The composition of insular and mainland biotas are often different.*
   'Oceanic islands are sometimes deficient in animals of certain whole classes, and their places are occupied by other classes; thus in the Galapagos Islands reptiles, and in New Zealand gigantic wingless birds, take, or recently took, the place of mammals.'
3. *Endemism is high among insular biotas.*
   'Although in oceanic islands the species are few in number, the proportion of endemic kinds (i.e., those found nowhere else in the world) is often extremely large.'
4. *Overwater dispersal is an important consideration in insular biology.*
   'Considering that these several means of transport [see above], and that other means which without doubt remain to be discovered, have been in

action year after year for tens of thousands of years, it would, I think, be a marvellous fact if many plants had not thus become widely transported. These means of transport are sometimes called accidental, but this is not strictly correct: the currents of the sea are not accidental, nor is the direction of prevalent gales of wind. It should be observed that scarcely any means of transport would carry seeds for very great distances; for seeds do not retain their vitality when exposed for a great length of time to the action of sea-water; not could they be long carried in the crops or intestines of birds. These means, however, would suffice for occasional transport across tracts of sea . . . or from island to island, or from a continent to a neighbouring island . . ..'

5. *Certain groups are absent from islands because of their inability to disperse across sea water barriers.*
   'This general absence of frogs, toads, and newts on so many true oceanic islands cannot be accounted for by their physical conditions: indeed it seems that islands are peculiarly fitted for these animals; for frogs have been introduced into Madeira, the Azores, and Mauritius, and have multiplied so as to become a nuisance. But as these animals and their spawn are immediately killed . . . by sea-water, there would be great difficulty in their transport across the sea, and therefore we can see why they do not exist on strictly oceanic islands.'
   'Why, it may be asked, has the supposed creative force produced bats and no other mammals on remote islands? In my view this question can easily be answered; for no terrestrial mammal can be transported across a wide space of sea, but bats can fly across.'
6. *A population isolated on an island may diverge under the pressures of natural selection from mainland populations and eventually speciate: the nature of the changes that occur on islands is adaptive.*
   '. . . species occasionally arriving after long intervals of time in the new and isolated district . . . would be eminently liable to modification and would often produce groups of modified descendants.'
   'In their new homes they will be exposed to new conditions, and will frequently undergo further modification and improvement . . . and will produce groups of modified descendants.'
7. *Insular taxa are often more closely related to those of a nearby island or mainland than to those from more distant regions. The degree of divergence will vary among taxa.*
   'If we look to the islands off the American shore, however much they may differ in geological structure, the inhabitants are essentially American, though they may be all peculiar species. We may look back to the past ages . . . and we find American types then prevailing on the American continent and in the American seas. We see in these facts some deep organic bond, throughout space and time, over the same areas of land and water, independently of physical conditions.'

'The same principle which governs the general character of the inhabitants of oceanic islands, namely, the relation to the source whence colonists could have been most easily derived, together with their subsequent modification is of the widest application throughout nature . . . It will, I believe, be found universally true, that wherever in two regions . . . many closely allied or representative species occur, there will likewise be found some identical species; and wherever many closely-allied species occur, there will be found many forms which some naturalists rank as distinct species, and others as mere varieties; these doubtful forms showing us the steps in the progress of modification.'

8. *Competitive interactions may influence the evolution of insular forms.* '. . . an herbaceous plant, which had no chance of successfully competing with the many fully developed trees growing on a continent, might, when established on an island, gain an advantage over other herbaceous plants by growing taller and taller and overtopping them. In this case, natural selection would tend to add to the stature of the plant . . . and thus first convert it into a bush and then into a tree.'

The best summary of Darwin's views is undoubtedly an excerpt of the summary he wrote himself for the Origin of Species.

'. . . all the grand leading facts of geographical distribution are explicable on the theory of migration, together with subsequent modification and the multiplication of new forms. We can thus understand the high importance of barriers, whether of land or water, in not only separating, but in apparently forming the several zoological and botanical provinces. We can thus understand the concentration of related species within the same areas . . . Bearing in mind that the mutual relation of organisms is of the highest importance, we can see why two areas having nearly the same physical conditions should often be inhabited by very different forms of life; for according to the length of time which has elapsed since the colonists entered one of the regions, or both; according to the nature of the communication which allowed certain forms and not others to enter, either in greater or lesser numbers; according or not, as those which entered happened to come into more or less direct competition with each other and with the aborigines; and according as the immigrants were capable of varying more or less rapidly, there would ensue in the two or more regions, independently of their physical conditions, infinitely diversified conditions of life – there would be an almost endless amount of organic action and reaction – and we should find some groups of beings greatly, and some only slightly modified – some developed in great force, some existing in scanty numbers – and this we do find in the several great geographical provinces of the world.

On these same principles we can understand, as I have endeavoured to show, why oceanic islands have few inhabitants, but that of these, a large proportion should be endemic or peculiar; and why, in relation to the means of migrating, one group of beings should have all its species peculiar, and another group, even within the same class, should have all its species the same with those in an adjoining quarter of the world. We can see why whole groups of organisms, as batrachians and terrestrial mammals, should be absent from oceanic islands, whilst the most isolated islands should possess their own peculiar species of aerial mammals or bats. We can see why, in islands, there should be some relation between the presence of mammals, in a more or less modified condition, and the depth of the sea between such islands and the mainland. We can clearly see why all the inhabitants of an archipelago, though specifically distinct on the several islets, should be closely related to each other; and should likewise be related, but less closely, to those of the nearest continent, or other source whence immigrants might have been derived. We can see why, if there exists very closely allied or representative species in two areas, however distant from each other, some identical species will almost always there be found.'

Alfred Wallace continued in the tradition of thought launched by the promulgation of the theory of evolution and applied it rigorously to the problems of island biology. He had acquired extensive field experience in the East Indian archipelago and he brought his wide knowledge to bear in writing the first modern monograph specifically dedicated to the topic of island biology. This was his 'Island Life' published in 1880 and as a second, revised edition in 1895. This book, now a classic, is a comprehensive treatment of patterns of distributions on islands, dispersal problems, and the role of evolution, climate and geographical and geological changes in influencing insular biogeography. There is also a detailed treatment of the specific biogeography of various major archipelagos and island groups of the world. Many of the concepts embodied in this book still have considerable relevance to current thinking.

Two quotations from 'Island Life' succinctly summarize Wallace's views and exemplify the similarity of his and Darwin's ideas.

'. . . the distribution of the various species and groups of living things over the earth's surface, and their aggregation in different assemblages in certain areas, is the direct result and outcome of a complex set of causes, which may be grouped as 'biological' and 'physical'. The biological causes are mainly of two kinds – firstly, the constant tendency of all organisms to increase in numbers and to occupy a wider area, and their various powers of dispersion and migration through which, when unchecked, they are enabled to spread widely over the globe; and secondly, those laws of evolution and extinction which determine the manner in which groups of organisms arise and grow, reach their maximum, and then dwindle away, often breaking up into separate portions which long survive in very remote regions. The physical causes are also mainly of two kinds. We have, first, the geographical changes which at one time isolate a whole fauna and flora, at another time lead to their dispersal and intermixture with adjacent faunas and floras – and it is here important to ascertain and define the exact nature and extent of these changes, and to determine the question of the general stability or instability of continents and oceans; in the second place, it was necessary to determine the exact nature, extent and frequency of the changes of climate which have occurred in various parts of the earth – because such changes are among the most powerful agents in causing the dispersal and extinction of plants and animals.'

'. . . the presence of certain types of plants or animals in one island rather than in another, are now shown to be dependent on the long series of past geological changes – on those marvellous astronomical revolutions which cause a periodic variation of terrestrial climates – on the apparently fortuitous action of storms and currents in the conveyance of germs – and on the endlessly varied actions and reactions of organised beings on each other.'

Since its inception, the Darwinian–Wallacean viewpoint has been modified in a variety of ways as new information came to hand; however, its broad outlines are still largely accepted today.

Two major advances which have shaped subsequent thinking have been (1) the unravelling of the mode of inheritance, and the development of an understanding of population genetics and (2) the accumulation of geological data relating to sea level changes and relative positions of land masses.

Although Darwin assumed that characteristics were transmitted from one generation to the next, the weakest link in his theory was lack of an understanding of the mechanisms of inheritance. Since Darwin there have been great strides in genetics. The variation in a population, so important to Darwin's theory, is now known to arise from recombination of existing genes, and from mutation to new forms. Not only is it now known how traits are inherited, but how populations can and do change, genetically and phenotypically, under the influence of isolation and selection of variants in the population by the natural environment. Modern advances in genetics have therefore provided a mechanism for evolution and support the basic tenets of the theory of evolution and speciation. The progress of knowledge and theory in these fields has been sign-posted by various books of synthesis along the way, notably 'The New Systematics' (Huxley 1940); 'Genetics and the Origin of Species' (Dobzhansky; three editions, 1937–1951) and 'Animal Species and Evolution' (Mayr 1963). Several reviews have related new advances specifically to the island situation; 'Island Biology' (Carlquist 1974) looks at dispersal problems and at adaptation on islands, and Baker and Stebbins (1965) relate advances in genetics to insular problems in 'The Genetics of Colonizing Species'.

Two of the most important post-Darwinian concepts that bear directly on insular biology were the concept of genetic drift (Wright 1921 and later papers) and the Founder Principle (Mayr 1942). The former indicates that in small, isolated populations (such as occur on some islands), random changes in frequency of particular traits may occur which potentially could alter the characteristics of the population and lead to divergence from other conspecific populations in ways independent of natural selection. Thus, in addition to the adaptive changes suggested by Darwin, it was postulated that adaptively neutral ones would take place as well.

The Founder Principle indicates that a newly isolated population may differ markedly from the one from which it was derived, because only a subset of the total genetic variation of the original population is represented in the founding individuals. For example, if only a few individuals from a mainland population managed to reach an island, only the limited range of genes they carry is all that is available to their descendants (except that subsequently arising from mutation).

Darwin believed in the permanence of the general locations of the continents and oceans. For someone holding that view and yet believing that the biota of islands was ultimately derived from that of mainlands, there were two possible, though not mutually exclusive, ways islands could have become populated, either by (1) organisms reaching them sometime in the past when there were 'land bridges' connecting island and mainland, or by (2) dispersal of organisms over sea-water barriers. Darwin and Wallace were well aware that both occurred and they followed geologists of their day in distinguishing between continental islands and oceanic islands. It was accepted that continental islands had received some of their biota when they were still connected to a mainland and that some present-day species were the descendants of those which originally had a widespread distribution including island and mainland. Similarly, islands of a given archipelago could exchange species when their component islands were interconnected. Later geological investigation has provided remarkably precise and detailed knowledge of eustatic sea-level changes and of changes in elevation and subsidence of land. It is now known quite accurately when various islands were interconnected or attached to a mainland. For example, during the Pleistocene glaciations New Guinea and Tasmania were both connected to and disconnected from the Australian mainland several times as a result of eustatic sea level changes; these changes resulted in faunal exchanges interrupted by isolation, both of which influenced patterns of speciation (Keast 1981). Similarly, most of the islands of the Puerto Rico–Virgin Island area were interconnected 10,000 years ago, but subsequently broke up, the last islands becoming isolated as recently as 6,000 years ago; time of isolation had an effect on inter-island faunal similarity (Heatwole and MacKenzie 1967). Many other examples could be given. There is little doubt that there has been exchange of species among islands of some archipelagos, or between continental islands and mainlands, and that these influences are reflected in the present-day biota.

However, oceanic islands pose quite a different problem. Because of their remoteness, and/or the great depth of water between them and a mainland, previous land bridges seem unlikely. Even in Darwin's day there were divergent ideas about the origins of the biota of continental islands. Some scientists believed they were populated by dispersal across sea-water barriers in the absence of any land connection, and others postulated land bridges, despite the intervening oceanic depths. This controversy has persisted almost to the present day (see discussion by Fosberg 1963).

If all of the land bridges that have been proposed to explain the distribution of particular plants and animals on oceanic islands, had in fact existed, a reconstruction of them through time would look like a map of inter-lacing freeways across the oceans of the world.

There were many sceptics about the ability of organisms to cross extensive sea barriers and they found it more credible to accept the theory of land bridges. No less an eminence than Thomas Barbour who devoted many years to the study of the biogeography of the Caribbean islands, stated that the patterns of evolution observed there 'would be impossible if individuals were being carried from one island to another by flotsam and jetsam dispersal' (1914) and in a more cynical vein he and Noble (1915) remarked: '. . . the many zoological groups which are represented upon so many of the islands alone refute the flotation theory. If they did not we could lie-to in the mid-Caribbean and watch the rafts go by, speculating as a pleasant game as to which bore Onycophora and earthworms and which cyprinodonts or amphibia . . ..'

However, geological evidence supporting most of the postulated land bridges has not been forthcoming, and in fact the bulk of geological evidence actually militates *against* the acceptance of such theories. By contrast, the evidence for over-water dispersal has steadily accumulated.

The classic works of Darwin (see above) Guppy (1906, 1917) and Ridley (1930) can leave little doubt that many plants are regularly dispersed across sea-water barriers. Similarly, Gressitt and Yoshimoto (1963a, b) found a remarkable number and variety of animals in the air over the ocean. The anecdotal accounts of flying insects found long distances from land is impressive. Finally, Heatwole and Levins (1972a), studying the area about which Barbour and Noble made their remarks, found that about 25% of drift picked up at sea contained at least one live terrestrial animal, and often many. The animals represented included 19 families of insects as well as pseudoscorpions, snails, spiders, mites, milipedes, isopods and worms. Perhaps had Barbour and Noble employed empiricism rather than cynicism they would have drawn different conclusions.

It seems that the era of oceanic bridge-building by biogeographers is almost over. There is overwhelming evidence for across-water dispersal of animals and plants to oceanic islands and little or no geological support for most of the postulated land bridges to them.

Another geological influence on island biogeography is the theory of continental drift. The hypothesis contends that the major continents were once part of a common land mass which subsequently broke up and the pieces (continents) drifted to new locations, carrying with them their plants and animals. Proponents of the view based their opinion on faunal and floral similarities of different regions and on the shapes of the continents. Only in the last few decades, however, has overwhelming geological support become available. The evidence is now convincing that continental drift has occurred and the theory of plate tectonics proposes a mechanism whereby it can occur. It is now necessary to consider the geological history of land masses and their movements in order to understand completely the distribution of plants and animals. This is extremely important for the larger land masses such as continents, and islands such as New Zealand and Madagascar, but may be less important for smaller ones, or those which are relatively young. However, Rosen (1975) has recently shown that distributional patterns in the Caribbean, including those of the major islands, can be interpreted in terms of the postulated plate tectonics of that region. As more of the details of plate tectonic movements in various areas become known, it is probable that they will have an increasing relevance to insular distribution patterns.

In summary, Darwin and Wallace established modern insular biogeography by pointing out the role of insular isolation in evolution and speciation and in emphasizing the importance of historical factors and dispersal capabilities in determining the nature of island biotas. These basic ideas have been mostly supported by later work. A better understanding has been achieved as a result of advances in heredity, population genetics, and geology. The concepts of genetic drift and the founder principle have been valuable theoretical modifications. Acceptance of continental drift and the theory of plate tectonics has required modification of Darwin's and Wallace's biogeographic view; their belief in oversea dispersal rather than land bridges as explanations for the biotas of oceanic islands has been largely vindicated.

4. The equilibrium theory

A turning point in modern insular biogeography and ecology was the publication by MacArthur and Wilson in 1963 of a paper entitled 'An Equilibrium Theory of Insular Zoogeography' and later a book (1967) called 'The Theory of Island Biogeography'. These two publications made several important contributions. They (1) provided a theory relating numbers of species on islands to island size and distance from mainlands, (2) reviewed the empirical data relevant to that theory, (3) put insular ecology into a dynamic rather than static perspective, and (4) pointed the way toward an experimental rather than solely observational approach to the topic; implications of the theory were amenable to empirical testing. As a result of all of the above they stimulated a massive amount of research which has furthered our understanding of insular organisms to the extent that this field has become one of the most active and exciting growing points of modern ecology.

The heart of the theory is that the number of species on an island is the result of an equilibrium between the immigration rate of species new to the biota of the island and the rate of extinction of species already present. It is argued that when an island is completely devoid of species, *any* propagule that arrives will be a new species for the island. The greater the number of species present (i.e., the greater the proportion of the mainland species-pool represented in the insular biota), the greater is the likelihood that an arriving propagule will be of a species already present and will not add a new species, only new individuals. Thus, a given island will have higher immigration rates of new species if it has few species than if it has many. Because of the increase in proportion of 'repeats' with increase in number of species present, the immigration curve would be expected to appear somewhat as in Fig. 1.

The rate of extinction of species also was postulated to be influenced by the number of species present. The more species present, the more there are to become extinct and the greater is the likelihood that at least a certain number will, just by random processes. Also, the greater the number of species present, the more likely it is that there will be species that have an adverse effect on others (e.g. through competition, predation, or combative interference) and raise extinction rate even higher. Consequently the extinction curve would be expected to rise with increasing numbers of species on an island (Fig. 1).

The number of species that an island will actually contain is indicated by the point of intersection of the immigration and extinction curves, and is known as the 'equilibrium number of species' or Ŝ. The equilibrium number of species is not the maximum number the island could hold when 'saturated'. For example, an increase in immigration rate could raise the number of species present (see shift in Fig. 2A). In reality Ŝ is probably an average value around which species numbers fluctuate.

Two major influences on immigration and extinction, and hence on the number of species an island has, are (1) isolation and (2) island size. Islands

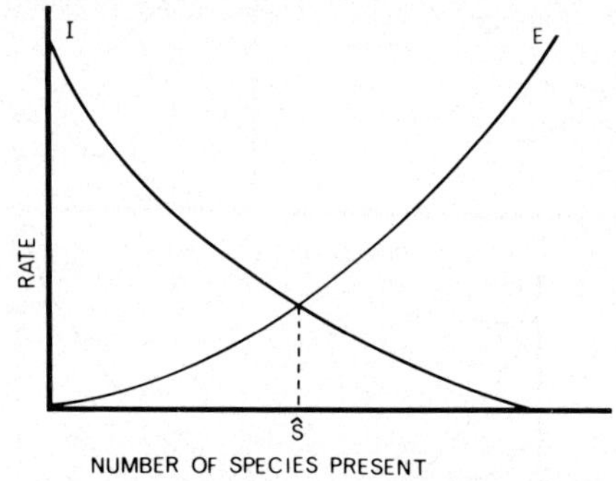

Fig. 1. Graphic portrayal of the Equilibrium Theory. Ŝ is the equilibrial number of species. Curve I indicates the immigration rates of species new to the island. Curve E indicates extinction rates. For explanation, see text.

that are far from a mainland will have lower immigration rates than a near island because of the greater difficulty for animals and plants to cross the extended sea water barrier. Such a lowering of the immigration curve would result in a shift to the left of the intersection point with the extinction curve and consequently a decrease in number of species (Fig. 2A). Conversely, raising the immigration curve to that predicted for a very near island would increase the species numbers. According to the theory then, the equilibrium number of species on an island will decrease with increasing distance of the island from a mainland as long as all other factors remain the same.

The larger an island is, the greater is the size of the 'target' it presents to a dispersing propagule and consequently immigration rates may be slightly enhanced. Of more likely importance, however, is an effect of island size on the extinction rate. Larger islands are usually more diverse ecologically and should thus provide a greater variety of niches with the result that a greater number of species might find suitable conditions and would be less likely to become extinct. Populations would be smaller on small islands and hence the chance of extinction by random events, would be greater. Species requiring large foraging ranges or large territories would be able to build up only small populations on a tiny island and would be particularly susceptible to extinction on small islands. A storm would be more likely to wipe out the small populations on a small island or to wash over the entire island than would be the case on a larger one. Competition might be more severe for the limited kinds of resources on a small island then on a larger, more diverse one. These theoretical considerations converge to predict that small islands

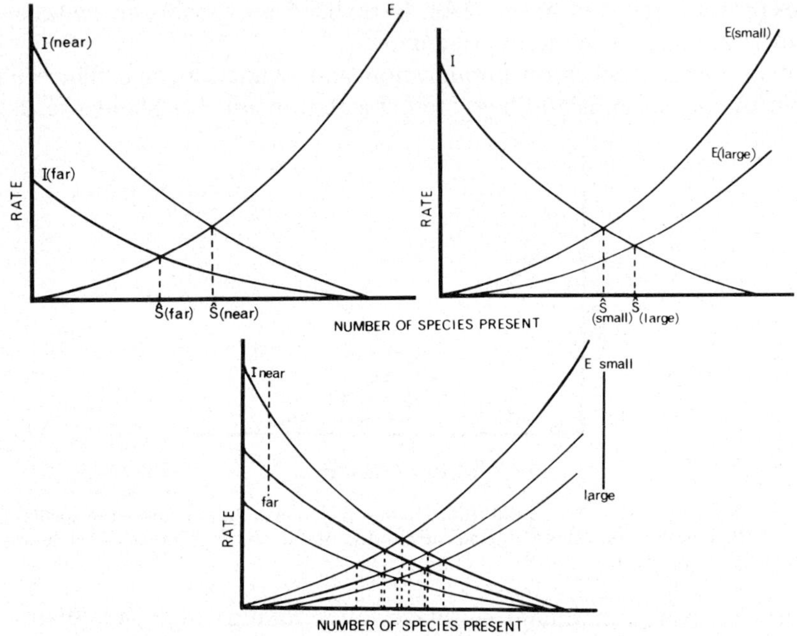

Fig. 2. Diagrams of immigration and extinction curves and equilibrial number of species on near and far islands (A) large and small islands (B) and a combination of islands of different sizes and distances from mainlands (C). Symbols as in Fig. 1.

will have higher extinction rates than a larger one with the same number of species present, and consequently small islands will tend to have fewer species than larger ones (Fig. 2B).

By way of recapitulation, the equilibrium theory of island biogeography can be stated as follows. The number of species on an island is determined by an equilibrium between immigration and extinction rates, is inversely related to distance from a mainland, and is directly related to island size (Fig. 2C).

There are a variety of factors which complicate the basic model. As MacArthur and Wilson (1967) indicate: 'Stating postulates for even the simplest models is a risky business, because we are unsure of the complex biological phenomena underlying them.' Also, physical differences among islands are important; not all islands of the same size and distance from a mainland are equal in other respects. By the time MacArthur and Wilson (1967) wrote their book, they were able to review material published by various investigators which assessed the theory and identified other factors relating to it. The important conclusions and predictions of the book, based on the literature and on their own additional theoretical constructs, can be summarized briefly.

1. There is a general relationship between island area and species number of the form:

$$S = CA^z$$

where S is the species number, C is a constant that varies widely among taxa and probably in different parts of the world, A is the insular area and z is a constant. z is usually between 0.20 and 0.35 and theoretically should be 0.27. Values above 0.27 can be explained as the outcome of the breaking up of biotas of large islands into semi-isolated communities due to their greater amount of environmental variation and larger number of topographic barriers. Low values indicate some degree of faunal exchange. Species counts on sample plots on mainlands often give z values of 0.12 to 0.17, probably because of flooding by transient species from nearby areas.

2. Areal effects are modified by, or correlated with, a number of factors including the following:

(a) General ecology – in some cases, small isolated islands with favourable habitat, such as rainforest, have more species of certain taxa than larger, less isolated ones in the same general region but with xerophytic vegetation.

(b) Island complexity or habitat diversity may alter extinction rates and influence species number.

(c) Elevation – part of the variation of the species number can be accounted for by differences in island elevation. High islands are more diverse ecologically than low ones.

(d) Geography and configuration of islands – clumping of islands into archipelagos or the interspersion of stepping stones between them and the mainland, influences immigration capabilities and hence species numbers. Even small stepping stones may have an effect and large ones close to the recipient island may increase flow of propagules by many orders of magnitude.

3. A model was developed predicting that the number of species should increase with area more rapidly on distant islands than on near ones, and should decrease more rapidly with distance on small islands than on large ones; the empirical data then available were consistent with the prediction.

4. Species-turnover on extremely small and unstable islands may be so high that extinction rates are not area-dependent within a certain range of small island sizes. [This is the 'small island effect'.]

5. A model of immigration and extinction on single islands was developed which led to the approximate equation:

$$X = \frac{1.16\hat{S}}{t_{0.90}}$$

where X is the extinction rate at equilibrium, $\hat{S}$ is the equilibrial number of species, and $t_{0.90}$ is the time required to reach 90% of the equilibrial species number. This equation allows a high turnover rate on islands that are colonized quickly. Methods for measurement or estimation of rates of extinction, immigration and colonization are presented.

6. A crude model of colonization was developed which took into account that small changes in birth and death rates have a great effect on colonizing success, that speeding developmental rate and increasing span over which reproduction occurs were usually more effective than increasing fertility, and that the most effective propagules are those at the age of maximum reproductive value. It was predicted that chances for initial success of a colonizing propagule should be high if the intrinsic rate of natural increase were large in comparison to the birth rate (death rate low). Species will persist for a long time when those same conditions obtain, if the carrying capacity of the environment is high (permits a large population). Other properties influence the intrinsic rate of increase. It will be reduced if colonists disperse so rapidly that location of mates is adversely affected. Presence of competitors may also reduce the ability of a population to realize its intrinsic rate of natural increase. If the recipient island is nearly saturated with species the intrinsic rate of increase may approach zero; under those conditions a safe population size is not likely to be achieved by any newly arriving species.

The carrying capacity of the island is also important in other respects. Small islands, or islands with many competing species, have only low carrying capacities (only small populations of a given species can be maintained) and established populations are not likely to persist for long periods of time, i.e., time to extinction is short and a species must reinvade frequently if it is to be maintained. Carrying capacity for a particular species is very low if the suitable habitat on an island is restricted. In such a case the effective island size for that species is small. If no suitable habitat occurs, both effective island size and carrying capacity are zero and establishment cannot take place.

A superior insular species is thus one which has good dispersal powers, has a high rate of natural increase, but yet can maintain large populations under insular conditions, i.e. can use a wide range of resources.

Empirical data reviewed by MacArthur and Wilson tended to support the model but also indicated that successful island species did not always combine all these attributes; some of these characteristics were prevalent in some species, others in others. Preference for unstable, scattered habitats seemed to be a preadaptation to successful establishment; species survive by the relatively quick and temporary occupancy of suitable new habitats. As the new habitats undergo succession, more specialized and competitively superior species arrive and displace the more generalized 'fugitive' ones in the new, successionally more mature environment. By that time, the fugitive species may have found new habitat elsewhere; they are only able to persist by moving from one place to another.

7. A species can be excluded from an island in many ways. For example, it was postulated that amongst species whose populations are limited by scarcity of resources there can be no more species than there are types of resources. The upper limit of species thus may be determined by the number of resources per habitat and the number of habitats on an island.

A species may competitively exclude another whose ecological requirements are closely similar.

Invasion of new species may be prevented because only a certain number of individuals can be accommodated on an island. There is a limit to the number of species among which those individuals can be apportioned, as each population needs a certain minimum number of individuals to remain viable.

Lack of predation may also limit insular diversity. In the absence of predators or disease one or a few species may become dominant and exclude others. Conversely, excessive predation may eliminate a species.

In summary, three conditions which reduce diversity may prevent invaders from establishing; these are (1) competitors too similar, (2) predators too few or too many, or (3) species too rare.

When a new species invades an island it usually encounters a different physical and biotic environment from the one it left on the mainland. Establishment may involve expansion or contraction of its niche (depending upon the range of resources it encounters and on whether it meets more or fewer competitors). When compression occurs it is more often a compression of habitat than of diet.

8. Evolution can change the number of species. Isolation of islands can lead to the formation of new, autochthonous species. In order for this to happen, islands must be relatively large and stable; otherwise populations will not persist long enough to undergo sufficient local adaptation and divergence from mainland populations. Near the outer limits of the dispersal range of a particular taxon, speciation on the islands of an archipelago and exchange of those species among islands, can result in a greater number of species on a given island than is generated by immigration from outside the archipelago alone.

MacArthur and Wilson indicated several weaknesses of their equilibrium theory. They cited in particular the lack of empirical knowledge about the precise shape of immigration and extinction curves, and the lack of precision of the terms 'immigration' and 'extinction' [actually a failure to distinguish between transients and established species]. They also admit that the assumption that the immigration and extinction curves are of similar shape for different biotas, different islands and different times, may not be true.

Terminology

As MacArthur and Wilson recognized (see above) one of the problems confronting island ecologists was the vagueness of the terminology relating to various aspects of insular dynamics. In terms of the equilibrium model it was necessary to define more sharply 'immigration' and 'extinction' and other terms relating to the degree of permanency of species on islands.

Since the equilibrium theory was propounded, various persons have addressed themselves to this task and concepts have been considerably clarified. The following basic terminology is suggested as a standard, and subsequently will be used throughout this book. It is an eclectic combination of the revised terminology of Whitehead and Jones (1969), Simberloff and Wilson (1969), Simberloff (1969), Heatwole and Levins (1973) and Lynch and Johnson (1974) as well as definitions presented here for the first time. These definitions should be considered as applying specifically to island ecology, not in a more general context.

Propagule:	the minimum number of individuals of a species potentially capable of successfully founding a new population.
Invasion:	the arrival of one or more propagules on an island, whether or not the island is already occupied by that species.
Immigration:	the process of arrival of a propagule on an island not occupied by that species.
Colonization:	the existence on an island of at least one propagule of a species.
Establishment:	colonization followed by reproductive success for at least one generation.
Extinction:	the total disappearance of an established species from an island.
Resident:	a member of an established species.
Transient:	an individual or a propagule occurring on an island but without reproductive success.
Species-turnover:	change in species composition as a result of establishment and extinction.

A special case must be made for all the above terms for species (such as many birds) that have long-distance migration involving regular seasonal departure from, and return to, an island. The seasonal migratory cycle is not considered as part of the establishment-extinction cycle.

5. Recent approaches

Since the formulation of the equilibrium theory, significant advances in insular biology have been made through (1) experimental approaches to colonization and island dynamics, and through new interpretations of the (2) shape of the island-area curve, (3) the role of trophic relations in structuring island communities, and (4) the effects of evolution on equilibrial number of species. Finally, (5) island theory has been extended to other kinds of isolated habitats in an attempt to provide a more general theory.

In addition, many studies have further documented various aspects of dispersal and competitive interactions, and have measured or estimated immigration, extinction or turnover rates, or have provided empirical data on the ecology of particular insular species, islands, or archipelagoes. These latter studies will be discussed where appropriate at various places in the book. In this chapter the topics 1–5 will be briefly discussed.

Experimental studies

The equilibrium theory makes predictions that can be tested experimentally. One of the most important tests of the theory was the defaunation by fumigation of small small mangrove cays[1] in Florida, U.S.A. These cays were surveyed for their fauna before fumigation and at various intervals afterwards. The islands were rapidly re-colonized, predominantly by passive and active aerial transport, and 250 days after defaunation all islands but the most distant one had regained numbers of species similar to those of untreated islands (Wilson and Simberloff 1969, Simberloff and Wilson 1969).

Observed immigration and extinction rates were usually between 0.05 and 0.50 species per day. Actual turnover rates were probably much higher

[1] We prefer the original English word 'cay' to the derived American one 'key'. Consequently, we use the former (except when 'key' is part of a proper noun) even when papers we refer to used the latter.

(because some immigrations and extinctions might have been missed); the lower limit was probably about 0.67 species per day.

By the end of the second year the species numbers had not greatly changed from those of the previous year and could thus be considered to be in equilibrium. The original species composition was approached more closely after the second year than after the first year (Simberloff and Wilson 1970).

This experimental study thus confirmed several of the most important aspects of the equilibrium theory. New concepts also arose from the study. The number of species rose to a peak and then dropped back to a slightly lower equilibrial level. The initial increase in species up to the peak of the curve occurred when population densities were low and was interpreted to represent a non-interactive phase of colonization whereas the later level after the peak was looked upon as an interactive phase associated with population growth and increased numbers of interactions (Simberloff 1969).

A second experimental approach has been the introduction of species onto islands and observation of their subsequent fates. Snails, frogs, *Drosophila*, anoline lizards and ants were introduced onto a small Caribbean sand cay (Levins and Heatwole 1973). The first three groups were from mesic habitats and become extinct on the island almost immediately, apparently because of lack of favourably moist microhabitats. Two species of lizards and one species of ant from physical habitats similar to those on the island persisted from several weeks to about one year. Their eventual extinction probably resulted from biotic interactions. The introduced lizard species were both known to be preyed upon by species of lizards native to the island. On mainlands some of the species, when sympatric, use more restricted habitat than they do when alone. On the island, patches of the restricted habitat type of one of the introduced species were small. The wider and more common habitat normally usable by the introduced species in the absence of other anoles, was occupied by a congener native to the island. The introduced species was probably driven to extinction by exclusion from most of the island because of interaction with the native species.

The introduced species of ant interacted aggressively with two local species (one diurnal, the other nocturnal). These interactions caused the introduced form to move its nests to new sites. Energy was expended in the frequent moves to new places and by hostile encounters at all of them, and extinction resulted because the colony's vital activities such as foraging for food were continually disrupted (Levins and Heatwole 1973, Levins *et al.* 1973).

Clearly a combination of factors including microclimate, predation and competition caused the extinction of the artificially introduced species.

The shape of the island-area curve

As indicated above, MacArthur and Wilson realized that for a range of sizes of very small islands, equilibrial species-number was independent of island area. They attributed this to the instability of such islands, i.e., the exceptionally high area-independent extinction rates and the consequent high turnover rates. For example, total extinction of the biota of all sizes of tiny islets up to a certain size, by storms or wave action, would be likely, whereas larger islands would be less subject to wholesale extinction.

An alternative explanation was provided by Whitehead and Jones (1969) for vascular plants. They suggested that the ecology of the island exerted a selective control over the establishment of arriving propagules. Very small islands lack some of the characteristics of larger islands that many species require for survival. For sand cays one such characteristic is the presence of a freshwater lens. Cays smaller than about 100 m in shortest diameter are incapable of holding a freshwater lens in the sand and consequently only strand plants (salt tolerant) can grow there. Above the size-threshold for water lenses non-strand species can grow and an upward inflection occurs in the species-area curve. Another threshold is the smallest island size capable of supporting man; humans influence species-number by introducing edible and ornamental plants, and accidentally introducing weeds.

Thus the species-area curve of vascular plants is a superimposition of the separate curves of strand plants, non-strand native plants and plants introduced by man. Each group alone provides a straight line on a semilogarithmic plot (species versus log area) but the three superimposed gives a line which curves upward with inflections at the thresholds. Similar considerations probably apply to animals.

Still another explanation for the 'small island effect' was suggested by Heatwole (1975). Most very small islands that have been studied are cays. Aside from the differences in soil water noted above, and the reduced exposure of the central areas of larger cays to salt spray, cays do not change in environmental diversity very much with size. All are flat and little differentiation is possible; increase in size merely increases the area of a given kind of habitat, not the number of habitats. Only on rocky islands, especially those above the upper size limit of sand cays, does the increase in ecological diversity with increasing size become important; rocky islands often have an altitudinal component to their diversity.

These three explanations of the 'small island effect' are not mutually exclusive and all could be operative to some extent. The small island effect may not hold equally for all taxa. Mühlenberg (1977a, b) found that on some of the Seychelles Islands, numbers of species of spiders were related to island area but numbers of species of ants were not. Gilpin and Diamond (1976) have refined our understanding of the shapes of immigration curves, extinction curves and species-area curves. They tested a quantitative and biologically reasonable model against data for birds from 52 islands, and accounted for 98% of the variation in species numbers. The immigration and extinction curves were very concave. Extinction rates varied almost exactly as the reciprocal of island area; the effect of area on immigration rates was slight. The species-area curve departed from strict linearity on both a semi-logarithmic and log-log plot. One of the important aspects of insular ecology which remains to be adequately assessed is the effect of such variables as latitude, or size of the mainland species pool on the species-area curve.

Trophic relations and community structure

It is obvious that trophic relations have an effect on establishment of immigrants and hence on the composition of insular communities. Clearly, herbivores cannot become established in the absence of plants, nor can predators or parasites become established without their prey or hosts. It is not surprising therefore, that different types of islands have different trophic structures (as indicated by the proportions of species in the various trophic categories). A number of Florida mangrove cays had a similar trophic structure, which differed markedly from that of a small vegetated sand cay, and even more markedly from that characteristic of sand cays lacking vascular plants (Heatwole and Levins 1973). On the small vegetated sand cay, trophic structure remained relatively stable in time despite rapid turnover of species and change in species composition.

On mangrove cays the largest trophic category was the herbivores, followed by predators and then entomophagous parasites; other categories contained relatively few species. On bare sand cays the largest categories (excluding the ectoparasites of sea birds) were scavengers and detritus feeders, followed by predators. The vegetated sand cay was intermediate between these two extremes, having high, and nearly equal proportions of herbivores, scavengers, and detritus feeders, followed by predators and omnivores (ants).

The trophic differences among these islands reflect differences in their types of energy base; plants in the case of the mangrove island; marine carrion and sea bird carrion and guano on the bare sand cay; a combination of all the above on the vegetated sand cay. The different trophic structures may also reflect successional stages in the development of islands. After Krakatau was defaunated by volcanic action, the first community to become established resembled that of bare sand cays, consisting mainly of scavengers and detritus feeders. Later (after plants had become re-established, conditions resembled those typical of vegetated islands; omnivores, herbivores, predators and parasites became established and altered the trophic structure (Dammerman 1948).

The trophic structure of mangrove cays probably does not represent the climax phase of cay development; although they are forested they lack dry land and consequently the ground layer of plants and animals characteristic of other forested cays is missing. One-Tree Island may be a more typical situation.

In an analysis of the trophic structure at different stages of recolonization of the mangrove cays defaunated by Wilson and Simberloff (1969), Heatwole and Levins (1973) concluded that trophic structure became re-established well before the original taxonomic structure did. Diamond (in press) found a similar phenomenon in birds recolonizing an island after a volcanic explosion. In the case of the mangrove cays, the equilibrial trophic structures of individual islands were not the same as that of all species in the region or of all census records. Heatwole and Levins (1973) suggested three phases in the process of establishment:

(1) Independent immigration of species. If immigration rates are high compared to the time it takes established species to reach demographic equilibrium, the number of species may temporarily exceed the equilibrial number (as in fact occurred). During this phase, changes in trophic structure are the passive consequences of independent species' establishment.

(2) Establishment of trophic structure equilibrium. Demographic interactions among species cause extinctions and either facilitate or prevent establishment of immigrating species.

(3) Replacement of species by trophically equivalent ones of other taxa. While trophic structure is maintained, continued immigration and extinction results in an approach to some taxonomic equilibrium. Species are replaced by ecological equivalents better adapted to the specific local conditions.

Depending on relative immigration rates of the various species, phases (2) and (3) may grade imperceptibly into one another.

Simberloff (1976b) objected to this re-interpretation of data on the grounds that the determinateness and importance of trophic structure may have been overemphasized by the pooling of data from separate islands and through the use of a statistic strongly dependent on species number. He correctly pointed out that the theory of trophic structure stability had not been rigorously proven. However, neither has it been disproven and many existing data are consistent with it. In a later, comprehensive analysis of the stochastic aspects of colonization (Simberloff 1978), he indicated that at present the hypothesis could not be rejected, but that 'If patterns of energy flow do constrain colonization, these patterns must first be worked out in detail for at least most of the species before a proof can be given.' This is indeed a formidable task, especially in view of the high species-turnover occurring on islands. Detailed empirical studies of the community ecology of islands are required. We will return to the topic of trophic structure stability in relation to the One-Tree data in Chapter 22. It is necessary to point out here, however, that in terms of the more restricted terminology employed in this book, trophic structure stability is more applicable to establishment than to colonization.

Evolution and the equilibrial number

Competitive exclusion may be one of the causes of extinction on islands and may influence equilibrial species-numbers. If there are two strongly competing species from a mainland pool and both are able to disperse to offshore islands, several situations may occur. One species may be so superior that it always eliminates the other. Consequently, the competitively inferior species can only become established on those islands which by chance have not been occupied by the superior one. Another possibility is that either species can become established on any island in the absence of the other; whichever species is the first to become established prevents occupancy by the other species. A third possibility is that conditions on some islands are suitable for one species and give it a competitive advantage, whereas on other islands the opposite species may be favoured. Finally, if the island is large enough it may be sufficiently diverse (resembles mainland conditions) to support both species.

A mosaic pattern of distribution in which islands containing one species are interspersed more or less randomly among those having the opposite species, but with no island containing both, is suggestive of competitive exclusion. Such distributions have been observed among ants of the Pacific (Wilson and Taylor 1967a, b) and Caribbean islands (Levins and Heatwole 1973).

Where competition is fierce among sympatric species one would expect selection to favour attributes that could reduce competition, that is, character displacement might occur, or there may be divergence in their choice of habitat. Over a long period of time evolution should result in co-adapted species, each with its own niche, and competition should be reduced. Species from such a co-adapted mainland pool would not be expected to compete on islands unless conditions there were sufficiently restricted to prevent ecological segregation. In any event, immigrant species from a co-adapted pool would be less likely to compete on islands than would species from several scattered sources. Given enough time, however, even random assemblages of species which contact each other frequently within an archipelago may evolve co-adaptations, and competition between them would become reduced. When that happens, competitive exclusion of species on specific islands would become less frequent, extinction rates would be lower, and the equilibrium number of species would increase. It follows that islands can hold greater numbers of species which are co-adapted than of those which are not. Evolution would thus tend to increase the equilibrial number of species (Wilson and Taylor 1967b).

There is evidence for this theory among Pacific ants. The fauna of western Polynesia is mostly truly native and presumably co-adapted, whereas that of the eastern and northern archipelagos consists of a synthetic aggregation of tramps[2] of Indo-Australian, African and New World origins, which have been accidentally introduced by man during the last 400 years; they are therefore

[2] Tramp species are those that are widespread as a result of frequent, accidental introductions by humans.

not co-adapted. Islands with a mostly native ant fauna have a greater number of species than those of equivalent size carrying only tramp species (Wilson and Taylor 1967a, b). This result is not an artifact of the size of the species-pool; 35 species of tramps are known from the archipelago, yet no island has the entire number and most contain fewer than a quarter of them.

Other isolated habitats as 'islands'

A variety of relatively isolated habitats have been considered as 'islands' and the equilibrium theory applied to them, including lakes and ponds (Lassen 1975, Keddy 1976), mountain tops (Johnson 1975, Brown 1978), caves (Culver 1970), wildlife preserves (May 1975), vacant urban lots (Crowe 1979), rocks on the sea bottom (Osman 1978), host plants (Janzen 1968), leaf stalks (Beaver 1979), carrion (Beaver 1977), primate social groups (Freeland 1979) and mammalian hosts of ectoparasites (Kuris *et al.* 1977) to select only a few references as examples.

In addition, valuable insights have been achieved experimentally by placing artificial substrates (plastic mesh sponges; fouling-panels) in the sea as sites for settling of marine animals, and treating such substrates as 'islands' (Schoener 1974a, b; Schoener *et al.* 1978).

All of the above types of studies have contributed to a better understanding of ecology. However, such isolated habitats are not islands. They can be demonstrated to share some of the characteristics of islands; but they also differ from true islands in many ways, as some of the above studies have emphasized. Reference to almost any partially isolated situation as insular has reached absurd proportions.

This book is not an appropriate place to review similarities and dissimilarities between different kinds of habitats as the scope of the book extends only to true islands. However, it is appropriate to plead for a satisfactory frame of reference for studies of islands in relation to other isolated habitats. There has developed a general theory of island ecology which is being stretched to cover the ecology of all isolated habitats. Its more precise applicability to real islands can only be distorted, with resultant confusion and lack of precision. What is needed is a comprehensive, comparative treatment of the ecology and biogeography of *isolates*, of which one kind consists of real islands. We suggest the term *isolate* be used as a generic term to include all isolated habitats, and that the term *island* be restricted to its original meaning, a small land mass completely surrounded by water. A clear terminology will avoid vague and nebulous thinking and promote frank comparisons of the attributes of different types of isolates.

Finally, there have been a number of advances in research on community structure generally, which have applicability to islands. Some studies have dealt specifically with insular communities (see Cody and Diamond 1975 for a review). All the relevant topics cannot be reviewed here. However, some of those most pertinent to the present study are discussed in Chapter 22.

PART THREE

The One-Tree Island community

6. General description and history of the island

One-Tree Island is a coral cay of 4.93 hectares located near the southern end of the Great Barrier Reef, almost exactly on the tropic of Capricorn (latitude 23° 31′ S, longitude 152° 08′ E) (Roberts 1978) and 80 km from the Queensland mainland. It is the easternmost of the nine islands which compose the Capricorn Group. Its nearest neighbour is Heron Island, 19 km to the WNW (Fig. 3). It is a Type 3 Coral Island according to the geomorphological classification of Fairbridge (1950).

The island is quite young. One-Tree Reef, on which it is located, grew upward from an old platform (now at 20–25 m depth) during the most recent eustatic rise in sea level. The reef reached present day sea level about 4,200 years ago and the cay was formed subsequently (see Heatwole, 1981, for a general review). The island consists of sand and coral rubble arising from weathering of the seaward reef face. Pieces of coral were ripped off the reef by storms and piled in ridges onto a sand cay which had developed earlier in the island's genesis. A series of concentric storm ridges attest to the action of successive storms in the formation of the island, the most recent ridges being those on the seaward edge.

One-Tree Island is situated on the southeastern end of One-Tree Reef, the longitudinal axis of which runs roughly east–west. Being located on the most exposed corner (facing prevailing winds) of the most seaward reef in the group, it is subjected to the greatest impact of wave and wind action. The island is roughly boomerang-shaped with the two arms pointing north and northwest and the outer curvature facing southeast (Figs. 4, 5 and 9).

The beach of coarse coral shingle surrounding the island is about 9 to 15 m wide, with variable slopes. The interior of the island is slightly concave and is dissected by the storm ridges previously mentioned. The island has a maximum elevation of about 2 m. The beach is periodically eroded by current or wave action and materials are frequently deposited, resulting in minor changes in the details of the beach's contour from time to time.

Close to the centre is a depression containing a shallow brackish water pond. Water level is affected by rainfall and by the tidal regimen and the pond sometimes dries completely.

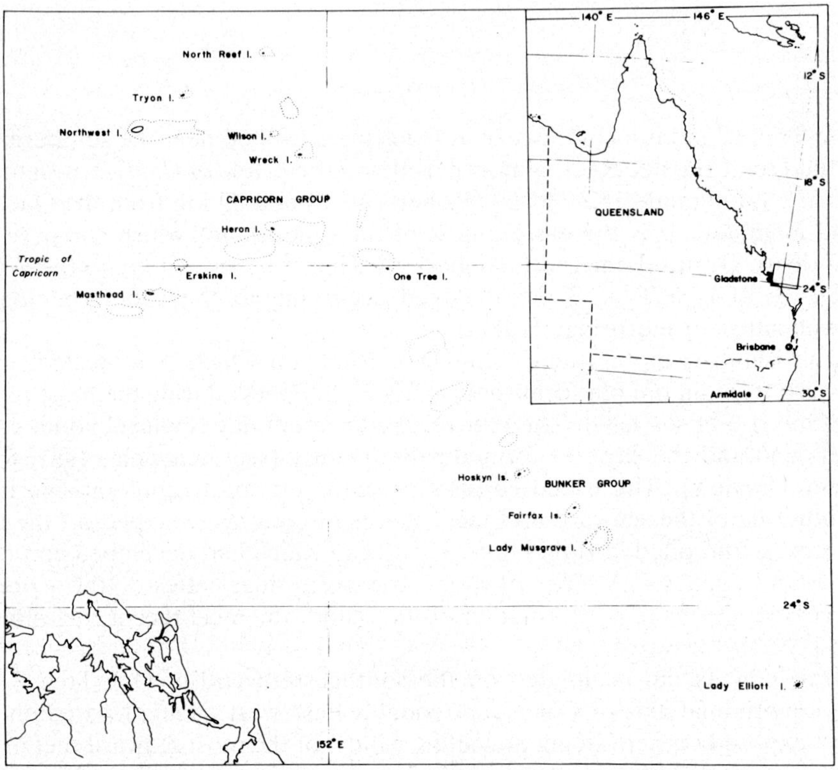

Fig. 3. Map of the Capricorn and Bunker Groups of islands, Queensland, Australia. Location of main map indicated by square in inset.

Fig. 4 (above). Aerial view of part of One-Tree Reef showing One-Tree Island.

Fig. 5 (below). Aerial view of One-Tree Island. Note different types of vegetation and central pond. The buildings of the marine station (not present during the initial part of the study) appear on the peninsula on the right.

Fig. 6. Vegetation types of One-Tree Island. 1. Strand or beach vegetation. 2. *Argusia* shrubs at edge of beach. 3. Herb flat in non-drought conditions with *Pisonia* grove in background. 4. Herb flat in drought condition with edge of *Pandanus* grove on right. 5. Succulent mat around pond (note crested terns, *Sterna bergii*). 6. View of One-Tree Island from the eastern beach, showing *Argusia* shrubs (centre, foreground), killed as a result of 'Cyclone David' in 1976.

One-Tree is vegetatively more diverse than its name suggests; there are quite a number of trees. However, in 1843 viewed from a distance from the British survey vessel H.M.S. Fly a clump of *Pandanus* trees was mistaken for a single tree, and the island was christened with the misnomer it still bears (Jukes 1847).

The vegetation of the upper beach is restricted to a few scattered unbellifers (*Trachymene* sp. aff. *procumbens*) and various transient species plus occasional individuals of the species of low plants also found in the interior of the island (Fig. 6). Just above the beach the island is encircled by a ring of shrubs (*Argusia argentea*), inside which is a rather homogenous field of herbs, vines and grasses (the herb flat; see Fig. 6). There are three large groves of *Pisonia* trees and one of *Pandanus*. These are small forests with nearly closed canopies on the tops and sides (Figs. 6, 7) and an open floor nearly devoid of herbaceous cover and richly endowed with leaf litter (Fig. 8). In addition some *Pisonia* trees are associated with the shrub ring on the eastern edge of the island and on the northern edge of the *Pandanus* clump. The pond is encompassed by a dense mat of low succulents, primarily *Sesuvium portulacastrum*. The distribution of these vegetation types on the island is shown in Figs. 5 and 9 and the area of each is indicated in Table 1.

Table 1. Area of different habitats on One-Tree Island.

Habitat	Area (m^2)	% of total island area
Main part of herb flat	25,799	52.3
Marginal part of herb flat with low vegetation cover	4,717	9.6
Pisonia forests[a]	3,468	7.0
Argusia shrubs[b]	2,389	4.8
Scaevola shrubs	28	0.06
Pandanus grove	750	1.5
Total herb flat	30,516	61.9
Total trees and shrubs	6,635	13.5
Succulent mat	1,635	3.3
Beach	10,247	20.8
Pond	288	0.6
Total (m^2)	49,321	100.1
(hectares)	4.93	—

[a] Canopy area 4,190 m^2; [b] Canopy area 3,181 m^2.

Jukes (1847) briefly described the vegetation observed at the time of the discovery of One-Tree Island (Jan. 10–11, 1843). In addition to the *Pandanus* grove, he noted pink-flowered succulents (*Sesuvium*) which he referred to as 'mesembryanthemum?', and low trailing bushes (probably *Melanthera*). Other subsequent, brief accounts of the flora have been presented by MacGillivray and Rodway (1931), Steers (1938) and Domm (1971).

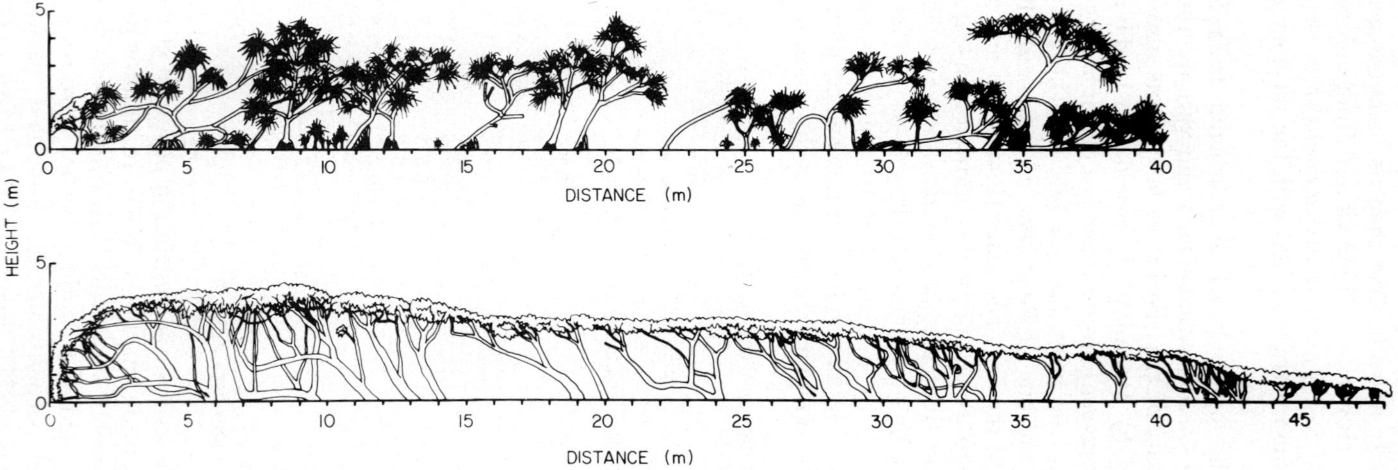

Fig. 7. Diagrammatic profile through a *Pandanus* (upper) and *Pisonia* (lower) grove on One-Tree Island. Note closed canopy, especially in the *Pisonia*.

Fig. 8. Views inside *Pandanus* (left) and *Pisonia* (right) clumps. Note abundance of leaf litter.

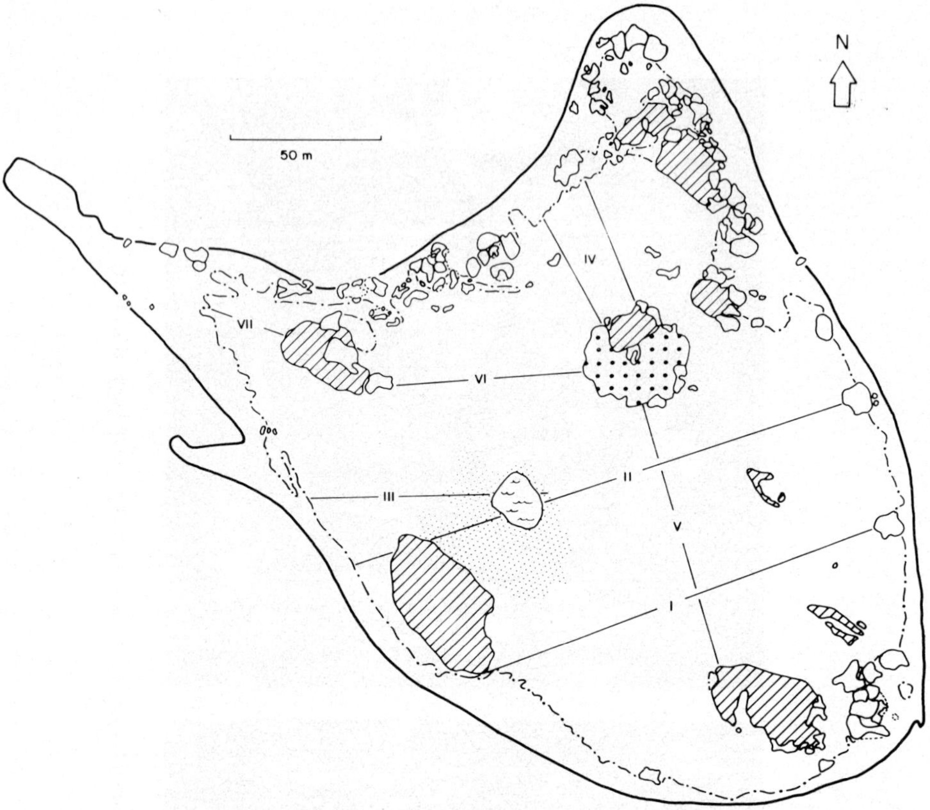

Fig. 9. Map of the vegetation zones of One-Tree Island. Hatched areas indicate *Pisonia* groves, areas of large dots a *Pandanus* grove, area of small dots the succulent mat, wavy lines the pond, and open outline *Argusia* shrubs. Straight lines and Roman numerals indicate the releveé transects. The line of dashes and dots delineate the herb flat.

The climate of the Capricorn island group is subtropical maritime. Long-term weather records are not available for One-Tree Island. However, Heron Island, only 19 km away has an official Australian Meteorological Bureau weather station, and weather data collected there were used to indicate the general levels and patterns of seasonal change during the present study. There were probably minor weather differences between the two sites. Temperature maxima would probably be somewhat higher for One-Tree than for Heron as the former has less forest cover.

The general thermal environment is mild, and characterized by regular seasonal fluctuations of moderate magnitude and by relatively small daily fluctuations. The lowest air temperature recorded at Heron Island from 1968 to 1970 was 10.6° C on 8th July 1970; the highest one was 33.5° C on 6th January 1969. Mean monthly minima ranged from 15° C to 24.4° C and mean monthly maxima from 20.6° C to 31.1° C (Fig. 10). Daily temperature range (maximum minus minimum) was not great and this span remained relatively constant throughout the year (Fig. 10). The seasonal pattern is a regular one, all three years being practically identical although 1968 was slightly cooler than the following two years. The hottest month is January and the coldest one July or August.

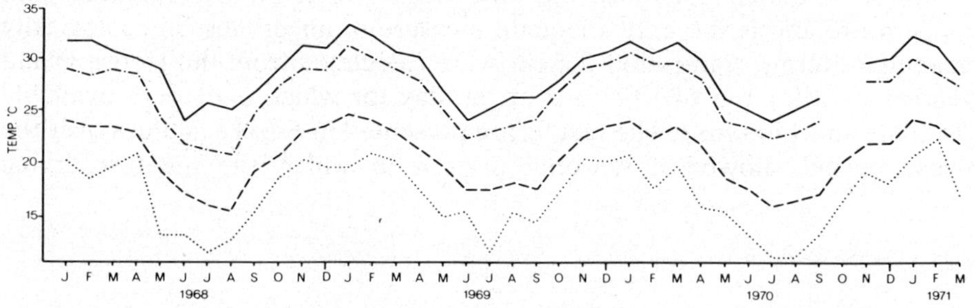

Fig. 10. Air temperature at Heron Island. Upper (solid) line indicates monthly maxima, second upper line the mean of daily maxima by months, the second bottom (dashed) line the mean of daily minima by months, and the bottom (dotted) line the monthly minima.

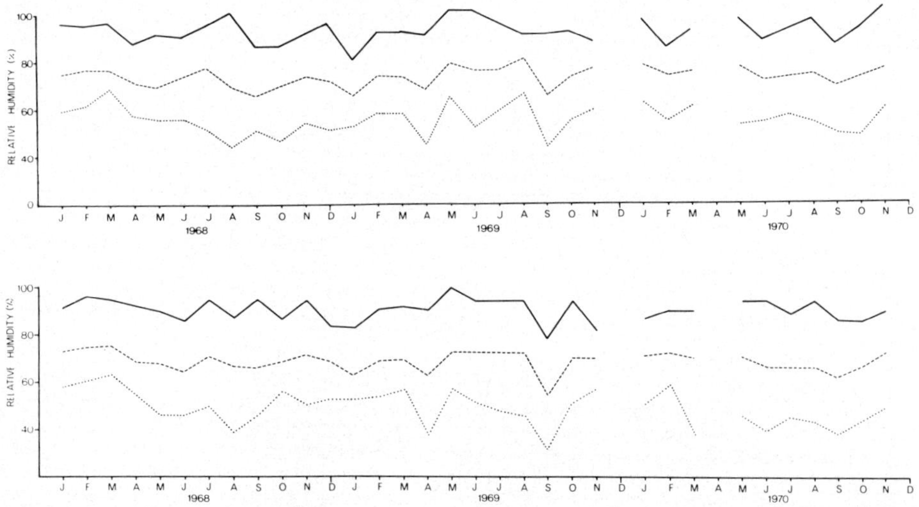

Fig. 11. Relative humidity of the air at Heron Island weather station. Upper: at 0900 hrs. Lower: at 1500 hrs. Solid line indicates monthly maxima, dotted line monthly minima and dashed line monthly mean.

Maximum-minimum temperatures recorded on One-Tree during one of the coolest periods for the general area (17–21 July 1968) were 14.5°–24° C in one of the *Pisonia* groves and 11.5°–25.5° C in the herb flat vegetation. Values from 13 to 17 May 1968 were 12.0°–29.0° C in the *Pisonia* except for one value of 43.0° C when the sun directly struck the maximum-minimum thermometer.

The relative humidity at the Heron Island station at 0900 hrs ranged from just above 40% to saturation, over a 3-year period. Mean values ranged from about 70% to 80%; seasonal variation was slight with no pronounced seasonal patterns (Fig. 11). Values at 1500 hours followed a similar pattern but in general were slightly lower.

Saturation deficit combines the influence of both relative humidity and temperature and is the only adequate measure of air dryness in ecologically meaningful terms. Saturation deficits were calculated from the Heron Island weather data for the two periods of the day for which data were available (0900 hrs and 1500 hrs). The results are presented in Fig. 12. Values over the 3-year period showed a seasonal pattern in which the greatest drying

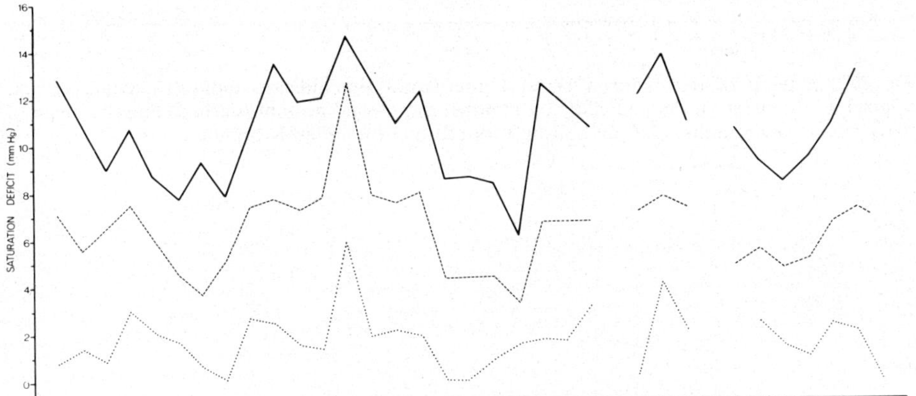

Fig. 12. Saturation deficit of the air at Heron Island weather station. Upper: at 0900 hrs. Lower: at 1500 hrs. Solid line indicates monthly maxima, dotted line monthly minima and dashed line monthly mean.

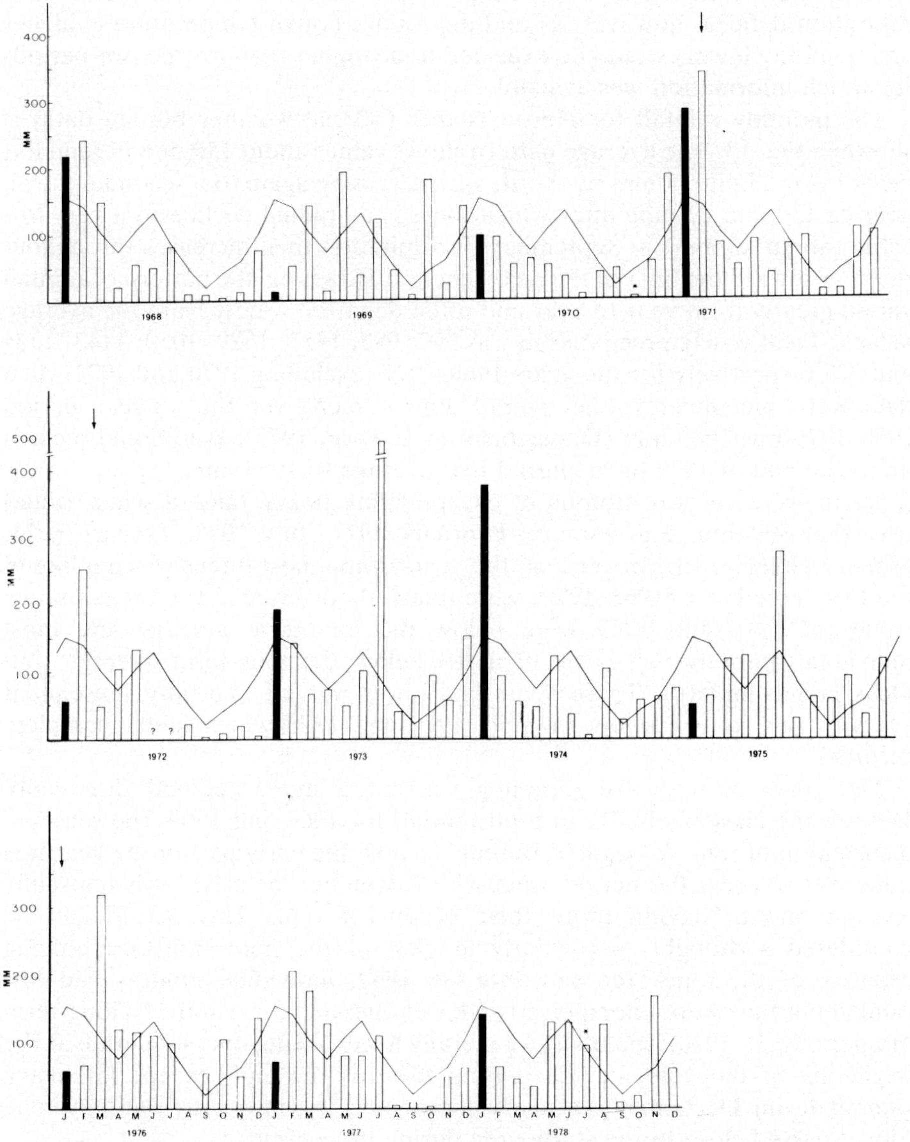

Fig. 13. Monthly rainfall at Heron Island, 1968–1978. Line indicates the average monthly rainfall for the period 1956–1974, taken from 'Climatic Averages Queensland' (Department of Science and Consumer Affairs Bureau of Meteorology 1975). *Indicates that rainfall data incomplete for that month. Arrows indicate severe storms. Values for January blacked in for easier reference.

potentials (highest saturation deficit) tended to occur during the austral summer months. Values tended to be lower (less potential for evaporation) at 0900 hrs than at 1500 hrs. The data presented do not indicate the maximum saturation deficits, however, as mid-day values (when temperature is higher and humidity lower) would be expected to be higher than for the two periods for which information was available.

The monthly rainfall for Heron Island[1] (official weather bureau data) is shown in Fig. 13. The average pattern shows values about 150 mm in January, declining to about 75 mm by April, and increasing again to a secondary peak nearing 130 mm in June after which there is a gradual decline to a very low value (about 20 mm) in September. Precipitation then increases throughout the year toward the high of the next January. However, the pattern of rainfall varied greatly from year to year and often departed widely from the average values. Total yearly precipitation was 972, 992, 1357, 1529, 1059, 1143, 1224 and 927 respectively for the years 1968–1977 (excluding 1970 and 1972 when data were incomplete); the overall annual mean for the 18-year period 1956–1974 was 1069 mm (Department of Science, 1975). Additional records up to the end of 1978 have altered that average to 1062 mm.

There were several months of exceptionally heavy rainfall when values exceeded 300 mm, e.g. January–February 1971, July 1973, January 1974, March 1976 (Fig. 13). In general, the years of the most intensive sampling of the One-Tree biota (1968–1970) were relatively dry years. The mean annual values of 1968 and 1969 were below the long-term average and most individual monthly values of 1970 fell below the long-term average (but October incomplete). These years had lower rainfall than any subsequent years of the study (except possibly 1972 and 1978 which had incomplete records).

The years of intensive sampling warrant a more detailed discussion. Despite the close similarity in total rainfall for 1968 and 1969, the seasonal distribution of rain was quite different. In 1968 the early part of the year was quite wet whereas the period August to November inclusive, was unusually dry, no month having more than 14 mm of rain. This period can be considered a drought, particularly in view of the poor moisture holding capacity of the One-Tree substrate. In 1969, individual months had low rainfall but they were interspersed with wet ones and there were no long-term dry periods. In 1970, rainfall was generally lower, being only moderate at the beginning of the year and decreasing towards the end, except for heavy rainfall during December. The early part of 1971 was unusually wet ('Cyclone Fiona' passed close to the study area during this period).

There were two other droughts during the period 1968–1978. One occurred

[1] Rainfall data not available from One-Tree Island during the period of intensive study. The values for One-Tree and Heron should be very close although some showers fell on one but not the other. Rainfall data from both islands, collected in 1978, showed close agreement (Pitman, personal communication).

in late 1972 and another from August through October 1977 (Fig. 13). It would appear that occasional droughts are a normal part of the environment of One-Tree Island; they occurred at 4 to 5-year intervals during the study.

Wind is an important ecological factor at One-Tree Island; it carries salt spray and organic materials (see Blanchard 1968), and especially during storms exerts damaging physical pressure. Some of the *Pisonia* groves exhibit very marked 'pruning' (Figs. 6, 7). It may also affect the immigration of arthropods to the island.

Wind direction is overwhelmingly from the southeastern quadrant (Fig. 14). No other quadrant even approaches that one in frequency of wind

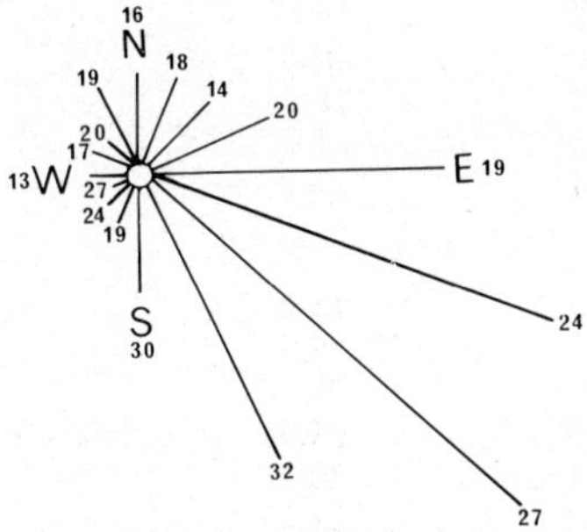

Fig. 14. Wind rose for Heron Island, 1970. Length of ray is proportional to frequency of winds from that direction. Numbers at end of ray indicates mean wind velocity (km/hr) from that direction.

direction, and the directions of rarest occurrences are those from the south-southwest to west-northwest.

Wind is rather constant. Of the 558 wind readings taken at Heron Island in 1970 (0900 and 1500 hrs daily for most days), 62 or 11.1% indicated calm conditions and 496 (88.9%) indicated wind of measureable velocity. During periods when the wind was blowing, the speed was usually brisk; monthly means for 1970 averaged 20.4 km/hr and ranged from 9.8 to 28.3 km/hr. There was little variation in conditions from one time of the year to another. This conclusion is also supported by the overall monthly means and ranges for 1968–1970. Values for all months were similar (Table 2). The strongest winds for a month varied from 28 to 61 km/hr (mean 43.9) and thus relatively strong winds occur at all times of the year. Southerly winds tend to be somewhat stronger than northerly ones (Fig. 14).

The above values refer to unexceptional weather conditions (no strong tropical storms in the vicinity in 1968–1970). However, tropical hurricanes or typhoons (called 'cyclones' in Australia) superimpose special conditions of heavy rain and high winds (mean wind speed of 63 km/hr or higher are part of the definition of such storm conditions).

Severe tropical storms in Australia generally increased in the time period

Table 2. Wind speeds at Heron Island, 1968–1970.

Year and month	Mean wind speed (km/hr)		Range in wind speed (km/hr)	
	0900 Hrs	1500 Hrs	0900 Hrs	1500 Hrs
1968				
January	12.0	16.5	4–28	0–37
February	15.4	17.4	4–46	2–37
March	14.3	15.9	0–28	4–37
April	12.6	9.8	2–43	0–39
May	21.8	18.7	9–43	7–44
June	26.8	21.5	0–48	6–52
July	28.3	23.7	6–46	0–46
August	20.5	16.5	0–44	0–56
September	17.0	19.8	0–52	4–37
October	15.9	18.3	0–41	4–32
November	15.5	21.3	0–43	0–37
December	19.1	22.2	2–61	2–33
1969				
January	14.6	15.9	0–35	2–32
February	22.2	27.0	0–39	2–46
March	22.4	24.6	7–32	7–46
April	25.9	23.5	13–46	0–37
May	24.1	20.2	0–52	0–46
June	26.8	21.7	11–46	0–46
July	26.5	27.0	9–50	13–48
August	22.8	19.8	0–44	0–37
September	25.4	18.0	0–57	0–59
October	18.1	19.1	0–46	0–41
November	11.7	14.4	0–28	0–32
December	20.4	—	11–33	—
1970				
January	18.1	18.3	0–33	0–41
February	20.0	23.7	0–41	0–46
March	19.4	20.7	0–59	0–46
April	25.2	—	0–50	—
May	27.8	22.8	0–50	0–48
June	18.7	16.7	0–46	0–52
July	26.3	23.7	0–50	0–43
August	22.2	17.2	0–52	0–39
September	23.5	20.2	0–56	0–37
October	22.4	20.4	9–54	9–46
November	23.1	23.1	0–46	0–59
December	19.1	—	0–37	—

1909–1975; the average incidence was 10 per decade in the general region of the southern Barrier Reef for the entire period, but was 15 per decade between 1959 and 1975. Most occur within the months of December through April although they can occur any time of year (Lourenz, 1977).

Several storms struck or came close to One-Tree Island in the course of the present study. 'Cyclone Fiona' passed near One-Tree on 22 February 1971 bringing winds of up to 102 km/hr and a high rainfall (Fig. 13). The direct effect of the wind on the island was not great. Many *Argusia* flowers on the eastern edge of the island were broken off and several branches of *Pisonia* were felled. On 2 April 1972, 'Cyclone Emily' neared One-Tree bringing winds of up to 129 km/hr. Part of the roofing was torn from a marine station building, small boats were sunk, a barge broke loose, and some experimental underwater 'reefs' destroyed; damage to the biota consisted of the killing of some shore birds and blowing leaves and some branches off of the trees and shrubs (Domm, pers. comm.). On 20 January 1976, 'Cyclone David' struck One-Tree Island. New rubble ridges were formed; the sea flooded over the eastern part of the island and reached a depth of about 75 cm around the building of the marine station. The sea water resulted in the death of some of the *Argusia* shrubs (Fig. 6) and caused mortality among young sea birds (Hulsman 1977a). The subsequent visit to the island in November of that year revealed no beach vegetation on the eastern beach, an area where *Trachymene* had previously occurred.

Other types of storms also occur. On 15 April 1975, a tornado (sea spout) crossed the northeastern arm of the island, loosened the roof of a building of the marine station and took a 10-metre patch of *Pisonia* trees and hurled it into the lagoon. The effect, though severe, was local, and did not influence the rest of the island.

The soils of One-Tree Island are a mixture of components derived principally from the island's rubble framework and its vegetation. With weathering, rubble acquires a black crust (Hedley 1925) and imparts the dark colour characteristic of the exposed surface on One-Tree Island. Rubble in the soil is light-coloured. The average depth of the litter layer was 8.4 cm (range = 5–13; N = 10) in the *Pandanus* grove and 2.2 cm (range = 1.0–3.5 cm; N = 10) in the *Pisonia* forest. Mean litter depth in the herb flat was 5.6 cm (range = 1.0–20.0; N = 81). Exploratory pits to 1 m showed a humus layer restricted to the top few cm, overlying unconsolidated rubble particles ranging in size from finger-diameter sticks to blocks 20–40 cm across (Fig. 15).

The top organic horizon was sampled systematically in the study of the soil arthropod community (see below, and Done 1975 for full details). Its depth varied in the three areas studied, as indicated in the frequency histogram (Fig. 16). In the herb flat the humus was usually in pockets less than 3 cm deep and rarely deeper than 5 cm. In the *Pisonia* grove there was a more continuous organic horizon which was rarely shallower than 5 cm. The organic horizon was intermediate under the succulents. The *Pandanus* grove (not studied quantitatively) had deeper accumulations of litter and humus than any other area.

A mechanical analysis of the organic horizon produced the three following fractions.

Fraction A: large coral fragments ('gravel') retained on a 2.4 mm sieve

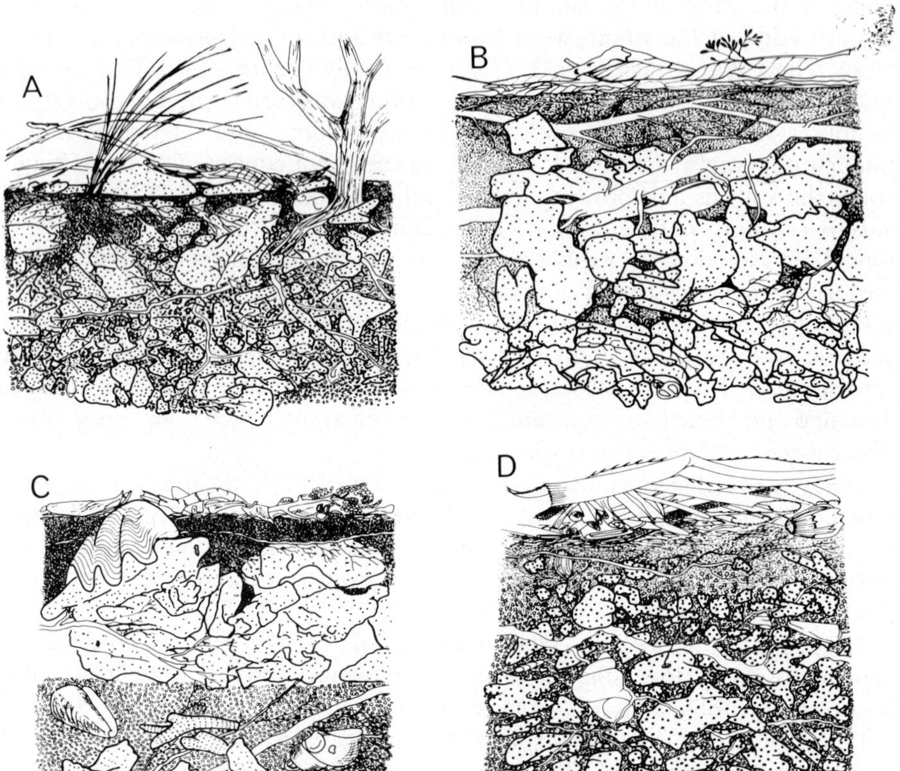

Fig. 15. Sketches of the upper layers of soil profiles in (A) the herb flat, (B) the *Pisonia* grove, (C) the *Argusia* area, (D) the *Pandanus* grove, (E) the succulent mat 3 m from pond and (F) the succulent mat 25 m from pond.

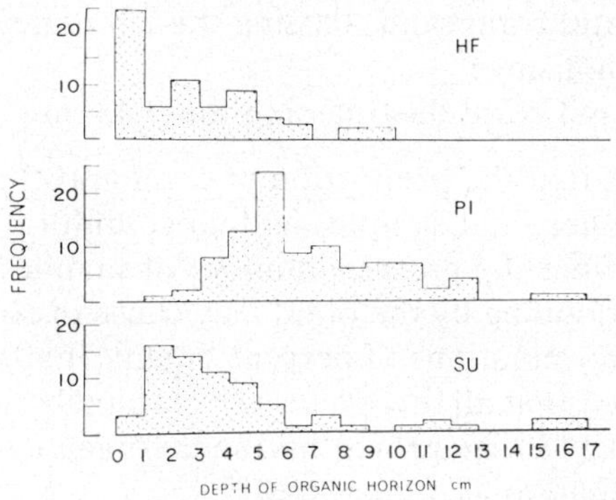

Fig. 16. Frequency of the depths of the organic horizon in the herb flat (HF), *Pisonia* grove (Pi), and succulent mat (Su).

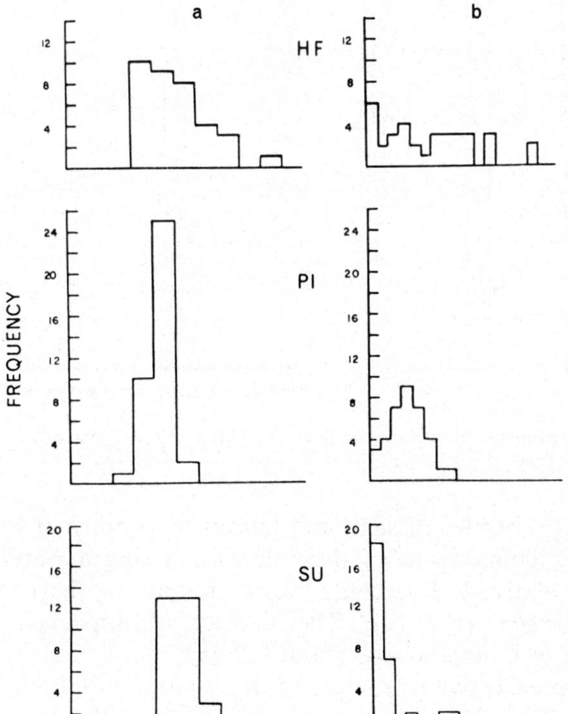

Fig. 17. Soil density (a) and per cent gravel (b) in soils of the herb flat (HF), *Pisonia* grove (Pi) and succulent mat (Su).

Fraction B: humus and coral sand, passing the 2.4 mm sieve but retained on a 0.43 mm sieve
Fraction C: humus and coral dust passing the 0.43 mm sieve.

Figure 17 indicates that the proportion of coral gravel tended to be highest and most variable in herb flat samples and most uniform in *Pisonia* samples. Under the succulent mat the greatest number of samples contained only 5% gravel. In this area, flooding by the pond may cause dissolution of the gravel.

Walkley-Black determinations of percent organic matter in fractions B and C combined, indicated that all three sites are extremely rich in organic matter (Fig. 18). Fraction A (*Pisonia* grove) had some organic matter that probably represented contamination and microbes.

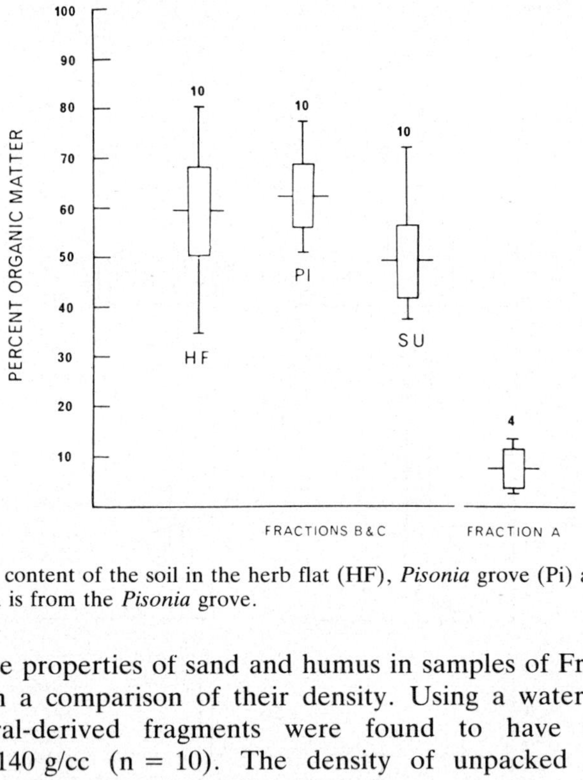

Fig. 18. Organic content of the soil in the herb flat (HF), *Pisonia* grove (Pi) and succulent mat (Su). Fraction A is from the *Pisonia* grove.

The relative properties of sand and humus in samples of Fraction B were inferred from a comparison of their density. Using a water displacement method, coral-derived fragments were found to have a density of 1.7142 ± 0.1140 g/cc ($n = 10$). The density of unpacked humus which appeared to be free of sand was about 0.2 g/cc.

When the three sites were compared, it was found that the herb flat had a higher content of coral sand (higher density) than did the *Pisonia* or succulent areas (Fig. 17). The samples from the succulent area were most uniform in amount of coral sand, with a mean value slightly higher than that of the *Pisonia* grove.

The foregoing data suggest that soil distribution and organic content are significantly more uniform in the *Pisonia* grove than in the other areas, a factor of considerable relevance to the resident soil arthropods (see Chapter 16).

Soil moisture was also considered primarily in relation to the soil arthropod communities. Soils whose water status is between field capacity and the permanent wilting percentage have an almost saturated soil atmosphere and evaporative losses from soil animals would be minimal. Free water can readily condense with a drop in temperature, but the atmosphere may become desiccating with a rise in temperature. In soils below the permanent wilting percentage, soil atmosphere is unsaturated and evaporation takes place; water cannot readily condense. Thus, in soils above the permanent wilting percentage free water for drinking is frequently available, below that point water intake must be principally from the food.

The field capacity and permanent witling percentage was determined for pooled samples from 10 points at each site, using a pressure membrane apparatus at pF 2.4 (1/3 atmosphere) and pF 4.2 (15 atmospheres) respectively.

A striking feature of the data is the high level of soil moisture content at which the permanent wilting percentage occurs (Table 3). Even in the driest sample containing organic matter (herb flat, Fraction B), the amount of water present at the permanent wilting percentage is equivalent to 55% by weight.

Table 3. Mean Field Capacity (FC) and Permanent Wilting Percentage (PWP) of soils from three habitats on One-Tree Island. All figures are of soil moisture in % of oven dry weight. Each mean based on two replicates each from samples pooled from ten cores. Fraction A measured only in the *Pisonia* grove but should be similar in all areas.

Location	Fraction A		Fraction B		Fraction C	
	FC	PWP	FC	PWP	FC	PWP
Herb flat	—	—	62.52	54.77	121.64	58.53
Pisonia grove	58.52	52.69	98.77	83.44	148.60	91.81
Succulent mat	—	—	116.29	102.31	120.21	65.51

Some of this hygroscopic water is presumably contained within organic fragments and is thus available to the detritivores which eat them.

Usually, gravels and sands contain only a relatively small amount of capillary water. However, the coral gravel (Fraction A) of the present study contained almost as much water at the permanent wilting percentage as it did at field capacity. The relatively high levels at both points are undoubtedly due to the chalky nature of coral fragments. Most of the water in coral sand is hygroscopic and thus unavailable to arthropods.

Fraction C is the most highly organic fraction and shows the greatest difference in soil moisture between field capacity and permanent wilting percentage, i.e., it contains the most capillary water, some of which would be available to arthropods. Fraction B, as a mixture of sand and organic matter, has intermediate properties; values both of field capacity and permanent wilting percentage lie between those for Fractions A and C.

During sampling for soil arthropods, soil moisture determinations were made on the cores from which animals were extracted. The raw values were corrected to the amounts of water in Fractions B and C combined, assuming Fraction A contained 55% of water by weight.

In both the herb flat and *Pisonia* grove the soil was often drier than the permanent wilting percentage (Fig. 19). Thus, soil air spaces are usually unsaturated, water will not condense readily, and arthropods must depend upon water from food to balance their water losses. This was surprising for the relatively shaded *Pisonia* grove. However, because water content at the permanent wilting percentage is so high there, animals living in organic soils should not go short of water. The plants did not usually wilt, as their roots penetrated to deeper moisture horizons.

The fact that the proportion of coral particles to organic matter is usually higher and more variable in the herb flat than in the *Pisonia* grove means that there will also be greater variability in soil water status at a given time in the former area. The soil of the herb flat is more likely to be exposed to sudden temperature changes which initiate condensation or desiccation (see below). All of these factors combine to make the herb flat a much more variable environment, in space and time, than the *Pisonia* grove, a difference important to the resident arthropod population.

In the central succulent area the slope is 1 in 35 near the brackish pond and 1 in 15 peripherally. There is a marked gradient of moisture (and presumably salinity)[2] particularly within the first 10 metres from the pond. From 10 to 30 metres the gradient is less marked. Samples close to the pond are frequently many times in excess of field capacity, whereas this is unusual beyond 15 metres (Fig. 19). Soils from within 5 metres of the pond rarely fall below the permanent wilting percentage but those beyond 5 metres frequently do.

The four open points in Fig. 19 mark samples which were taken the day following flooding of the area up to the 45 cm contour. Such floods occur only twice annually under the influence of the highest spring tides, and must represent important ecological events.

The thermal environments in several soil microhabitats were periodically monitored by Grant Multipoint thermistors and recorded during visits to One-Tree.

The microclimate near the soil was more extreme and fluctuated more widely in the succulent area and herb flat than in the more shaded *Pisonia* groves (Fig. 20). On a hot day air temperatures 5 cm above the ground peaked at about 37° C at mid-day in the succulent area but only just above 28° C in the *Pisonia*, with respective minima being 8–13° C and 11–15° C. Litter temperatures peaked at slightly lower values but peaks were similar for the succulent and herb flat, both of which were about 6° C higher than peaks in the *Pisonia*. Similar relationships held at increasing depths in the soil

[2] Pitman (personal communication) indicates that salt levels in the soil (distant from the succulent area but near the beach) on One-Tree were quite low.

except that (1) oscillations were generally damped, the magnitude of differences between sites accordingly being reduced, and (2) the herb flat had lower maxima than the succulent area. Minima in all areas did not differ greatly from those of the litter but maxima were lower. At a depth of 4 cm the total span from maximum to minimum in all areas was only about 7 centrigrade degrees (Fig. 20).

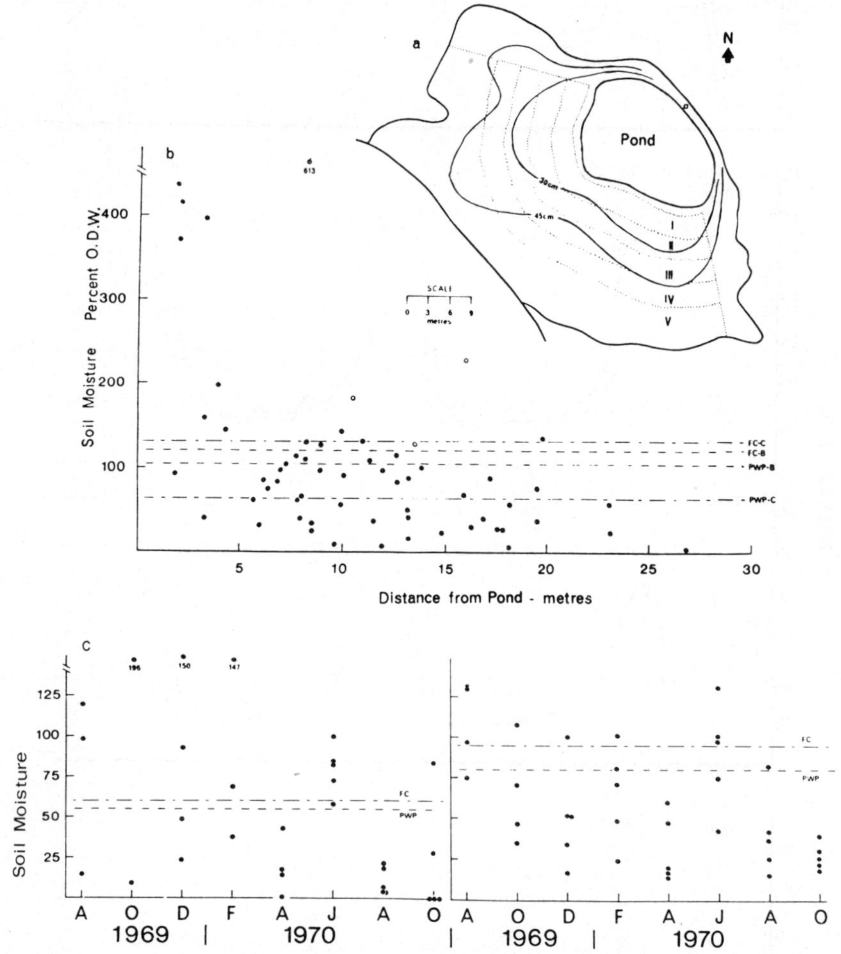

Fig. 19. Map of the pond and succulent mat showing: (a) sampling areas I–V; (b) soil moisture content of soil at different distances from the pond. The 30 and 45 cm contour lines of vertical height above pond indicate a slope of 1 in 35 and 1 in 15 respectively. The dotted lines indicate the limits of concentric zones from which humus cores were taken for arthropod extractions and soil-moisture determinations. FC–C and FC–B refer to the field capacity of fractions C and B respectively and PWP–C and PWP–B to the permanent wilting percentage of those fractions. (C) Soil moisture in % dry weight of the herb flat (left) and *Pisonia* grove (right) at different sampling times. FC and PWP refer to fraction B.

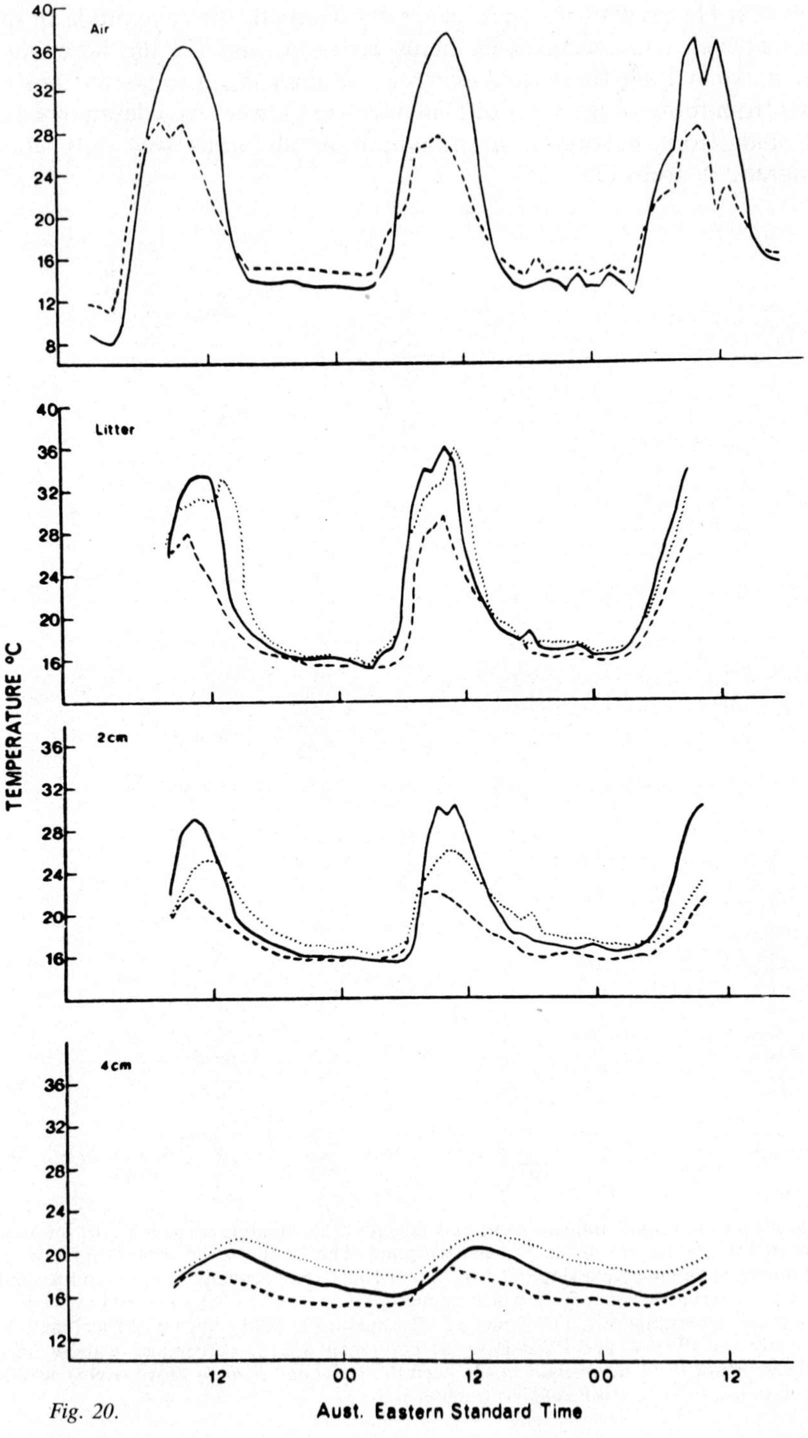
Air
Litter
2cm
4cm
TEMPERATURE °C
40
36
32
28
24
20
16
12
8
12
00
12
00
12
Aust. Eastern Standard Time

Fig. 20.

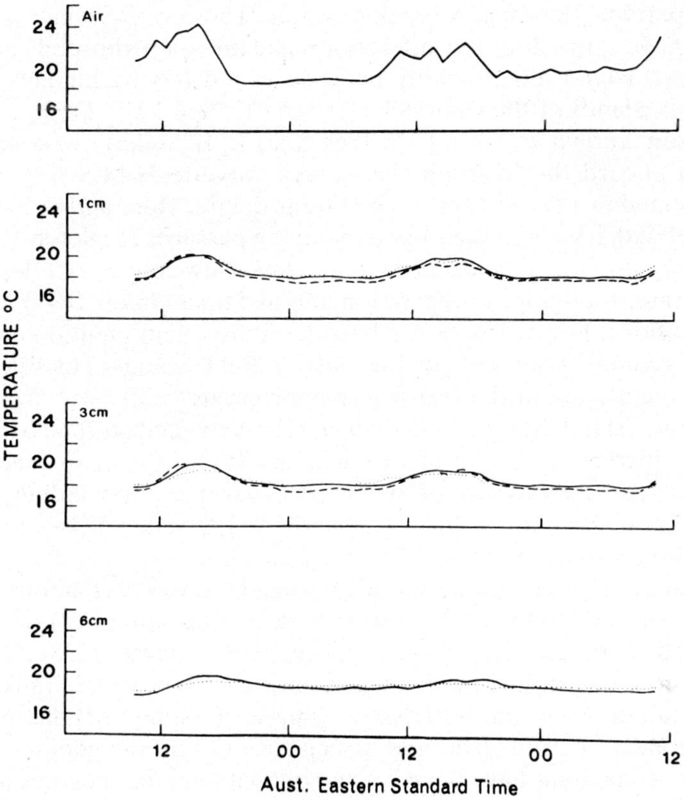

Fig. 20. Thermal microclimate at One-Tree Island on 5–7 October 1970 (left) in the air 5 cm above the ground, in the litter, and 2 cm and 4 cm deep in the soil, and on 26–27 May 1970 (right) in the air 5 cm above the soil, and 1 cm, 3 cm and 6 cm deep in the soil. Dotted line indicates the herb flat, dashed line the *Pisonia* grove, and solid line the succulent mat.

On a cool day oscillations were greatly damped, the span of air temperatures 5 cm above the succulent surface only being slightly more than 6 centigrade degrees. Oscillations progressively decreased with greater soil depth, at 6 cm amounting to only about 2 centigrade degrees. In the soil, temperatures seldom differed among sites by more than 1 centrigrade degree (Fig. 20).

In summary, the various habitats differ in the degree and frequency to which they are subjected to desiccation and high temperature. The herb flat has soils that are heterogeneous spatially and variable temporally, and are often subject to fluctuations of temperature and moisture that reach levels likely to be unfavourable to a wide range of soil arthropods. The soil under the *Pisonia* grove is less variable and reaches unfavourable levels less frequently. The succulent area has a gradient of environmental conditions away from the edge of the pond, seasonal stress occurring near the pond as a result of periodic flooding by brackish water. These various environments are likely to have quite different effects on resident soil arthropods and plants.

One-Tree Island has probably been disturbed less by humans than have most other islands of the Capricorn Group (McNeil 1955, Domm 1971). The first person known to visit One-Tree was J. B. Jukes, who served as a naturalist aboard the hydrographic survey corvette H.M.S. Fly, which was commissioned in 1841 to explore the Great Barrier Reef and to look for gaps in the reef with a view to their use as shipping passages (Ingleton 1944). Jukes went ashore on One-Tree on 11 January 1843 and wrote a brief description of the substrate, topography, vegetation and avifauna (Jukes 1847) which, with a few additions, has served as the basis for subsequent mention of One-Tree Island in general treatments of the Barrier Reef Islands. His brief observations are mentioned in the text where appropriate.

One-Tree Island has remained in a relatively undisturbed state, a fact which has increased its appeal to scientists. It is difficult to reach by boat because of the shallowness of the water over the surrounding reef. This difficulty of access and its lack of a sandy beach makes it less attractive to bathers and campers than other nearby islands.

The first attempt at a biotic survey of One-Tree was carried out in 1927 by MacGillivray and Rodway (1931) who made a collection of the plants; Kemp constructed a rough vegetation map in 1936 (Steers 1938). The island attracted little scientific attention again until 1965 when Dr. Frank Talbot of the Australian Museum initiated a long-term study of the ecology of One-Tree Reef. This project, and associated ones on the geology and water chemistry of the area, have been carried out with varying degrees of intensity up to the present time. A general account of all aspects of the research carried out at One-Tree is provided by Heatwole (1981).

The island is currently leased by the University of Sydney and is reserved exclusively for scientific research. A small building was erected in 1971 and a resident staff of two people has been present nearly continuously from that time until the present. These developments are likely to result in considerable alteration in the terrestrial ecology of the island; some changes have already taken place (e.g. roaches became established and by 1976 had become extremely abundant) and are mentioned in the text where appropriate. However, except for brief visits in 1972, 1976 and 1978, the present project was carried out primarily between 1968 and early 1971, before buildings or permanent residents were present and before human disturbance had any major impact. Attempts were made to minimize as much as possible our own effect on the island while carrying out our study (see next Chapter and Appendix I).

7. Materials and methods

General procedures

Investigators may exert a considerable direct influence on the community they are studying, particularly on a small, circumscribed area such as One-Tree Island, where total plant and animal populations are small. Consequently, special precautions were taken to influence the island as little as possible.

Potential extraneous energy inputs were counteracted by exercising care in food transport and preparation, and in disposal of garbage and sewage. Human-induced mortality of island biota (other than that occasioned by sampling) was minimized. As a guide to future investigators of island communities, these procedures are expanded upon in detail in Appendix I. We believe our precautions were effective although some minor exceptions are noted in the Appendix. The known and assumed effects of other visitors to the island are also noted.

For the purposes of this study, One-Tree Island was visited from 1968 to 1978. The island was chosen on the basis of a visit by the senior author in May 1968 and the first sampling took place in that year. Routine sampling took place at 2-monthly intervals during 1969 and 1970 and into early 1971, when the authors, usually accompanied by two volunteer assistants, spent about 1 week on the island at each visit. Brief visits for specific purposes or as a 'follow-up' to the more intensive periods of the study occurred in April 1972, November 1976, and September 1978.

Sampling procedures were standardized as far as possible during the study. However, refinements and, in particular, expansion of sampling took place well into 1969. Routine sampling of soil fauna began in August, 1969.

There is a dilemma associated with long-term studies of a small area. On the one hand, if sufficient samples are taken for statistical treatment to be carried out, such as calculation of confidence limits, the study must be a short-term one, otherwise the habitat is sampled away and the sampling itself becomes an important influence. On the other hand, if sample sizes are kept low in order to permit long-term study, confidence in the data is reduced.

We have tried to compromise between these two extremes. We have used rather small sample sizes and have consequently sacrificed some precision in order to be able to continue our work over several years. We have attempted to compensate by using two independent methods for estimation of as many of the parameters as possible. When feasible, a non-destructive method was selected to complement a more destructive one. As an alternative to detailed statistical testing, degree of agreement between methods was used to assess probable validity of the results. Laughlin (1976) has stressed the increased reliability of conclusions based on two independent approaches.

Cage plot sampling

Ground vegetation and its associated arthropods were sampled using a removal method. We used a sampler consisting of a nylon mesh sleeve surrounding a cylindrical frame, 70 cm in diameter and 100 cm high, made of wooden dowels and bamboo hoops. The sleeve (mesh size 0.5 mm) extended above the frame and could be closed with a draw string (Fig. 21). The investigator inserted his arms and upper torso into the samples and an assistant pulled tight the drawstring of the extended sleeve, around his torso. The sampler was then placed quickly over the area to be sampled, and pressed down. A pair of secateurs was used to cut the vegetation straddled by the outer edge of the sampler, permitting the cage to settle firmly on the ground. From inside the sampler all insects seen on the vegetation or nylon mesh of the sampler were collected with an aspirator usually containing a small amount of alcohol. A releveé (see below) of the enclosed vegetation was then made, and vegetation height measured, an assistant outside the sampler recording the data. Plants were then collected, one species at a time. Each leaf was pulled off and examined for insects and placed in a plastic bag. Stems, fruits, flowers and buds, and roots were examined and separately bagged by species. Some large, exceptionally leafy plants were examined for insects, shaken vigorously and then sealed in a plastic bag to be examined more thoroughly for insects later when leaves, stems and roots were separated inside an insect-proof tent. Dead standing stems were examined in the sampler for insects and then bagged. Finally, depth of litter was measured and the litter collected into a plastic bag. All bags were sealed and a final check made for insects on the mesh of the sampler before it was removed from the plot. The depth of the humus was then measured and pieces of coral lifted and notes made on abundance and type of macroscopic fauna (e.g. ants, silverfish, etc.) under them.

At each visit to the island 10 such plots, at 5 m intervals along a north-south

Fig. 21. Using the sampler for cage-plot sampling on One-Tree Island.

line across the herb flat were processed. Each trip's transect line was parallel to, and west of, the preceding line.

For biomass determinations, the plant materials collected from the removal plots were transported to Heron Island or Armidale and oven dried at 60°–65° C to constant weight. All dry weights referred to in this paper resulted from this procedure.

Sweep sampling

Epigeal arthropods were also sampled in the herb flat by the quantitative sweeping method of Menhinick (1963). This method involves sweeping an area a number of consecutive times with an insect net, each time covering the entire area with non-overlapping sweeps. The numbers of insects captured in the consecutive sweeps are plotted against the total number of insects previously captured, and the total present in the area is estimated by extrapolating the resulting curve to the x-axis. There are problems with this technique for some shapes of curves (see results section) but in general it proved to be successful in the present study. At each 2-monthly visit, an area 10 m × 10 m was swept 10 consecutive times. A steel-framed insect net with a nylon bag of 0.5 mm mesh was used. The sweeping damages vegetation so each plot swept was treated in all following trips as though it had been subjected to removal sampling and it was not subsequently re-sampled by any method.

Releveés

The destructive cage-plot samples mentioned above were necessarily small in number and in size because of the limited size of the island and the long duration of the study. To ascertain whether such small numbers of samples could provide reliable data, it was also necessary to assess the homogeneity of the vegetation by a nondestructive technique. Seven transect lines were designated which crossed the herb flat in different directions and traversed ecotonal boundaries between the herb flat and each other vegetation type on the island (Fig. 9); they were located by metal stakes and/or conspicuous landmarks at each end. At 5 m intervals along them, releveés of 1 m^2 quadrats were analyzed for the following characteristics: height of live vegetation, height of dead standing stems, numbers of buds, flowers and fruits of each species, and estimates of total plant cover and per cent cover of each species present. Occasionally other observations were made and are discussed under appropriate results sections.

By running these transects throughout the study, it was possible to assess temporal changes in heterogeneity and to describe the phenology of the various species. The releveés also provided a continual check on the biomass sampling as they constituted an independent measure of the change in relative abundances of the various species of plants.

Phenology of flowering and fruiting

The numbers of flowers and fruits of low plant species on the island were estimated by counting their numbers in the releveés described above and adjusting to an areal basis for each vegetation type (i.e., herb flat, succulent mat). The *Argusia* inflorescences were censused directly, separate tallies being made of these containing buds, buds and flowers, flowers, flowers and fruits, and fruits. Direct counts were made for the entire island by people walking around each tree, but the results are probably underestimates as it was difficult to see all flower-heads on the tops of the taller shrubs; buds were probably missed more than other categories. This source of error would not affect the seasonal patterns which were observed, however.

Pisonia inflorescences were also censused sporadically during the study. Those on the tops of the tree clumps could not be easily seen and would be missed during counting. However, this error is not as serious as would first appear, as the tops of the clumps were wind-pruned and fruits were only produced where there was protection from the wind, yet exposure to sun. Most clumps lacked fruits on the top and wind-pruned sides and under the shaded canopy, but carried them in abundance on the lee edge. Direct counts of all *Pandanus* fruits were made each trip.

Leaf and new stem production

Ten growing twigs each of *Pisonia* and *Argusia* were individually labelled with a numbered tag. All old leaf scars were painted with plastic non-toxic paint. The leaves on each twig were then counted and recorded. On subsequent visits, the number of leaves which had fallen in the intervening period was ascertained by counting the unpainted leaf scars terminal to the painted series. Similarly, the number of new leaves produced during the period was obtained by subtracting the previous number of leaves on the twig from the sum of current leaves plus new leaf scars. New leaf scars were then painted with a different colour and the process repeated on each successive visit. The increase in twig length between visits was determined by measuring the distance between the apex of the twig and a reference paint mark.

A slight modification in the above technique was necessary for *Pandanus* because of its different growth form. On each of ten trees, one branch was individually numbered and labelled. The leaves of the first parastichy were each pierced and string through the hole used to attach a numbered tag to each leaf. Number of leaves becoming dry and/or dropping between visits to the island was ascertained by noting the tag number of the lowest basal green leaf on the parastichy. As there are three separate parastichies in each branch, the total number of leaves lost was approximately three times the number of labelled ones lost. Number of new leaves produced per parastichy was obtained by observing the number of green leaves in the parastichy terminal to the previously labelled ones. This value was tripled to indicate total production for the branch.

Litter fall

In addition to the estimation of leaf litter fall from marked leaves of the tree and shrub species discussed above, an independent method of estimating litter production was used. Shallow leaf-trap trays were constructed from plastic window screen (1.5 mm mesh) and placed on the ground under the canopies of the different species of trees and shrubs. Three large trays (60 × 145 cm with sides 8 cm high) were placed in the *Pandanus* grove and two of the same dimensions in the *Pisonia* forest on the western edge of the island. Smaller trays (35 × 60 cm with sides 8 cm high) were placed under individual shrubs, five under *Argusia* shrubs and one under a *Scaevola* shrub. Five additional trays of the smaller size were sited under the vegetation layer in the herb flat, at 5 m intervals along a north-south transect line leading south from the *Pandanus* grove. At each visit to the island, all plant matter was collected from the trays, sorted into leaves, sticks and fruits, and then oven-dried and weighed. Any object lying across the edge of the tray was cut and only the portion on the inside included in the sample.

The leaf traps permitted good drainage and collection of samples was frequent (every two months); no evidence of rotting or destruction by detritus feeders was noted in the samples although some breakdown of materials may have occurred. Any departures from the above procedures for a given trip are indicated with the results.

Sampling arthropods associated with *Argusia* flowers

Arthropods in the herb flat were sampled by the removal plot and sweep methods. In addition there were a large number of nectarivores and other arthropods associated with *Argusia* flowers. These were sampled separately. Two independent methods were used.

The first procedure involved marking inflorescences and counting the numbers of each species of arthropod on them at hourly intervals over a 24-hour period. Numbers of individuals were plotted against time and activity peaks obtained for each species. It was assumed that at the activity peak for a given species, all individuals of that species were actively visiting flowers. An estimate of the total numbers on the island was obtained by multiplying the number of individuals of each species per inflorescence at that species' activity peak by the total number of inflorescences.

The second method was the 'plot removal' method of Hanson (1968). Each 'plot' consisted of an *Argusia* flower head. A total of 23–35 inflorescences were marked and individually numbered. The marked sample consisted of buds, half-buds/half-flowers, flowers, half-flowers/half-fruit, and fruit, each represented in approximately the proportion as they were recorded in the whole island census. The sample on each bimonthly visit was located in trees in the northwestern corner of the island.

Each plot was examined carefully twice daily, as close as possible to 0900 and 1600 hours, respectively. Each time a species was found on a plot on which it had not previously been recorded, it was counted and recorded for that plot at that time. Once a particular species had been found on a plot, it was ignored in future censuses of that plot. Sampling continued for at least 2 days. The population estimate for each species was determined using Hanson's formula:

$$\hat{K} = X_1 \Big/ \left[1 - \left(\sum_{i=2}^{n} X_i \Big/ \sum_{i=1}^{n-1} X_i \right)^{\frac{1}{2}} \right]$$

where $\hat{K}$ is the estimate of population on the marked inflorescences, X_i is the total number of individuals sighted at time i, X_1 is the number sighted during the first period, and n is the number of times the plots were censused; in this study n is the number of times after and including the time a particular species was first recorded.

This formula is based on the assumption that the probability of finding an individual on a plot is unknown and not constant. The formula is not appropriate where $X_n > X_1$ or where a species occurred only once. In these cases, $\hat{K}$ was set equal to $\sum_{i=1}^{n} X_i$.

The whole island estimate for each species was the product of the sample estimate and the proportionality constant relating total inflorescences on the island to the number of inflorescences in the sample. The sample was thus assumed to provide representative estimates for all arthropod populations concerned. An unknown level of variation existed due to microhabitat differences of *Argusia* on different parts of the island.

Sampling foliage fauna

The fauna inhabiting the foliage of *Argusia* and *Pisonia* was sampled by examining 10 quadrats (each 1 m^2) of foliage. The canopy of both species on One-Tree is very thin, averaging about 20 cm in depth for *Argusia* and 70 cm for *Pisonia*. Thus a given quadrat represented a volume of about 0.2 m^3 for the former and 0.7 m^3 for the latter. The quadrats were located at 5 m intervals along a line at waist height along the *Argusia* shrub ring or along the edge of *Pisonia* clumps. The quadrat was delimited by a meter stick and then quickly examined for flying insects leaving the area. These were counted by species and then the leaves of each twig in the quadrat examined carefully for insects and spiders, separate tallies being kept for each species. Finally the leafy twigs were counted and the thickness of the canopy measured. The dimension and height of each individual or clump of trees and shrubs were measured and the shape and canopy form noted. The total canopy area was then calculated and used to convert values per m^2 to totals for the entire island.

Ant baiting

On the last visit to the Island, September 1978, after it had been occupied by humans for some time and the stringent precautions taken earlier in the study not to disturb the terrestrial community were no longer in force, a series of sardine baits were laid for ants in order to get a better assessment of the relative importance of this group. The baits were placed at one-metre intervals along a transect line through the herb flat. Baits were laid at 0810 hours on 16 September 1978, and were checked at 1030 hours, 1510 hours, and 2020 hours on the same date and at 0830 hours on 17 September 1978. The transect was dominated by *Melanthera* with *Abutilon, Ipomoea, Achyranthes* and *Plumbago* also being present. The vegetation was 50–80 cm high. On the day of sampling the sky was cloudless and there was a brisk southeast breeze.

Four bait stations were laid out in the succulent mat, one at the edge of the pond, one near the centre of the mat in the open, another nearby but under a piece of coral, and the fourth near the ecotone between the succulents and herb flat. Under the *Pisonia* forest four baits were laid on the ground, two one metre up on the trunks of trees, and one in the canopy. In the *Argusia* two baits were laid on the ground and two one metre up in shrubs. These stations were checked at the same time as those in the herb flat. In all cases baits were replaced when necessary.

Sampling soil arthropods

Soil arthropods were sampled quantitatively by the conventional manner of removing a soil sample and extracting the arthropods from it using a heat gradient. Each soil sample was taken with a soil-corer constructed from 50 mm diameter polyvinylchloride tubing with 2 mm nylon mesh gauze and 1 mm cheese cloth at its upper end, and its lower end filed sharp for insertion into the ground. Preliminary studies had indicated that samples from the upper 5 centimetres of the organic horizon maximized the estimates of abundance and species richness of soil arthropods and also provided a consistent and convenient comparative method. Samples from the underlying coral rubble were very poor in soil arthropods.

Three habitats were studied, the herb flat, the southern *Pisonia* grove and the succulent mat. Each habitat was divided into 5 sub-areas (e.g. see Fig. 19a). On each visit, random map coordinates were chosen to locate one sampling site within each area. The approximate location was found in the field and a sample taken where the soil-corer fell when thrown over the shoulder. First the litter from a circle 10 cm in diameter was collected and placed in a plastic bag, and was used later for extraction of animals and determination of oven dry weight. The soil-corer was then inserted into the soil laid bare by the removal of litter, dug out using a trowel, and closed at its open end with a lid, labelled, wrapped tightly in a plastic bag and placed in a carrying bag in the shade. Where the corer was obstructed by chunks of coral, a second site less than 1 m from the first was chosen. Since it was desirable to get the core into the extractor as soon as possible, sampling was always done on the last day of each visit to the island.

There were eight visits to the island (Aug., Oct., Dec., 1969; Feb., March, May, Aug., Dec., 1970) during which soil arthropods were sampled, which with three areas each with five sub-areas gives a total of 120 samples (reduced to 117 because 3 samples were spoilt or lost) included in the analysis.[1]

Extraction of arthropods from the cores was carried out at the laboratory at Heron Island, using a modified Macfadyen high-gradient cylinder apparatus, taking 50 simultaneous cores (Macfadyen 1962; for modification see Done, 1975). The extraction took place over 4 days. The heat was left off for the first few hours, a practice recommended by Macfadyen (1962). Surface temperature was increased from about 20° to 45° C over the 4 days, whereas the temperature at the bottom of the core increased from 20° to 40° C.

In the laboratory the samples were cleaned, the animals sorted to species with the aid of a micro-aspirator, cleared in lactic acid when necessary, dimensions of the body (without appendages) and life history stages noted (larvae, small nymphs, large nymphs, adults) and the data recorded on computer coding sheets. For the purpose of estimating biomass, all animals were assumed to be ellipsoidal in shape; their volume was calculated by the formula:

$$V = \frac{4}{3}\pi \frac{L^1 1^1 1_2}{8}$$

where L is the length of the animal, 1_1 the breadth and 1_2 its depth. Assuming a density of 1 for live arthropod tissue, the volume could be directly converted to weight for making biomass estimates.

Based on the literature, species were assigned a trophic designation as primary consumer, carnivore, omnivore, or unknown. The methods and apparatus, computer techniques, statistical procedures, and the difficulties in measuring and assigning life history stages and trophic levels are described in full detail by Done (1975).

For each sampling period, a classification of the samples was undertaken to assess both the similarities between them, and in particular, whether or not samples grouped according to the areas from which they were taken. Details of the classificatory procedure are as follows: raw data were subjected to a log (n + 1) transformation and standardized as a proportion of 2_x variance (after Burr 1968). Seventy-six species were used as attributes in calculation of Lance

[1] Actually, 800 samples were collected and extracted but as each one required 6–20 hours for sorting, counting and measuring, the additional 680 have been preserved but not examined further. They are available to anyone who would like to examine them.

and Williams' (1967) Canberra Metric dissimilarity index; double zero matches were excluded. Samples were classified by using the group average strategy in the TAXAN programme of Burr (1968) and dendrograms were prepared.

For readers wishing to explore these techniques in more depth, a detailed discussion of the methods and procedures, including the rationale behind them and their strengths and weaknesses, has been provided by Done (1975).

Leaf damage

Proportion of the standing crop of low vegetation consumed by insect herbivores was estimated by tracing onto paper the outline of the damaged portions of each leaf from the ten cage plots described above. The total damage for each species in each plot was cut out and weighed. A large piece of identical paper of known area was also weighed and the result used to convert the weight of the paper cut-outs representing damage, to surface area. Complete outlines of samples of the uneaten parts of leaves were also traced and the surface area calculated in the same way as for damaged parts. The actual leaves were then oven-dried and weighed, and the leaf weight per unit surface area ascertained. This figure was used to convert the area of leaf damage to weight eaten by insects. It was also used to calculate biomass data from measured surface areas, or vice versa, in other contexts. Inasmuch as new leaves might have a different weight to surface area ratio from older ones (seasonal differences might affect leaf thickness etc.), this calibration procedure was carried out during each sampling period.

The *Argusia* and *Pisonia* insect damage was measured by a different method. Yellow leaves about to absciss were considered as having already sustained all the herbivore damage they would ever suffer. Hence, measurement of the damage to these leaves would provide a good estimate of the total damage by herbivores to a leaf during its life. A meter stick was used to measure off 1 m intervals at waist height along the line of *Argusia* shrubs on the eastern beach and around the western clump of *Pisonia*. In each case the proximal yellow leaf on the twig nearest the end of the meter stick was collected and the surface area of eaten and uneaten parts measured, and oven dry weight of the leaf, determined. These data combined with those on leaf longevity and turnover permitted a calculation of the rate of leaf consumption by herbivores. Supplementary data on *Pisonia* leaves were obtained from treating the 'canopy leaf samples' in the same manner described for the plants of the cage plots in the herb flat.

A similar approach was used on *Pandanus* except that leaves were taken at 5 m intervals. In each case the terminal (i.e., most recently dried) brown leaf on the whorl was collected and damage assessed as indicated for the previously described samples.

Sample sizes were 50 for *Argusia*, 25 for *Pisonia* and 10 for *Pandanus*. The last was sampled only infrequently.

Inherent in the methods used are several sources of error which have been reviewed by Nielsen (1978). First of all the apparent loss of leaf surface area to chewing insects includes not only the leafy material consumed but that wasted by insects during feeding; the latter may be as much as 20% of the total amount removed. There is often necrosis of the leaf tissue along the border of feeding tracks with a concomitant increase in size of the hole. Finally, the size of the hole may increase with further leaf growth, the extent of this effect varying with position of the hole in the leaf. Obviously if most grazing takes place during early growth of the leaf, the error due to the growth-effect is much larger than if it occurs primarily after leaves have nearly reached their final size. We do not know at what stage of leaf growth most grazing occurred for the various plants on One-Tree Island, nor do we know if there are interspecific differences in pattern or extent of hole-enlargement during growth. Furthermore, we do not know the amount of herbivore wastage or the extent of necrosis following attack. All of these factors tend to overestimate the amount of material eaten by insects and since we cannot correct for it, the reader should be aware that our values are probably overestimates and the extent of overestimation may vary among species. Nielsen (1978) suggested that leaf hole area may give leaf consumption values 2–3 times greater than actual values and applied an average correction of 54% to his own data. However, if the leaf-hole areas are looked upon from the standpoint of the plant they are probably much more realistic as they indicate extent of effective leaf area loss. The decrease in surface area between damaged leaves and whole ones represent loss in potential photosynthesizing ability, regardless of whether that decrease is due directly to consumption or wastage of leafy material by insects, or indirectly to subsequent necrosis or hole expansion during growth.

Special sampling techniques and general observations

In addition to the quantitative sampling described above a number of qualitative or semi-quantitative observations were routinely made. The island's perimeter was walked each trip and a census made of items washed up on the beach. The areas in which birds were nesting or roosting were mapped and counts made of each species at various times of day.

Notes were made on the presence of any insect species not included in the quantitative sampling. All trophic interactions which were observed were recorded as were various observations on life history and reproduction. When immature and adult stages had different feeding habits, the trophic level of the stage believed to have the greatest impact on the community was usually the trophic designation assigned. For example, hymenopterans which have parasitic entomophagous larvae were listed as parasites, although the adults in some cases may sometimes feed on nectar. In a few cases, where both stages were known to be important, the larvae and adults were treated as different 'species' for analysis of trophic structure, e.g. *Utetheisa* larvae were listed as chewing herbivores and the adults as nectarivores.

Several species were not adequately sampled by the usual quantitative methods and had to be specially treated. Moths of the genus *Utetheisa* inhabited the *Argusia* shrubs. When a person closely approached the shrub the adults would fly up and remain aloft for a considerable time before settling again. These were sampled by approaching each shrub on the island, waving a handkerchief attached to a stick around each side of it and directly counting the numbers of adults as the flew up. The caterpillars of this species were included in the sampling of *Argusia* foliage fauna. Each caterpillar in the sample was measured.

Adult grasshoppers were sampled by dragging a rope 30 m long between two investigators as they traversed the herb flat and counting the number of adult grasshoppers as they flew up, and then adjusting to a unit-area basis.

Individuals of the most common species of plant restricted to the beach (*Trachymene*) were counted individually and on some occasions their dimensions measured and reproductive condition noted. On each trip, transient species of plants were examined individually and notes made on their general health and reproductive status.

Periodic observations were made on the pond in the centre of the island. Occasional collections of the fauna and flora were assembled and on each trip general notes on water level and clarity were made. Physical parameters were also measured from time to time.

8. The flora

We recorded a total of 31 species of vascular plants from One-Tree Island between 1968 and 1972 (Table 4), of which 21 were permanent residents, i.e., were present at all visits and were known to set seed on the island. In addition, there were four species which were temporarily established, i.e., set seed which germinated into new plants but subsequently became locally extinct, and/or which immigrated to the island during the study. Finally, there were six transient species which grew briefly on the island but either died before setting seed, or if fruits were produced, the resulting seeds did not produce new plants (but see discussion of individual species below).

Table 4. The vascular flora of One-Tree Island, 1968–1972.

Trees	*Tribulus cistoides* (R)
Pisonia grandis (R)	*Abutilon indicum* (R)
Rhizophora stylosa (T)	*Cucumis* sp. (TH)
Pandanus sp. (R)	*Cucurbita pepo* (TH)
	Trachymene sp. aff. *procumbens* (R)
Shrubs	*Plumbago zeylanica* (R)
Argusia argentea (R)	*Ipomoea pes-caprae* subsp. *brasiliensis* (I)
Sophora tomentosa (T)	*Ipomoea macrantha* (R)
Scaevola taccada (R)	*Lycopersicum esculentum* (TH)
	Solanum nodiflorum subsp. *nutans* (IE)
Forbs and vines	*Melanthera biflora* (R)
Boerhavia diffusa (R)	*Sonchus* sp. (TH)
Commicarpus insularum (R)	*Spathoglottis paulinae* (IE)
Sesuvium portulacastrum (R)	
Tetragonia tetragonioides (R)	**Grasses**
Portulaca oleracea (R)	*Lepturus repens* (R)
Achyranthes aspera (R)	*Thuarea involuta* (R)
Cakile edentula (IE)	*Stenotaphrum micranthum* (R)
Canavalia rosea (R)	*Digitaria ciliaris* (T)

R = Permanent resident; T = Transient; TH = Transient probably introduced by humans; I = Species which immigrated during the study and became established; IE = Species which immigrated, became established, and subsequently became extinct, all within the period of the study.

In addition to the 31 species in Table 4, there were several others known to have been brought by humans after the establishment of the marine station (see section on Miscellaneous Cultivated or Human-associated Plants).

Of the total established flora, the majority of species were forbs, with grasses (four species) being the second largest category (Table 4). There were three species each of trees and shrubs. Three of the herbaceous species were succulents, and 12 were either trailing or creeping species or had tendencies toward a prostrate or straggling habit.

The general biogeographical and ecological characteristics of each species is discussed in turn. Much of the information reviewed under the species accounts appears in the literature under synonyms. The nomenclature employed in this book is that currently followed by the Queensland Herbarium.

Nyctaginaceae

Pisonia grandis R. Br.

Pisonia grandis is a tree with spongy, sappy wood and large (up to 25 cm long) ovate, yellow-green leaves; the bole is large and smooth and the canopy spreading. Maximum height is about 30 m. It is usually evergreen although prolonged dry periods may render some trees nearly leafless; this is almost a yearly phenomenon in some localities (Stemmerik 1964, Cribb 1969).

The latitudinal boundaries of this species coincide almost exactly with the 24° parallels in both the north (Marianas, Taiwan) and south (New Caledonia, Austral Islands, Pitcairn Island) and hence with the boundaries of the tropics. Within these latitudes it is found from reefs just off the East African coast, eastward through the islands of the Indian Ocean, East Indies, some coastal Australian islands and the Pacific Islands to Pitcairn Island (St. John 1951, Airy Shaw 1952, Hunt 1967–68). It is only rarely found on mainlands or on high islands within its geographic range, generally occurring on low coralline islands. It frequently forms dense stands with a closed canopy and would appear to represent the final stage in succession on coral cays. However, Gillham (1963) indicates that it is shade intolerant and where taller species are present they replace *Pisonia* which then becomes a seral rather than a climax stage. It seems to prefer leeward sides of islands; the wood is brittle and easily broken by storms. Groves of *Pisonia* are prone to wind-pruning.

Exposure to sea water reduces the success of germination of *Pisonia* seeds (Hnatiuk 1979). They are not buoyant (Wickens 1979b).

A characteristic soil type, the Jemo Series, often develops under mature stands of *P. grandis* (Fosberg 1954); it is of organic origin, dark brown to black, and spongy or peaty in texture. Sometimes it lies directly on unaltered sand but at other times the upper layers of the inorganic material are cemented into a dark brown, strongly phosphatic hardpan. The organic matter is probably derived largely from leaves and twigs of *Pisonia* with contributions of guano from the many birds that habitually nest in the trees.

This species appears to have developed a mutualistic relationship with sea birds. Certain species (e.g., the White-capped Noddy, *Anous minutus*) nest in *Pisonia* trees in large numbers, using old *Pisonia* leaves as the main constructional material (Gilbert 1925–27, Cribb 1969). Conversely, the plant may be dependent upon sea birds as in some localities it seems unable to regenerate without phosphate-enriched soil (Airy Shaw 1952, Stemmerik 1964). It grows well on coral islands with large bird populations but if bird colonies abandon a site, *P. grandis* often eventually disappears or becomes rare. However, St. John (1951) mentions that *P. grandis* does occur on some high islands and suggests that dispersal factors may be implicated more than edaphic ones in the distribution of this species.[1] Large sea bird colonies are usually established on small islands uninhabited by humans and hence it is to such places that the bird-dispersed fruits are carried. Regardless of which factors are crucial in determining the distribution of *P. grandis*, it is clearly evident that sea birds are important in disseminating the seeds to new localities. The fruits of *Pisonia grandis* are clavate-cylindrical, slightly longer than 1 cm, and bear 5 longitudinal rows of dark brown papillae (Fig. 22); at maturity these become covered with a sticky, resinous exudate which retains its adhesive properties for an extended period of time (Cribb 1969). There are numerous reports of these fruits or even entire inflorescences adhering to sea birds, sometimes to the extent of binding the wings to the body and incapacitating the birds (Longman 1914, Musgrave 1925–27, Gilbert 1925–27, Cribb 1969; also our personal observations) and in fact the Balinese use the sticky fruits to catch birds (Stemmerik 1964). The frequency with which birds are completely immobilized by *Pisonia* fruits and subsequently die led Mather (pers. comm.) to suggest that the sticky fruits may also be advantageous to *Pisonia* by trapping and killing birds whose bodies then provide fertilizer for the tree.

Insects also get stuck on the sticky fruits sometimes (Hahn 1962). The frequency with which this occurs has led to the suggestion that the seeds may need the presence of animal matter to assist their germination and that the sticky seeds may be an adaptation in that context (Longman 1914).

Boerhavia diffusa L.

Boerhavia diffusa is an erect, creeping or climbing herb up to 1 m high, with cymose panicles of white, red, or purple flowers (Fig. 23). It is distributed pantropically in open dry places, secondary forest and rocky or sandy places

[1] Note also that Spicer and Newbery (1979) report that not all sites inhabited by *Pisonia* are rich in phosphates.

Fig. 22. Fruits and seeds of various plants of One-Tree Island. 1. Seeds of *Pandanus* sp. on leaf litter. Note the insect emergence holes. 2. Fruit of *Pandanus* sp. 3. Inflorescence of *Pisonia grandis*. 4. Fruit of *Tribulus cistoides* (scale graduated in mm). 5. Fruit of *Pisonia grandis* (scale is 1 cm). 6. Fruit of *Plumbago zeylanica* showing viscous glands (scale as in no. 5). 7. Fruit of *Boerhavia diffusa*; note sand grains adhering to sticky glands (scale graduated in mm). 8. Fruit of *Commicarpus insularum*; note sticky glands around the top (scale as in no. 7). 9. Fruiting spike of *Achyranthes aspera*; note spiny nature of fruits. 10. Umbels of fruits of *Trachymene*. 11. Ripe and green fruits of *Solanum nodiflorum*.

Fig. 23. Some of the plants of One-Tree Island. 1. *Melanthera biflora.* 2. *Ipomoea macrantha.* 3. *Sesuvium portulacastrum.* 4. *Boerhavia diffusa.* 5. *Canavalia rosea.* 6. *Tetragonia tetragonioides.* 7. *Plumbago zeylanica.* 8. *Tribulus cistoides.* 9. *Abutilon indicus.* 10. *Achyranthes aspera.* 11. *Scaevola taccada.* 12. *Pandanus* sp. 13, 14. *Trachymene.* 15. *Argusia argentea.*

(Stemmerik 1964). It is frequently found back of the seashore and on small coral islands (Merrill 1946). The fruits are club-shaped with numerous stalked glands scattered over the surface (Fig. 22). The glutinous secretions from the glands are sticky and cause the fruits to adhere to surfaces with which they come in contact (Ivens 1967). This species is almost certainly bird-dispersed. Similar fruits of a related species have been observed in down preened by a red-footed booby (*Sula piscatrix* = *S. sula*) from its feathers (Guppy 1906).

Commicarpus insularum Meikle

Commicarpus insularum is an erect to creeping herb with white flowers in axillary umbels. This species was only recently described (Meikle 1979) and material from One-Tree Island collected during the present study was examined in connection with that description.

The clavate-truncate fruits have sticky glands which are most prominent towards the apex of the fruit (Fig. 22). This species is undoubtedly dispersed by adhering to birds (see *Boerhavia diffusa*).

It is one of the few One-Tree plant species with a restricted distribution. It occurs in littoral areas of eastern Queensland and islands of the Great Barrier Reef; it has been recorded from One-Tree Island, Heron Island, Tryon Island and Lady Elliot Island (Meikle 1979).

Aizoaceae

Sesuvium portulacastrum L.

Sesuvium portulacastrum is a perennial, prostrate herb up to 80 cm long; the leaves are linear or lanceolate and are thick, fleshy and succulent (Fig. 23) (Merrill 1946).

The species is tropicopolitan. It often occurs in dense, nearly monospecific mats. It is a halophyte, preferring saline clays, sands or muds in coastal areas, along tidal creeks or on islands (Backer 1951), and is a common weed on tropical beaches (Docters van Leeuwen 1936). Davis lists it as the most frequent and constant plant on the strand of Florida cays. It may even inhabit sites overflowed by sea water at high tide (Guppy 1891). It may often be found on very small, remote islands.

The means whereby this species is dispersed so widely and to such remote places, has not been directly observed. The small, black seeds are non-buoyant (Guppy 1917) and hence are not directly dispersed by sea. However, seeds of this species falling into crevices in driftwood or pumice may be transported when such objects subsequently become afloat (Ridley 1930). Ridley further suggested sea dispersal of floating portions of the green plant and Howard (1950) observed this to have taken place. This species is quite salt tolerant and inhabits areas of beach where it is sometimes drenched by sea water (Walter 1964); it could probably withstand a considerable time in sea water. Walter (1964) provided data indicating that the osmotic pressure of the cell sap is variable, depending on the salt content and dryness of the habitat but that quite high values can be sustained (up to 52.6 atm.). As the solid, succulent stems root from the nodes (Backer 1951), the green plants would be effective propagules; Muir (1937) indicated that in his experiences pieces of stems were transported by sea at least over short distances. Longevity of plants in sea water should be empirically tested.

Ridley (1930) also suggested bird dispersal. He noted that on Christmas Island, Boobies (*Sula sula*) carried succulent portions of the plant and used them in nest construction. It seems unlikely that birds would carry nesting material greater distances than from one island to a closely neighbouring one, and it is doubtful if such dispersal could account for the worldwide distribution of the species. The small seeds could perhaps be transported long distances in mud on birds' feet.

It was the main component of the succulent mat on One-Tree Island but was not observed elsewhere until September 1978 when a small new patch was found on the southeastern tip of the island. It was healthy and flowering; it may have resulted from a new invasion.

Tetragonia tetragonioides (Pall.) Kuntze

Tetragonia tetragonioides is a fleshy, erect, annual herb with small yellow flowers (Fig. 23). It occurs wild in coastal regions of Australia, New Zealand, Japan and the Pacific Islands although it is cultivated as a potherb over a wider area (Backer 1951). Its habitat is sea coasts and margins of salt marshes (Beadle *et al.* 1962). The seeds have short spines and cling to animals; it is also possible that they are sea-dispersed. Its use by man makes human dispersal a probability also.

Portulacaceae

Portulaca oleracea L.

Portulaca oleracea is an often prostrate succulent herbaceous annual (Merrill 1946) up to 40 cm high; the leaves are obovate to spatulate and fleshy. The flowers are yellow and the small seeds occur in dehiscent capsules.

The species is found in the temperate and tropical areas of the world (Wilson 1932, Ivens 1967); it is ruderal, occurring in waysides, deforested areas, cultivated fields and waste places, as well as along sandy shores and on exposed rocks (Geesink 1971, Holm *et al.* 1977), and in sandy openings in littoral woodland (Schimper 1903). It is found on a number of remote islands.

Coquillat (1951) lists it as the eighth most common vascular plant in the world and Holm *et al.* (1977) indicate it to be the ninth most troublesome weed of the world in terms of man's agriculture.

The fleshy stems root on contact with soil and it can be spread by fragmentation (Holm *et al.* 1977). The seeds are small (0.5–1.5 mm) and drought resistant but are not buoyant (Guppy 1906, Geesink 1971, Wickens 1979b). According to Geesink, *P. oleracea* seeds can be transported in cargo and have been dispersed by man to such an extent that the native range can no longer be ascertained with certainty. This species is used as a vegetable and medicinally, and was intentionally dispersed by European explorers and travellers. Goats can disperse it locally (whole seeds have been found in their droppings) although long distance dispersal by this means would be unlikely without the aid of man (Guppy 1906). The seeds are also known to pass through the digestive tracts of sheep and pigs (Muenscher 1955).

In freshwater, submerged seeds of *P. oleracea* have been reported to germinate and the seedlings float; they may thus be dispersed by freshwater (Ridley 1930).[2] Guppy (1891) suggests that seeds may be carried 'in the seed vessels attached to the floating plant or branch' and sea dispersion by flotation of branches has been suggested by Ridley (1930); Muenscher (1955) has indicated that fleshy stems will take root and mature seeds. However, tolerance of seedlings, or adult plants to sea water is not known. Seeds of *P. oleracea* have been found in grooves in floating debris in freshwater rivers and Guppy (1917) suggests that this species may also be transported at sea by pumice or driftwood.

In reviewing the dispersal of *P. oleracea* Ridley (1930) noted that (1) it has been intentionally widely dispersed by man, (2) its non-intentional occurrence on remote islands often, though not always, occurs after human settlement and/or is concurrent with the first appearance of human-dispersed weeds, and (3) it is absent from some apparently suitable islands near mainlands which support this species. He concluded that this species may naturally disperse by sea occasionally but that its side distribution is mainly attributable to dispersal by humans.

Ridley also reviewed bird-dispersal of *P. oleracea* and noted that some terrestrial and semi-aquatic birds eat the seeds and thus may disperse them in their digestive tracts.

Zimmerman (1977) fed seeds of three species of *Portulaca* to the house sparrow (*Passer domesticus*) and found that *P. oleracea* seeds had a viability of 62% after passing through the birds' guts, a value considerably higher than for the other two congeners and indeed higher than reported for any other weeds.

Dispersal to One-Tree Island may have been by land birds since it is not far from the mainland and a number of Australian land birds are included among

[2] Hnatiuk (1979) was unable to duplicate the underwater germination reported by Ridley (1930).

those species known to contain *P. oleracea* seeds in their stomachs (Field Wren, *Calamanthus fuliginosus;* Chestnut-rumped Thornbill, *Acanthiza uropygialis;* Cinnamon Quail-thrush, *Cinclosoma cinnamomeum*; Black-backed Blue Wren, *Malurus melanotus*, and several warblers; Ridley 1930). Since a variety of land birds rather frequently reach One-Tree (see Chapter 17) it is quite possible that *P. oleracea* originally reached the island by this means.

Zimmerman (1977) carried out an experimental study of the attributes of *P. oleracea* which adapt it to a colonizing mode of life. He found that covered capsules yielded large numbers of seeds and hence the species is self-compatible; he suggested that self-compatibility, a tendency toward cleistogamy and strict environmental control of floral opening encourages selfing and that such a breeding system enables *P. oleracea* to exploit newly inhabited sites and begin a new population. Lack of an after-ripening requirement means that a new generation may start immediately if conditions are favourable. The specific thermal and photic conditions required for germination are narrow and ensures that the seeds will germinate only in areas where there is a good chance that seedlings will reach reproductive maturity. If such requirements are not met, the seeds can remain viable for long periods (up to 40 years according to Muenscher 1955 and Kozlowski 1972) and are thus available whenever favourable conditions do arise. Once a population is established seed longevity would permit development of high numbers of reserve seeds and enhance the persistence of the species over unfavourable periods. Zimmerman noted that once established, populations are difficult to eradicate.

Zimmerman summarized his findings by indicating that self-compatibility, the seed characteristics of longevity, resistance to digestion (see above), lack of an after-ripening requirement, relatively narrow germination response to temperature and light, and overwintering proficiency 'combine to give the plant quick-response capability when conditions surrounding the seed meet germination requirements'.

Thus, the characteristics of *P. oleracea* are such that (1) dispersal by birds and humans is facilitated, (2) a quick-response to immediately favourable conditions can be made and (3) unfavourable conditions endured in the seed stage. The first favours immigration, the second enhances establishment, and the third reduces the danger of extinction, a suite of attributes of obvious value in an insular situation.

Amaranthaceae

Achyranthes aspera L.

Achyranthes aspera is an erect to spreading herbaceous annual to biennial up to 1.25 m tall with terminal spikes of flowers (Backer 1949, Beadle *et al.* 1962,

Kleinschmidt and Johnson 1977) (Fig. 23). It is pantropical (St. John and Smith 1971) and has often been introduced by man. It is found in open places such as roadsides and wastelands and in hedges and thickets (Backer 1949, Ivens 1967). In the Pacific it was a ruderal weed before the advent of white explorers (Guppy 1906). It is particularly common and luxuriant in areas where sea birds nest (Diamond 1979).

The flowers are stiffly reflexed against the rachis of the inflorescence. The bracts and bracteoles have spiny tips and when the fruit is ripe, the perianth also becomes rigid and spiny (Fig. 22). The long, projecting inflorescences readily catch against an object moving through the vegetation and the fruit enclosed in its envelope of perianth, bracts and bracteoles, is detached and persistently adheres (Ridley 1930, Ivens 1967). These fruits were a nuisance on One-Tree Island because of their ready attachment to clothing, an observation previously made by Musgrave (1925–27). *A. aspera* is probably frequently and widely transported by birds (Diamond 1979). Bullock and Primack (1977) note that the fruits are brushed off of carriers passing through vegetation but that they persisted for greater distances in the open; both characteristics are important for dispersal for not only must the desmochore adhere sufficiently to be transported but it must also be released from the carrier in an appropriate place.

Brassicaceae

Cakile edentula (Bigelow) Hooker

Cakile edentula is a succulent herbaceous annual (or fortuitous perennial). It is usually erect, up to 80 cm tall; the flowers are white to pale lavender (Rodman 1974).

This species (including its various subspecies) inhabits the shores of the Great Lakes and St. Lawrence River and the eastern seaboard of the United States and Canada. It has also become naturalized after introduction onto the Pacific coast of North America (Barbour and Rodman 1970) and in coastal Australia. It arrived in Australia near Melbourne sometime before 1863 and possibly as early as 1840, probably in the sand ballast of American sailing ships. It subsequently spread eastward and northward along the coast into New South Wales and Queensland and westward into South Australia and Western Australia. It also reached Tasmania, Lord Howe Island and New Zealand. In parts of this range it is now being displaced by the more recent European arrival, *Cakile maritima* (Rodman 1974).

C. edentula grows on sandy beaches just beyond reach of the high tides in the pioneer zone of beach vegetation; it is often the species of vascular plant growing closest to the sea (Rodman 1974). The likely physiological responses of *C. edentula* may be inferred from those of the taxonomically and ecologically allied *C. maritima,* studied by Barbour (1970a, b). *C. maritima* is a halophyte and grows in areas where topsoil salinity is 450–1800 ppm. It can withstand greater amounts of salt spray than can grassland plants (survival good at 0.1–0.3 mg salt/cm^2/day but poor at 1.0 mg/cm^2/day). In the laboratory, germination and root growth were not inhibited by 0.1 or 1.0% NaCl and overall seedling growth was not inhibited by 0.8% salt. However, a minimum of about two weeks of rainy weather is required for germination and good seedling survival.

The fruit of *Cakile* is an indehiscent, hard-corky two-segmented silique. Each segment contains one or more seeds; the outer segment is deciduous whereas the proximal one remains attached to the plant (Rodman 1974).

It is probable that the following summary of features relevant to sea dispersal of *C. maritima* would apply equally well to *C. edentula*. *C. maritima* is undoubtedly sea-dispersed as the fruits are common in beach drift and seedlings have been seen germinating from such drift. The seeds themselves are too heavy to float but the deciduous segment of the fruit contains a light, spongy tissue which provides buoyancy for the enclosed seed. Flotation time is short for a sea-dispersed plant (only 7–10 days); drying of the pods had very little effect. The cause of sinking is the entry of water at the base of the pod through the sutures of the lateral halves. The seeds are salt tolerant and capable of germination after immersion in sea water for at least 45 days; they do not germinate while in sea water, however (Guppy 1906, Ridley 1930). Barbour (1970a) found that in *C. maritima,* the deciduous part of the fruit was subject to being blown about on the surface (and hence to being set adrift) whereas the lower, attached part of the fruit is often buried by sand along with the parent plant. Thus the two-parted fruit, embraces two dispersal strategies: colonization of distant places and re-seeding of the parental locality. Because of its similar fruit morphology, *C. edentula* almost certainly behaves in a similar way.

Howard (1950) and Heatwole and Levins (1972a) mention whole green plants of *C. lanceolata* being carried by sea currents and it is probable that *C. edentula* is dispersed in the same way.

Sea currents are not the only means whereby *C. edentula* has become dispersed, however. It has been transported in ship's ballast a number of times, probably having reached the western coast of North America, the Azores, Australia and New Zealand by this means (Rodman 1974 and see above). Beadle *et al.* (1962) indicate that *C. edentula* is dispersed by wind as well as by ocean currents. However, they may not have intended wind dispersal to refer to transport across sea barriers but perhaps merely along shores. After introduction onto the western coast of North America, *C. edentula* extended its range to new localities along that coast at the phenomenal rate of about 65 km per year for 50 years (Barbour and Rodman 1970).

C. edentula was not a part of the permanent flora of One-Tree. It was first observed (as three dead plants) on the beach in May 1969; on the subsequent trip (July–August) one live plant was flowering. None were subsequently found until March–April 1970 when thirty seedlings and three large healthy plants were present on the beach (Fig. 24) and by May there was one plant fruiting. None were present again until February 1971 when some seedlings were observed. Given the dispersal capabilities of this species and the fact that in all cases the plants were on exposed parts of beaches, it is probable that several immigrations occurred with temporary establishment and subsequent extinction, and we have consequently considered it as such for analyses of species-turnover.

Fabaceae

Canavalia rosea (Sw.) D.C.

All references to *Canavalia maritima* in the literature refer to *Canavalia rosea. Canavalia rosea* is a trailing or prostrate herb with trifoliate leaves of oblongate to orbicular leaflets and with pink flowers (Fig. 23) (Beadle *et al.* 1962). The stems may reach to 30 m or more in length (Backer 1929, Docters van Leeuwen 1936). This species is pantropical, usually occurring in sandy places along sea shores especially on islands, often in thickets back of the beach (Merrill 1946), or high on the beach (Wilson 1977), but also occasionally occurring a short distance inland.

In contrast to many strand species, *Canavalia* is not a true halophyte; although it lives in salt-rich localities, it does not have a high salt content of the tissues (Guppy 1906).

C. rosea is a sea-dispersed species. Its seeds have been found in beach drift in a variety of places including Fiji and Ecuador and in the sea off New Guinea, Christmas Island and Cocos-Keeling (Ridley 1930). Flotation is achieved as a result of an air cavity between the cotyledons (Docters van Leeuwen 1936); the integrity of the seed covering is necessary as without it the kernel swells and sinks. The time that seeds of this species remain afloat in sea water varies widely. About 10% of fresh seeds sink immediately, 50% remain afloat up to 3 weeks, and 10% up to 3 months. Of seeds which had been kept for 3 years and then tested, 50% still floated after 11 weeks. Seeds that sank in the early weeks did so because of swelling and abortive initiation of germination (Guppy 1906). The vegetative parts of plants may also be sea-dispersed (Howard 1950).

Sophora tomentosa L.

Sophora tomentosa is a shrub with yellow flowers and long moniliform pods (Ridley 1930, Merrill 1946) bearing 1–10 seeds (van Steenis 1948). It is

Fig. 24. Some transient plants on One-Tree Island. 1. *Cakile edentula.* 2. *Sophora tomentosa.* 3. *Rhizophora stylosa.* 4. *Sonchus* sp.

circumtropical in distribution and inhabits beaches and sand banks that are rich in lime and deficient in humus (Backer 1929, Muir 1937), especially on islands; Schimper (1903) also records it from littoral woodland in Java. Although it often lives in a salt-rich environment it is unlike many strand plants in that it does not have a high concentration of chloride in its tissues (Guppy 1906).

The pods of this species float in sea water for a few weeks. The seeds, liberated when the pod breaks down, float even longer. When fresh seeds were placed in sea water, four-fifths of them remained afloat after 3 months. Of seeds that had been kept for 3 years and allowed to dry before exposing them to sea water, half were still afloat after a year and were still sound (Guppy 1906). Muir (1937) found that 48% of the seeds placed in fresh water were still afloat after 104 days. Buoyancy results from the low specific gravity of the kernel itself. When the seed coat is removed, the kernel floats initially but soon absorbs water and sinks in a day or so (Guppy 1906). Hence, persistent flotation in sea water required the integrity of the seed coat. Dispersal of this species is via ocean currents (Sykes and Godley 1968); its seeds are frequent components of beach drift (Guppy 1906). However, germination success is decreased by exposure to sea water (Hnatiuk 1979).

This species was not a permanent part of the flora of One-Tree. It was absent throughout the study until March 1970 when one small plant (18.5 cm high) was observed on the beach (Fig. 24). It continued to grow and was still healthy in March 1971. However, by 1976 it had disappeared, and it must be considered a transient.

Zygophyllaceae

Tribulus cistoides L.

Tribulus cistoides is a creeping to erect perennial herb (Fig. 23) with large yellow flowers. Although of tropical American origin (Holm *et al.* 1977) it is now pantropical. It occurs on sandy beaches and coastal dunes (van Steenis 1949a).

The fruits do not float (Guppy 1906)[3] and hence this species is unlikely to be sea-dispersed. They are probably animal dispersed (Wickens 1979a). The mericarp is heavily armed with stout spines (Fig. 22) that penetrate the feet of animals treading on them. Certain species of this genus are nuisances because of their seeds clinging to the feet and wool of sheep. On mainlands, dispersal by mammals and even by automobiles (the fruits clinging to tyres) seems to be

[3] However, Holm *et al.* (1977) believed it to have spread to Hawaii by 'ocean currents and animals'.

important (Ridley 1930). Agents for dispersal over oceans are probably birds, especially large species. There is one direct observation of *Tribulus* attached to the foot of an albatross (Ridley 1930).

Malvaceae

Abutilon indicum G. Don.

Abutilon indicum is a woody, almost shrubby biennial to perennial herb with large, showy yellow flowers (Fig. 23). It is widespread over tropical Africa, Asia and Australia (Bailey 1899). It is not typically a strand plant but usually occurs well back from the high water mark; it is more common in open forested parts of Heron Island than in the seral scrub (Gillham 1963). On One-Tree Island it was only found in the herb flat until September 1978 when a few small plants were found on the beach.

In order to evaluate whether this species is likely to have floated to One-Tree Island we ran an experiment to ascertain the length of time the seeds and capsules would remain afloat in sea water. Individual brown mature and partly mature seeds sank immediately in still sea water. Green immature seeds floated initially but sank when the small bubbles trapped in indentations on their downward side were manually displaced. It is concluded that sea transport is not possible for individual seeds at any stage of maturity.

Sixty entire capsules were tested in 12 groups of 5. Group one consisted of green capsules containing immature seeds; groups 2–11 contained mature seeds in capsules of progressively later stages of dehiscence. Each group was placed separately in a large beaker of sea water which was then allowed to stand quietly until all carpels had sunk. Group 12 consisted of 5 dry capsules; treatment differed in that the water was agitated, the capsules repeatedly submerged and held under the surface for several seconds, and one of them squeezed flat, forcing gas bubbles from it. All capsules in beakers 1–11 floated high in the water initially, those in number 12 floated after treatment, but lower in the water. The external surfaces of the latter appeared to have taken up water in contrast to those in the other beakers. The squeezed capsule floated with one side barely above the water surface and sank within a day.

The non-squeezed capsules in beaker 12 all sank between three and four weeks after the experiment was begun. In the beakers of quiet water, the capsules sank between 3 weeks and 5½ months. It would appear that entire capsules (or individual carpels) would probably float for sufficient time to be transported short distances across sea barriers. Agitation of the water, such as occurs in the sea, reduced flotation time but still allows several weeks. However, the fact that *A. indicum* usually grows back from the beach means it would not be set adrift very often and although sea transport is a possibility, and may account for the presence of this species on One-Tree Island, it is doubtful if it has reached all the islands in its range by this means. We have found it on 23 widely scattered islands of the Barrier Reef–Coral Sea area. It also occurs on remote Christmas Island where it was believed to have been inadvertently introduced by man (Ridley 1930). Human transport is also a probable means of dispersal for this species elsewhere. Seeds of a variety of species are known to be transported intact and viable in the digestive tracts of land birds (Ridley 1930) and it is likely that *A. indicum* has been transported by this means in the Great Barrier Reef area. Its small seeds are hard and would probably withstand considerable mechanical and chemical digestive action. Kozlowski (1972) indicated seeds of this species germinated after 17 years of dry storage.

Rhizophoraceae

Rhizophora stylosa Griffith

Rhizophora stylosa is a pioneer tree of seashores. It has long aerial prop roots descending from the lower branches and is supported at the base by several stilt roots; it has long, looping pneumatophores (Jones 1971).

It is distributed throughout the islands of the southwestern Pacific, through the East Indies, and along the coasts of southeast Asia and northern and eastern Australia (Walsh 1974). It often occurs on the outer fringes of mangrove forests (Jones 1971). It inhabits stiff clays, sandy beaches and coral reefs (Hou and van Steenis 1958) and is obviously salt tolerant.

The entire genus is sea-dispersed, as floating seedlings which germinate in the ovary, remain there and develop for a number of weeks, eventually dropping from the tree into the water and floating away (Ridley 1930). The fruits and young seedlings of some species of the genus at least are not buoyant but by the time they normally drop from the tree, about 95% float in sea water (Guppy 1906). Guppy also mentions that these propagules may drift for several months in sea water and retain their viability; they can survive dry and stranded on a beach for at least 9 weeks.

R. stylosa was not a permanent member of the One-Tree flora. It was represented on the island by only one individual which was first seen as a 3-leaved seedling on the beach in May 1969 (Fig. 24). It subsequently grew to a height of 25 cm by May 1970 but by August 1970 had died. Consequently, *R. stylosa* was classed as a transient species. Such ephemeral existence in unsuitable habitat may be of common occurrence in mangroves. Davis (1942) stated, in regard to the Tortugas sand cays, that seeds and seedlings of mangroves 'float there by thousands, and some of them become rooted on the beach, only to die from excessive exposure'.

Apiaceae

Trachymene procumbens vel. aff.

The genus *Trachymene* is currently under revision and the nomenclature of the species found at One-Tree has not yet been settled. It has been referred to as *T. procumbens* (probably incorrectly) in the past (Everist, pers. comm.); in the absence of a more correct name at the present time, this nomenclature will be used here. However, until a name can be assigned with confidence, literature on distribution and habits cannot be evaluated properly. The genus is chiefly Australian but also occurs in New Guinea, the Philippines and some of the Indonesian islands (Buwalda 1949).

The species is an annual herb which grows on the more exposed parts of the beach (we have also observed it in similar habitat on Lady Elliot Island). The plants are compact and dome-shaped, the clump growing up to more than 60 cm in diameter (Fig. 23). Umbels of small white flowers are numerous, with the fruits occurring in conspicuous, dark compact clusters several centimetres above the general level of the foliage (Figs. 22, 23). The fact that it thrives on the more exposed parts of beaches, even on the windward side of the island, suggests that it is salt tolerant. Mode of dispersal is unknown.

Plumbaginaceae

Plumbago zeylanica L.

Plumbago zeylanica is a straggling herb reaching up to 2.5 m in height (Fig. 23). This species was probably of African origin but it is now widely distributed as a weed throughout the Old World tropics, extending to Hawaii. It inhabits open, rather dry places, often those modified by human activities, such as roadsides (Ridley 1930, van Steenis 1949b).

The fruits are oblong with five furrows (Fig. 22); they are covered by a viscous secretion which causes them to readily adhere to animals or clothing. Part of their dispersal has probably been by man and his domestic animals; for example this species arrived in Australia as early as 1802, perhaps brought in on sheep's fleeces (Ridley 1930). Its distribution to small islands has probably been by adherence to birds.

Convolvulaceae

Ipomoea pes-caprae (L.) subsp. *brasiliensis* (L.) Ooststr.

This subspecies is sometimes elevated to specific status in the literature. *Ipomoea pes-caprae* is a perennial, trailing herb with large purple flowers; the stems root at the nodes. It is pollinated by a variety of Hymenoptera, Lepidoptera and Coleoptera (Kugler 1973).

This species is circumtropical in distribution, usually occurring on or immediately behind sandy beaches although it is occasionally found inland along roadsides, ditches or canals, in openings in littoral woodland, or on the rim of a volcanic crater; it is often a pioneer species on sandy areas and is useful as a sand-binder (Schimper 1903, van Ooststroom 1953, St. John 1970, Sykes 1970). Wilson (1977) suggests it may be competitively superior in habitats with shifting substrates. It often occurs on remote islands (Hemsley 1892). It was a colonist of the Taal volcano, Philippines after it was sterilized during eruption (Brown *et al.* 1917). It is salt tolerant (Fosberg 1974), growing in places where ground water has salt concentrations of over 5%; the cell sap sometimes reaches osmotic pressures up to 12.7 atmospheres (Walter 1964).

The seeds of *I. pes-caprae* were considered by Guppy (1906) to be 'amongst the most typical of floating seeds'. Flotation results from air spaces left by the incomplete filling of the seed cavity by the embryo because of its shrinking during early development. Variation in the relative extent of this air space results in inter- and intraspecific variation in buoyancy and in duration of time seeds remain afloat (Guppy 1906). Duration of flotation also depends on various environmental factors. Guppy (1906) found that about 1/3 of the seeds began to swell and then sank and died within the first 3 months as a result of the initiation of germination while still in the water. High temperatures reduce flotation time. The vegetative parts may also be dispersed by sea (Howard 1950).

The properties of the seeds of *I. pes-caprae* predispose it to dispersal by ocean currents, its seeds being characteristic components of beach drift in tropical Asia and America and the tropical islands of the Pacific and Indian Oceans (Guppy 1891, 1906). Viable seeds were found on beach drift on Aldabra (Wickens 1979a). However, even species highly adapted for water dispersal may also be dispersed by other means occasionally; for example, seeds of *I. pes-caprae* have been found in the stomach of a tern (Wickens 1979a).

This species was not part of the permanent flora of One-Tree. It first appeared in November 1969 as one plant on the beach which persisted throughout the study. In its vicinity 3 seedlings appeared in mid to late 1970. It was considered as an immigrant which became established.

Ipomoea macrantha Roem and Schultes

This species is widely known in the literature under the names *I. tuba, I. grandiflora* and *I. violacea. I. macrantha* is a twining, perennial herb with large white flowers (Fig. 23). It is pantropical and occurs on beaches and other seaside habitats (van Ooststroom 1953), including tidal forests where brackish water influence is not so great (Rao and Sastry 1970).

The seeds are buoyant (see discussion of *I. brasiliensis*) and are extremely salt tolerant. Guppy (1906) found that they germinated after a year of floating on sea water. Clearly this species is sea-dispersed. It was a prominent part of the permanent flora of the herb flat.

Boraginaceae

Argusia argentea (L.f.) Heine

Argusia argentea is a shrub to small tree with small white flowers in scorpioid cymes (Fig. 23). It occurs on the islands of the Indo-Pacific region from East Africa to the East Indies, Australia, Taiwan, the Ryukyus, the Philippines, and most of Polynesia (Ridley 1930, Heine 1976). It is a typical beach and strand plant (Guppy 1906) which can be established on open beaches only slightly above the high tide line (Lesko and Walker 1969), although on some islands old trees may be found further inland (Fosberg 1974). Guppy (1891) pointed out that this species tends to grow on the more exposed edges of islands, only because that is where the sea deposits its seeds (see below), and because dispersal of seeds via birds or crabs does not usually occur. Occasional plants that do reach more sheltered habitats grow better than they do in the more exposed places they usually inhabit. *A. argentea* is characteristic of coral islands and usually absent from mainlands (Ridley 1930). Ducie, considered to be the most isolated atoll in the world (Fosberg 1974) has *Argusia argentea* on it, but no other species of vascular plant.[4] Over 100 individuals up to 4 m in height occur on the perimeter of One-Tree Island.

The tissues of *Argusia* are often high in chloride (Guppy 1906) and Schimper (1903), Fosberg (1974) and Heine (1976) list it as halophytic.

The fruit is a dry berry with four loculi which at maturity contain one seed each. The mesocarp of the fruit is thick and cork-like and provides buoyancy (Ridley 1930, Lesko and Walker 1969). Flotation in sea water up to a year has been recorded; the fruits are components of beach drift (Guppy 1891, 1906; Muir 1937). The seeds are salt tolerant, no loss of germination occurrring after 120 days exposure to sea water. Sea water inhibits germination while the seeds are exposed to it but pre-treats them so that they germinate more rapidly when subsequently exposed to fresh water (Lesko and Walker 1969). The above attributes are adaptations for transport by sea currents, which is probably its most common means of dispersal (Heine 1976), although Guppy's assertion that bird dispersal does not occur may not be entirely correct as White and MacGillivray (1925–27) reported the silvereye using *Argusia* fruit as food.

[4] In 1827 and again in 1922 a grass (*Lepturus*) was also present although uncommon. At the latter date a vine (*Epigaea*) was present. Only *Argusia* was present throughout and only that species has been observed there from 1936 onwards (Rehder and Randall 1975).

Drift logs and pumice may be accessory means of dispersal as seeds of this species have been found in holes bored by teredo in logs and sprouting from pumice on Cocos-Keeling Atoll (Guppy 1891).

Solanaceae

Solanum nodiflorum Jacq. subsp. *nutans* R. J. Henderson

Solanum nodiflorum subsp. *nutans* is a short-lived erect perennial herb up to 1.5 m tall; the leaves are green to purplish and the flowers are white, sometimes suffused with purple. The mature fruits are fleshy and are purple in colour (Fig. 22) (Henderson 1974).

The species is widely distributed, occurring in the Western Indian Ocean, Australia, parts of Central and South America, the Caribbean, the United States and perhaps other areas as well. The subspecies *nutans* is native to coastal and near coastal eastern Australia and closely adjacent islands, but also occurs on Lord Howe and Norfolk Islands, in Fiji, New Zealand and New Guinea and perhaps in the north-western Pacific region and western coast of North America (Henderson 1974).

This species is extremely eurytopic and occupies a number of ruderal habitats. Henderson (1974) in treating the species complex generally, lists habitats such as cultivated ground, roadsides, garbage tips, vicinity of buildings and water outlets and other areas disturbed by man as well as the margins of rainforest and coastal sand dunes.

The seeds probably could be easily transported by man, and this agency may account for part of its widespread distribution. However, the fleshy fruits are eaten by gulls and silvereyes (MacGillivray and Rodway 1931) and the entire seeds are found in the feces of such animals; it is probably primarily bird-dispersed (Longman 1914). It is almost certain that it reached One-Tree as seeds in the digestive tract of birds. This species is treated in more detail in Chapter 18.

Goodeniaceae

Scaevola taccada (Gaertn. Roxb.)

All literature on *S. sericea* refer to *S. taccada*. *S. taccada* is an erect to spreading shrub up to 7 m tall; it has shiny, herbaceous to thin-fleshy leaves spirally arranged near the ends of branches (Fig. 23). It occurs from Madagascar to southeastern Asia, tropical Australia and in the Pacific Islands to Hawaii, in which places it is restricted to seashores and islands, usually on sandy beaches or coastal cliffs. It is a pioneer of newly deposited littoral sands, and of bare gravel and rock and often fringes the upper beach (Leenhouts 1957); sometimes it becomes established on open beaches only slightly above high tide line where it is subjected to salt spray and storm waves (Lesko and Walker 1969, Newbery and Spicer 1979) although it may also occur in littoral woodland (Schimper 1903). Like *A. argentea* his species grows best in more sheltered situations than the ones it occupies by force of its mode of dispersal (Guppy 1891).

The tissues of this species often have a high chloride content (Guppy 1906), and Schimper (1903) and Fosberg (1974) consider it a halophyte.

The fruits are highly buoyant, white drupes which are capable of floating for at least a year, and are often found in the beach drift of tropical coasts and islands (Guppy 1891, 1906). Buoyancy is due to a thick, corky mesocarp around the fruit (Lesko and Walker 1969).

High salinity of sea water inhibits seed germination in floating fruits but subsequent exposure to fresh water induces germination. Thus, seeds are somewhat protected from germinating under the unfavourable conditions of a sea voyage, but subsequent exposure to rain on a beach results in rapid germination. Soaking in sea water for 120 days did not reduce eventual germination success when later exposed to fresh water (Lesko and Walker 1969), and in fact Hnatiuk (1979) found improved germination after longer times in salt water. It was the only species out of 81 tested from Aldabra Atoll that had that characteristic.

It would appear that long-distance dispersal is achieved in the main by flotation on sea water. Ridley (1930) mentions that this species has been observed sprouting from pumice on beaches and consequently floating pumice may be an auxiliary dispersal agent. Ridley also noted that the thin epicarp is pulpy and may be attractive to birds so this species could be bird-dispersed as well. *S. taccada* was represented on One-Tree by only 3 shrubs, all near the northeastern corner of the island.

Asteraceae

Melanthera biflora (L.) H. Wild.

Melanthera biflora is a straggling perennial herb up to 1 m in height (Fig. 23) (Beadle *et al.* 1962). It has yellow composite flowers. Its geographic range includes the Old World tropics from East Africa through coastal India, southeastern Asia, northern Australia and islands of the Indian Ocean, southwestern Pacific and Polynesia to Japan; it inhabits muddy and sandy shores (Ridley 1930, Wild 1965), and is characteristic of the vegetation behind the beach of many shores (Merrill 1946) including islands. Spicer and Newbery (1979) and Newbery and Spicer (1979) indicated that in the Maldives, this species is found mainly on disturbed cleared ground.

This species is unlikely to be wind-dispersed for any great distance as the achenes lack a pappus (Longman 1914, Ridley 1930: figured by Wickens 1979b). Rather it would seem to be water-dispersed (Brown *et al.* 1917, van der Pijl 1969). Ridley (1896) refers to the fruits being drifted along the coast in the sea. The achenes are corky, having beneath the epidermis a layer of buoyant tissue which keeps them afloat for months (Guppy 1906). Backer (1929) indicates they floated for more than twenty weeks in a 3% solution of NaCl.

We determined mean number of achenes per composite flower as 65 (range 41–98; N = 10). *M. biflora* is a conspicuous member of the herb flat flora.

Poaceae

Lepturus repens (Forst. R.Br.)

Lepturus repens is a grass of variable habit, occurring in erect caespitose tufts or as creeper; jointed fruiting spikes 3–20 cm long are produced. It is one of the commonest and most widespread strand plants of the tropical Pacific and Indian Oceans. It inhabits sand or rocks, sometimes in dense mats (Ridley 1930, Fosberg 1955). Fosberg (1974) considers it a halophyte.

Various workers have suggested this species to be sea-dispersed (Wickens 1979b). The disarticulated mature rachis joints with their attached fruits act in dispersal; they float on sea water because of their spongy internal tissue. It is also possible that the sharp glume which is appressed elastically against the joint may be instrumental in attachment to birds (Fosberg 1955, 1974).

L. repens is the dominant among the grasses of One-Tree.

Thuarea involuta (Forst. f.)

Thuarea involuta is a prostrate creeping grass with short spikes. It is found in coastal tropical Asia and East Indies eastward throughout Polynesia and westward through the islands of the Indian Ocean; it occurs on mainland coasts but is most abundant on islands (Ridley 1930). It frequents sandy beaches (St. John and Smith 1971), where it has a binding effect on sand (Longman 1914). In the Maldives it is found chiefly in heavily trampled places (Newbery & Spicer 1979). It roots at the nodes (Docters van Leeuwen 1936).

The spike is short with a female spikelet at the base and male spikelets distally. At ripening, the male spikelets fall off and the rachis curves over the female spikelet to enclose it and then becomes thickened and hardened (van der Pijl 1969; figured by Ridley 1930). During maturation of the seed, an 'air-tissue' develops in the axis and serves as a float for the fruit (Arber 1934). The reproductive structures detach from the plant and are easily swept into the sea by wind. When placed in artificial sea water about 50% floated and half of these remained afloat for at least 2½ months. Thus, this species usually probably disperses by ocean currents, and has actually been observed to do so (Arber 1934); its seeds have been recorded from beach drift (Muir

1937). Flotation is reduced in fresh water, only 4% remaining afloat after 2 months (Ridley 1930).

Stenotaphrum micranthum (Desv.) C. E. Hubbard

Stenotaphrum micranthum is a slender, creeping annual grass with mostly terminal inflorescences. It is distributed among the islands of the Indian and Pacific Oceans and on some of the East Indian islands. It occurs primarily on small islets and in some regions is confined to them and missing on neighbouring larger ones and on mainland coasts; it is seldom found inland (Sauer 1972).

It is a pioneer strand species, often inhabiting open sandy beaches or the salt spray zone of coralline limestone. It also occurs around human habitations and plantings (Sauer 1972).

Ridley (1930) considered the maritime species of this genus to be primarily sea-dispersed; the spikelets are provided with somewhat coriaceous glumes which probably make them buoyant; Sauer (1972) found that inflorescences would float 7–10 days in sea water, a time Wickens (1979b) considered sufficient to account for *Stenotaphrum* to reach Aldabra by island-hopping across the Indian Ocean. Both Ridley (1930) and Sauer (1972) suggest that auxiliary dispersal may occur through man's agency.

Digitaria ciliaris (Retz.) Koeler

Digitaria ciliaris is a pantropical, weedy annual grass. The plants are usually decumbent with numerous inflorescences; rooting occurs at the nodes.

It was a transient species at One-Tree as only on individual was found near the campsite in March 1971; it was probably inadvertently brought in by humans. It is an important weed in Queensland (Kleinschmidt and Johnson 1977).

Pandanaceae

Pandanus sp.

The species in the Bunker–Capricorn group of islands was designated as *Pandanus pedunculatus* in older publications (e.g. Longman 1914, White and MacGillivray 1925–27). However, more modern authors have not committed themselves to a specific epithet (e.g. Cribb 1972) as the taxonomy of this group is far from resolved and species identification of the One-Tree form cannot be made with confidence.

The genus is distributed widely on islands and mainlands throughout tropical Asia, Polynesia and Africa (Ridley 1930). There are two ecological groups of *Pandanus*, the inland ones that even attain considerable altitudinal range in some regions, and the littoral ones. The two groups have different histories and dispersal mechanisms (Guppy 1906). Only the littoral group is discussed further as only it is relevant to One-Tree Island.

The littoral *Pandanus* have large fruits consisting of large woody drupes packed into a cylindrical or globose head (Figs. 22, 23). At maturity, the drupes separate. The pericarp is light and fibrous and provides buoyancy; its outer layer is smooth and polished and resists decay and the action of sea water (Ridley 1930). Consequently the drupes float in the sea for months and form a common component of beach drift (Guppy 1906, Ridley 1930). Sea dispersal of the drupes is probably the chief means of dispersal. Vegetative parts of plants are also found in beach drift (Guppy 1906) and possibly contribute to dispersal. Also, these plants are (or at least were) used by island people in a variety of ways and it is possible that they assisted in its dispersal (Guppy 1906, Ridley 1930). Longman (1914) indicates that *Pandanus* is among the first arrivals brought to coral beach drift by sea currents.

Orchidaceae

Spathoglottis paulinae F. Muell.

Spathoglottis paulinae is a terrestrial orchid with the pseudobulb just under the surface of the soil. The leaves are narrow and pleated and may reach almost 2 m in length; the flower stalk reaches to slightly more than 1 m tall and has a head of closely packed purplish to white flowers (Dockrill 1969).

Its recorded distribution is northeast Queensland from Rockingham Bay to the Endeavour River. Its occurrence on One-Tree is unexpected as it is not often found at an altitude below about 300 m and occurs in grassland or open forest where it prefers rather damp soils, and often occurs in rather poor clay (Dockrill 1969).

The seeds of orchids in general are very small and could be transported by wind or perhaps by insects or other animals (Guppy 1906, Ridley 1930). The seed of *Spathoglottis plicata* (a closely related widespread species in southeast Asia including islands; Luer 1975) is tiny but with rather extensive wings which prompted Ridley (1930) to figure it as an example of a wind-dispersed seed. *S. paulinae* may also be dispersed in this way; One-Tree is well beyond its previously recorded range.

This species was not a permanent member of the One-Tree flora. It was present on only one visit (Nov.–Dec., 1970). At that time it was represented by two plants but there were three old flower-stalks also present. This species had probably reproduced on the island and thus had become temporarily established, i.e., it immigrated, became established, and then may have become extinct, all during the study.

The only previous record of an orchid in the Bunker–Capricorn islands was of the terrestrial *Geodorum pictum* from Hoskyn Island; it also occurred in *Pandanus* groves (Cribb 1972).

Miscellaneous cultivated or human-associated plants

There were four species in this category. A melon plant (*Cucumis* sp.) was found near the tent site in January 1970, and another in November 1970; neither set fruit. Several pumpkin plants (*Cucurbita pepo*) were present from March to October 1970 and were known to flower but not to set fruit. A milk thistle (*Sonchus* sp; Fig. 24) was present, but not flowering in September 1969. The most common cultivated plant was the tomato (*Lycopersicum esculentum*). Three plants were present near the tent site in September 1969. One was again present in March 1970 and from one to three plants were present at each visit from then on through October 1970; they set fruit in July–August, 1970, but it is not known whether the resulting seeds germinated, and consequently the tomatoes, like the other human associates were considered transients. All of the above species were found only on the exposed beach near the tent site and were assumed to have been transported by humans. They were probably brought in by other scientific expeditions as the species involved were never included in our stores.

At the time of the 1978 visit three additional species had grown up near the caretaker's residential hut but had spread no further. They were believed to have sprouted from bird seed thrown out for birds by the caretaker: one unidentified and two mustards, *Brassica rapa* and ? *Sisymbrium*.

The caretakers also tended a small garden which in 1978 contained parsley, basil, sage, thyme, capsicum, chives, celery, pawpaw, cassava, nasturtiams and 4 species of ornamental succulents. There was also wandering jew planted in pots. None of these had escaped from captivity. However, should they be discovered in the wild in the future, their origin will be known.

Non-flowering plants

Non-flowering plants were not studied in detail. There was an unidentified moss that covered part of the rubble in the rubble flat. It was usually dry but greened up briefly after rains. Unicellular algae occurred on coral rubble and on dead *Pandanus*.

Several mushrooms were observed in December 1970, an especially wet period. *Agaricus campestris* was seen growing on humus in the *Pisonia* grove and five other species were on *Argusia* and *Pisonia* wood and on dead *Melanthera* stems.

A few other types of fungi were observed from time to time, including a species of Clavariaceae in the *Pandanus* grove and a yellow one of the Dacrymycetales on dead wood.

Lichens (mostly still unidentified) were abundant on the trunks of many trees and shrubs. They were sampled by wrapping a string around a trunk of a tree and recording the lengths of the string intercepted by each species. Three heights above the ground (25 cm, 1 m, 1.5 m) were employed on each of 10 *Pisonia* trees and 10 *Argusia* trees.

The average total lichen cover on *Pisonia* trees was 32% at heights of 25 cm and 1 m but decreased to 25% at 1.5 m. There were 5 species involved, all crustose. The main one, a Script Lichen, accounted for most of the area covered by lichens (76%, 80% and 62% from the lower to the higher zones respectively). It occurred on 90% of the trees. The second most common species occurred at all heights and was found on 80% of the trees. It accounted for much less of the lichen area than the graphidaceous one (6%, 19% and 38% in the zones from low to high, respectively). The remaining three species were relatively unimportant, each occurring on only 10% of the trees, and then found in only one zone (two in the lowest zone and one in the middle zone). They accounted for 1–13% of the area covered by lichens in the zones they occupied.

On *Argusia* trunks there were seven species, none of which were the same as those on *Pisonia*, and most of which were of the foliose type. Lichen cover on *Argusia* was much sparser than on *Pisonia.* The average total lichen cover was 21% of 25 cm, 6% of 1 m and 8% of 1.5 m. The most abundant species occurred on 90% of the trees and accounted for 77%, 57% and 12% of the lichen-covered area at 25 cm, 1 m and 1.5 m respectively. The second most common species occurred on 60% of the trees and never accounted for more than 20% of the lichen-covered area at any height. The five less common species were on 20% or less of the trees, were usually at 1 m or higher and none ever accounted for more than 3% of the total trunk area and only twice (42%, 30%) for more than 18% of the lichen-covered area.

The *Pandanus* were nearly lichen free. Many trunks had none; those with lichens had very low densities. Only two species were involved, perhaps two of the species on *Argusia*. Some dead stems had greater lichen densities.

General features of the flora

Dispersal

The One-Tree Island flora is largely sea-dispersed, almost half of the species whose major dispersal means is known, falling into this category. The next most prevalent means of dispersal is by birds, especially as diaspores carried externally. Man's activities dispersed a number of species and wind only one (Table 5). There was a great difference in dispersal means between the established species and the transient ones. Over 3/4 of the established species were either sea- or bird-dispersed whereas less than a fifth of the transient species were. By contrast wind and human transport accounted for over 3/4 of the transient species but none of the established ones (Table 5). Most of the human-transported transients were domestic plants believed to have reached the island in the stores of expeditions (e.g. tomatoes, pumpkins, melons). Of the species whose major mode of dispersal could not be ascertained with certainty, most appear to disperse by combinations of sea, bird and human means; the 'uncertain' category may be realistically viewed as containing species with a mixed mode of dispersal.

The relative frequency of dispersal mode of the One-Tree flora is similar to that of some other islands in that sea dispersal predominates. Guppy (1891) stated for the Cocos-Keeling Atoll 'in comparison with the westerly drift [of sea currents], birds have played a secondary part in stocking these islands with plants'. Howard (1950) found that 42% of the species in the flora[5] of the Bimini Group in the Bahamas were sea-dispersed; man accounted for 29%, bird-dispersal for 16% and wind for 12%. Davis (1942) noted that sea-dispersal was the most common dispersal mode for the strand and dune plants of the Florida sand cays. Brown *et al.* (1917) summarized previous data from Krakatau as indicating that 39% of flora was certainly transported by sea, 10% certainly by birds and 16% certainly by wind, with maximum possible percentages being 72, 19, and 32 respectively. Docters van Leeuwen (1936) reviewed results of subsequent expeditions and traced the changes in proportions of the flora arriving by different dispersal means. He noted that by 25 years after eruption sea dispersal was the greatest single category but then reached an asymptote, relatively few additional species arriving by that means in later years, whereas wind-dispersed species continued to arrive in rather large numbers. Animal-dispersed species didn't appear until several years after eruption, after which the number continuously increased. It was about 25 years after eruption before human-dispersed species arrived; their number fluctuated but was never very large. By 1933, 50 years after eruption, the tally was 41% of the vascular flora (including ferns) wind-dispersed, 28% sea-dispersed, 25% dispersed by birds or bats, and the remaining 6% human-dispersed or dispersed by unknown means, i.e. a considerable change from the earlier pattern.

Dispersal mode on One-Tree Island differs considerably from that of some other islands. Wicken (1979b) reports that the largest proportion of the native flora on Aldabra Atoll is believed to have been carried by birds, mainly internally. Seed dispersal was high, as indicated by the large number of viable diaspores washed up annually but was reduced in importance because of failure of most to become established. Cultivated plants and weeds brought by man were extremely successful. Wind transport (as at One-Tree Island) was the least effective.

The original flora of the Galapagos Island has only 9% of its species believed to have resulted from oceanic drift, the majority having been brought by birds (60%); there was a high proportion brought by wind (31%). Many more recent introductions have been made by man (Porter 1976).

Carlquest (1974) reviewed the relative frequency of dispersal modes

[5] Calculated from Howard's data, excluding species that were cultivated, were endemic or whose dispersal mode was unknown, as dispersal mode was not listed for any of these categories in Howard's tables. Cultivated species are obviously transported by man but constitute a special case.

Table 5. General characteristics of the flora of One-Tree Island.

	All species		Established species only		Transient species only	
	Number of species	%	Number of species	%	Number of species	%
Major means of dispersal						
Sea currents	13	42	12	48	1	17
Birds (externally)	6	19	6	24	0	0
Birds (internally)	1	3	1	4	0	0
Wind	1	3	1	4	0	0
Human	5	16	0	0	5	83
Uncertain	5	16	5	20	0	0
Distribution						
Widespread strand species	12	39	12	48	0	0
Widespread ruderals	4	13	4	16	0	0
Species with either more restricted distribution or with other type of ecology	15	48	9	36	6	100
Salt tolerance						
Halophytes	14	45	13	52	1	17
Non-halophytes	14	45	9	36	5	83
Uncertain	3	10	3	12	0	0
Type of life history						
Perennial	17	55	16	64	1	17
Biennial	2	7	2	8	0	0
Annual	12	39	7	28	5	83

Leaf size category						
Nanophyll	1	3	1	4	0	0
Microphyll	20	65	15	60	5	83
Mesophyll	8	26	7	28	1	17
Macrophyll	2	7	2	8	0	0
Megaphyll	0	0	0	0	0	0
Perennating bud						
Phanerophytes	9	29 (46)[a][26][b]	8	32	1	17
Chamaephytes	6	19 (9) [7]	6	24	0	0
Hemicryptophytes	0	0 (26) [18]	0	0	0	0
Cryptophytes	3	10 (6) [7]	3	12	0	0
Therophytes	13	42 (13) [42]	8	32	5	83
Types of flowers						
Bisexual	28	90	24	96	4	67
Unisexual	3	10	1	4	2	33
Vegetative reproduction only	0	0	0	0	0	0
Mode of pollination						
Wind	4	13	3	12	1	17
Insect	22	71	18	72	4	67
Unknown	5	16	4	16	1	17
Capable of vegetative reproduction						
Yes	5	16	4	16	1	17
No	20	65	16	64	4	67
Uncertain	6	19	5	20	1	17

[a] Raunkiaer normal spectrum;
[b] Spectrum for Death Valley, California; data from Oosting (1956, p. 76).

believed to have brought the ancestors of the present-day plant species to a variety of Pacific islands. The most important mode varied from island to island. The two most often of prime importance were sea currents and internal transport by birds.

One of the few cases where wind was believed to be the most dominant mode of dispersal to an island was in the study of Davies *et al.* (1938). They estimated that of the 18 species they believed to have reached the Five Islands by oversea dispersal (as opposed to those left there at the time the island was separated from the mainland) 50% were carried by wind, 28% by sea, 17% by birds and 6% by man.

Newly formed volcanic islands in freshwater lakes seem to be revegetated primarily by seeds that are bird-dispersed. Ball and Glucksman (1975) implicate ducks as the prime dispersal of plants to Motmot Island (Lake Wisdom, New Guinea) and Brown *et al.* (1917) estimate that 54% of the species of plants recolonizing Volcano Island (Lake Taal, Philippines) after eruption were bird dispersed, 21% wind dispersed, 9% water dispersed and 20% dispersed by man or his domestic animals (total is more than 100% because some species were included in more than 1 category).

Mode of dispersal may be related to island size. Nilsson and Nilsson (1978) in a study of islands in a Swedish freshwater lake, found that animal-dispersed species of plants were proportionally fewer (in comparison to water-dispersed and wind-dispersed plants) with decreasing island size. They suggested the following explanation: (1) animal-dispersed species have more difficulty in reaching the islands than water-dispersed or wind-dispersed ones; (2) proportionally more species become extinct and turnover is more rapid on small islands than on larger ones and consequently the former tend to accumulate the more easily dispersed species.

Most plants that reached One-Tree Island by water probably had their origin on islands or coasts to the north or northwest as the sea currents are predominantly from that direction most of the year; the rest of the time the current comes more from the east (Endean *et al.* 1956, Maxwell 1968) and transport from distant localities from that direction would be rare.

Ecological type

Two ecological categories of plants were conspicuous in the One-Tree flora, strand plants and ruderals (Table 5). Together they made up slightly more than half the flora. The remainder contained the transient species, including introduced domestic plants. When established species only were considered, strand species accounted for almost half (48%) and ruderals for 16% of the species.

Like strand and ruderal species, many of the remainder are eurytopic and widely distributed. For example, four of the One-Tree species (*Ipomoea pes-caprae*, *Melanthera biflora*, *Digitaria ciliaris* and *Portulaca oleracea* were among the colonizers of Volcano Island in Lake Taal (fresh water), Philippines, after eruption in 1911 (Brown *et al.* 1917). Thus, the flora of One-Tree Island overall is dominated by widespread eurytopic species. Similarly, Porter (1976) has referred to the Galapagos Islands as a jumble of open pioneer habitats inhabited by a weedy flora.

Germination

Data on germination success under favourable conditions are available for eight of the One-Tree species (Hnatiuk 1979). Percent germination was as follows: *Achyranthes aspera* 45%, *Ipomoea pes-caprae* 55%, *Pisonia grandis* 100%, *Portulaca oleracea* 25%, *Scaevola taccada* 80%, *Sesuvium portulacastrum* 20%, *Sophora tomentosa* 90%, and *Argusia argentea* 5%. For several of the species scarification was either a requirement for, or an aid to, germination. None of the *Argusia* seeds germinated without scarification,[6] and when the seeds were not scarified, germination success dropped to 20% in *Ipomoea*, to 75% in *Scaevola* and to 85% in *Sophora*. Other species were not scarified.

The time required for germination varied widely among species, and were as follows: *Achyranthes* 16–144 days, *Ipomoea* began at 13 days but upper limit not known, *Pisonia* 8–13 days, *Portulaca* 6–15 days, *Scaevola* began at 14 days, but upper limit not known, *Sesuvium* 11–40 days, *Sophora* began at 4 days but upper limit not known, and *Argusia* seeds germinated after 94 days. Holm *et al.* (1977) report that results of laboratory attempts to germinate *Tribulus* have been highly erratic.

As Hnatiuk (1979) has commented on these (and other) species in the context of Aldabra Atoll, a wide variety of germination strategies are represented. Some germinate rapidly after receiving water and complete germination rapidly and appear able to exploit short spells of wet weather. Others show staggered germination which ensures steady recruitment of young plants to the population such that even if a spell of adverse conditions wipes out existing plants, ungerminated seed would still remain to germinate later and maintain the species (for a discussion in relation to *P. oleracea*, see a previous part of this chapter). Harper (1965) and Mayr (1965) have suggested polyphenism as a mechanism assuring different seeds germinate in different years. This phenomenon is more common in colonizing, weedy species than in others.

Halophytism

Characterization of a flora in terms of its proportion of halophytes is difficult because of the variety of criteria used to define halophytism. Ecological

[6] The poor germination of *Argusia* in these experiments is in contrast to the statement by Guppy (1891) that this species, along with *Scaevola* and *Cordia subcordata*, were the 3 species most ready to germinate during the first month in the ground, of the 9 species he tested on Cocos-Keeling Atoll.

definitions range from those encompassing only species growing exclusively on salty soil to those which include species tolerant of salty conditions, but which grow elsewhere as well. In addition to the problem of obligate versus facultative halophytism, definition of the salt levels required for a soil to be considered 'salty' poses additional problems of ecological categorization (Duncan 1974, Caldwell 1974). A physiological definition might encompass those species which can accumulate high levels of salts in the tissues; however, as noted above, some species (especially legumes) growing in quite saline habitats do not do this.

For characterizing the One-Tree flora, the criteria used were the following: all plants capable of accumulating high levels of chlorides in the tissues were considered halophytes as well as those which characteristically or frequently occupy soils of high salt content or are commonly subjected to high levels of salt spray; all species listed by Duncan (1974) as halophytes were so considered for the present analysis.

Obviously all species inhabiting One-Tree on a long-term basis must be tolerant of a certain amount of salt. Yet, salt content of the substrate may decrease markedly just behind the upper beach (Barbour 1970a) and not all island plants are necessarily halophytes even by the broad criteria outlined above.

Less than half of One-Tree's total flora were in fact halophytic (Table 5). Transient species were mostly non-halophytic and their ephemeral existence may indeed be related to their inability to cope with local levels of salt. More than half of the established species are halophytic, and of these, three (10%) are known to tolerate high salt loads in their tissues; others probably do so also.

Life history

More than half of the One-Tree plant species are perennial. The proportion among established species is even higher (64%) as the annual category was swelled by the transient domestic plants (Table 5). Several species can adopt either an annual or perennial life history (the distinction is not always sharp in tropical and subtropical species) and the assignment of species to a given category was somewhat arbitrary. However, all possible classifications led to the conclusion that the majority of the flora does not rely entirely on seeds to carry the population from one year to the next.

Hart (1977) has suggested that the advantage of particular life-histories depends on the probabilities of survival during the 1st, 2nd and subsequent years. He indicated, on the basis of a mathematical model, that perennials would be expected in stable, closed environments where survival from the seed stage to the end of the 1st growing season was low but the fraction of first year plants that survived to the second season was high. As examples of habitats favouring perennials he cited climax grasslands and forests. He predicted that annuals should prevail in sites subject to seasonal disturbance, i.e. where survival of first year plants into the second year was low, and listed vernal pools and cultivated fields as examples. Biennials should be found in situations where survival is likely for at least a few years, such as mid-successional stages or forest openings. Observed proportions of herbaceous species in the three life-history categories coincided with expectations. In groupings of habitats into five categories from 'closed' to 'open' and with varying degrees of permanency, he found a progressive increase in proportion of annuals, from closed to open habitats with a corresponding decrease in perennials; biennials reached their maximum proportion in intermediate habitats, especially intermittent woodland ones such as tree falls or burns.

It is difficult to relate the One-Tree plants to such a spectrum. First of all, the lack of well-defined seasonality in the maritime subtropics makes this distinction between the various life-history categories less meaningful than in temperate regions. Secondly, there are several different habitats on One-Tree Island, each with its own vegetation. If only the herb flat is considered (Hart's spectrum does not include aborescent vegetation) a more direct comparison can be made. Of the 20 established species of plants in the herb flat, 12 (60%) were perennial 2 (10%) biennial and 6 (30%) were annual. This agrees with Hart's observation that in general, biennials are rather uncommon. The closest of Hart's categories (26.1% annuals, 26.1% biennials, 47.8% perennials) to the herb flat was the one in the centre of his list of five; he designated it as 'semi-open' and gave short turf and partially colonized areas subject to grazing as examples. The numbers of herb flat species in the three life-history categories were tested by χ^2 analysis against values expected on the basis of Hart's semi-open habitats. The difference was not significant ($0.3 > P > 0.2$).

Life form

Most species are small-leaved (microphyllous), a characteristic of floras from relatively dry regions. This probably reflects functionally dry conditions on One-Tree. Although rainfall is relatively abundant, the substrate is porous and facilitates rapid drainage. Davis (1942) has previously commented on the xeromorphic nature of strand vegetation generally.

The adaptation of the flora to physiological dryness is also reflected in the location of the perennating bud. Therophytes comprise the largest category (42% as compared to the Raunkiaer Normal Spectrum of 13%, see Table 5). A flora heavily endowed with therophytes is characteristic of arid regions (Cain 1950) and the life-form spectrum of One-Tree is, in fact, similar in some respects to that of Death Valley, California, a rather extreme desert (Table 5).

Hnatiuk and Merton (1979b) have reviewed the available data from atolls and tropical islands relative to life-form spectra. They note that Aldabra has a high proportion of phanerophytes (48%), chamaephytes (23%) and therophytes (13%), or in comparison to the Raunkiaer normal spectrum the flora was shifted toward chamaephytes at the expense of cryptophytes and hemicryptophytes, with about the same proportion of phanerophytes. The sand cays of the Hawaiian Islands also had high proportions of chamaephytes and phanerophytes, the combination which seems to be characteristic of atolls and sand cays, as other types of tropical islands they compared did not have a dominance of chamaephytes, and floras dominated by chamaephytes but without a high proportion of phanerophytes tend to occur in cold areas. Davis (1942) also found high proportions of phanerophytes, chamaephytes and therophytes on Florida sand cays.

In simplistic terms, 'deserts have therophytic climates, the tropics are phenerophytic (*sic*), Alpine and Arctic climates chamaephytic, and most temperate regions are hemicryptophytic' (Oosting 1956). Hnatiuk and Merton (1979b) interpret their comparison to indicate that chamaephytism on atolls is a response not to cold but to unfavourable water supply owing to either low rainfall or porous soil.

One-Tree Island shares the feature of chamaephytism accompanied by a relatively high proportion of phanerophytes, but differed in having a high proportion of therophytes.

Thus, the flora of One-Tree Island seems to have a spectrum typical of sand cays and atolls except that it moves further towards therophytism, probably an expression of its drier conditions imposed by the nature of its substrate and periodic droughts. Its life-form spectrum is as might be expected for the flora of a dry tropical cay.

Brynes *et al.* (1977) have analyzed the vegetation of Lizard Island, a continental island off the eastern coast of Queensland, Australia. It is much larger and ecologically more diverse than One-Tree Island. The majority of vegetation types on Lizard Island has phanerophytes as the largest single category (rainforest, low closed forest, closed scrub, *Thryptomene* closed scrub, woodland, heathland, mangroves, and strand), with the proportion of this group being 28–89% of the plant species.[7] The exceptions were herbland and grassland with high representation of chamaephytes (38% and 42% respectively) and therophytes (35% and 35% respectively), *Pandanus* swamp with 56% hemicryptophytes, and coastal dunes which was the only habitat with a spectrum similar to that of One-Tree Island. The coastal dunes had 21% chamaephytes, 18% hemicryptophytes, 3% geophytes and 36% therophytes. The major difference between the life-form spectrum of this vegetation and that of One-Tree Island is that the latter had no hemicryptophytes.

The semi-arid islands of the Nuyt's Archipelago off of South Australia have

[7] Calculated from tabular data provided by Byrnes *et al.* (1977).

high proportions of phanerotypes (14–49%), chamaephytes (22–38%) and therophytes (20–47%) with no other categories ever exceeding 7% (Osborn 1922, 1925) and consequently have a spectrum resembling that of One-Tree Island despite being in the temperate region. Osborn (1922) attributed the high proportion of therophytes to aridity.

Davis *et al.* (1938) presented the life-form spectrum of the vegetation of Five Islands, off of Port Kembla, N.S.W., Australia. It differed from that of One-Tree Island by having relatively few phanerophytes (11%), more chamaephytes (31%) and hemicryptophytes (17%), but had a similar proportion of therophytes (39%).

The combination of high proportions of chamaephytes and therophytes on such latitudinally dispersed and varied islands as One-Tree Island, Lizard Island, the Nuyt's Archipelago and the Five Islands raises the issue as to whether this combination may not have to do with insularity. Further study is required to resolve this problem. Ehrendorfer (1965) suggested that perennial chamaephytes and hemicryptophytes and annual therophytes represent adaptations to different types of pioneer habitats.

Sobey and Kenworthy (1979) have shown that in Scotland gull-affected sites had relatively high numbers of therophytes in comparison to those unaffected by gulls. The high proportion of therophytes on One-Three Island may be in part because it is a bird-influenced community. However, other similarities in the life-form spectrum between the two areas did not exist, and the high representation of therophytes in the two areas may be fortuitous.

In contrast to small islands off of southwestern Western Australia on which a high proportion of the common species were succulent-leaved (Abbott 1980), only three species of plants on One-Tree Island were succulents.

Breeding system

There are two major considerations regarding the relative advantages of autogamous verses exogamous breeding systems among insular plants.

One factor discussed by Baker (1955), is that species that can reproduce either sexually or asexually from a single individual would seem to be more likely to become established on islands than those that require the simultaneous arrival of two or more individuals. For this reason, hermaphroditic, self-compatible species might be expected to be more characteristic of insular floras than dioecious, self-incompatible, or obligately cross-pollinating species, or those with characteristics tending to favour outcrossing such as monoecism, heterostyly and dichogamy, unless species in the exogamous categories reproduced vegetatively.

Studies of the Plumbaginaceae (Baker 1953) and of some animals as well (notostracan crustaceans; Longhurst 1955), suggested that self-compatibility and hermaphroditism were related to establishment after long-distance dispersal. Looked at in this way, autogamy could be considered an insular adaptation.

The other consideration is that in isolated populations, founded by a few individuals, the variability of the mainland relatives is no longer accessible, and that variability which remains, or which can be developed, should be circulated throughout the entire population if the species is to be successful on a long-term basis (Carlquist 1966, 1974).

Viewed from this standpoint, a mechanism favouring outcrossing could be construed as an adaptation to insular life, and may outweigh the disadvantages of problems in establishment attendant to exogamy. Woodell (1979) points out that propagules are often carried in large numbers together, thereby increasing the likelihood of even dioecious species colonizing an island.

The data from One-Tree Island are pertinent to these ideas. All species except the dioecious *Pandanus* in the One-Tree flora were hermaphroditic or monoecious, and all except *Pandanus* and two transient domestic plants had bisexual flowers (Table 5). Only five species (16%) reproduced vegetatively and none of them do so exclusively. Finally, a number of the species are dispersed by fruits which either float or adhere to birds, each fruit containing a number of seeds. Consequently the transport capabilities emphasize dispersal of a number of seeds simultaneously.

Carlquist (1966) has suggested anemophily as an adaptation permitting out-crossing. On One-Tree Island four species are wind-pollinated. The rest are probably insect pollinated (Table 5), thereby necessitating presence of appropriate insects for establishment to occur. On One-Tree there is an abundance of general pollinators among the insects and immigration rates of such species are high (see Chapters 14 and 21); consequently obligate insect-pollination would not be a limiting factor in the establishment of a species unless it relied on specific pollinators.

Woodell (1979) in a study of Aldabra found that at least two general pollinators occurred on the island, a nectar-feeding sunbird (*Nectarinia sovimanga*) and a cetoniid beetle (*Mausoleopsis aldabrensis*). A number of genera (in some cases the same species) of plants also found on One-Tree Island were visited by one or both of these pollinators. The genera were: *Scaevola, Argusia, Ipomoea, Pandanus* and *Sophora.* The beetle was especially important in that it visited many species of flowers including those with a wide variety of geographic affinities, native and introduced, and including flowers of many colours and floral structures. Woodell drew attention to the similar, generalized pollinating role that a carpenter bee (*Xylocopa darwinii*) is reported by Linsley *et al.* (1966) to play in the Galapagos,[8] and concluded that the 'presence of an abundant non-discriminatory generalist pollinator

[8] It is interesting to note that both on the Galapagos Islands and Aldabra *Portulaca oleracea,* though common, was not visited by the generalist pollinators nor were other prostrate herbs in the latter locality; Woodell (1979) suggested it was because the insects couldn't take flight by dropping from below flowers.

. . . must be an important factor favouring the chances of new immigrant species becoming successfully established . . .'

Woodell also felt that generalist pollinators and reproductive system of the plant were related. With generalist pollinators available, out-crossing would be feasible, and he suggested that it would be an advantage, as a plant with a great number of flowers would have an added attractiveness to pollinators and thereby gain in competition for pollinators in short supply.

The One-Tree situation can be compared to that of Aldabra at least to a certain extent. One-Tree Island has a predominance of out-crossing species of plants and has a number of generalist pollinators. Whether these two characteristics are related in the way Woodell hypothesized must await further research. In this context it would be essential to ascertain whether the proportion of out-crossing species of plants and of generalized pollinators differs from those of the mainland.

Combining the information on pollination and breeding systems is necessary for an understanding of the insular adaptation of plants. The theories of both Baker and Carlquist (see above) contain merit. We feel than an eclectic view is probably the most realistic as the two views refer to different aspects of insularity. Baker's hypothesis that inbreeding is an advantage for long distance dispersal is concerned with the process of establishment on an island, whereas Carlquist's ideas are more related to the advantages of out-breeding in an isolated population after it has become established. The relative proportions of in-breeding versus out-breeding species on an island will depend on the characteristic of the island under consideration. For example, a remote island is less likely to experience multiple invasions of a given species simultaneously than is an island near the source of the propagules. Consequently, the more remote the island the more likely is self-compatibility to be critical for establishment, and the more accessible the island the less likely it is to be important.

Another consideration is ease of dispersal. A species that is easily transported by birds or sea currents will reach a given island more frequently (and hence have multiple invasions more often) than one that experiences great difficulties in crossing water gaps, i.e. an island is functionally more remote for the latter type than for the former.

Two predictions arise from these considerations: (1) insular plants dispersed over water with difficulty will have a lower proportion of dioecious species, or self-incompatible species than those characteristically dispersed by sea birds and water currents. (2) For a dispersal type, the proportion of the species which are dioecious or self-incompatible will be smaller on very remote islands than on less remote ones. Appropriate confirmation of these predictions should be sought.

One factor which complicates interpretation is the evolution of breeding systems that may occur after establishment on an island, and this is where Carlquist's theory becomes especially important. A species may become established on a remote island because it can begin a population from one seed. After establishment, however, the advantages of out-crossing may be sufficiently great that mechanisms affecting it will become fixed. Thus, the predominant reproductive system in the flora of a remote island may be quite different from that of the species which originally colonized. Analysis of the present day condition might indicate a high proportion of out-crossing species, and support Carlquist's theory, but ignore the historical aspects and unrealistically ignore the importance of Baker's hypothesis in the early stages of establishment. It is important to recognize that Baker's theory is relevant to colonization and establishment, Carlquist's to established isolated populations. Both Baker (1967) and Carlquist (1974) recognized this fact. The former gave an example of changes from self-compatibility accompanying a dispersive phase in the evolution of plants, to a later development of out-crossing. He points out that 'a high proportion of dioecism in a flora is what would be expected if there is a selective pressure favouring out-crossing but a shortage of self-incompatibility to enforce it.' Self-incompatibility is a rare and unlikely event and can break down easily; he suggests that dioecism is an evolutionarily easier alternative, and may account for the high proportion of dioecious plants in some remote archipelagos. Both authors agreed that the advantages of out-crossing to survival did not rule out the potential advantage of autogamy for establishment.

In contrast to the One-Tree plants of which probably 90% or more are autogamous, Carlquist (1974) indicated that the Hawaiian native flora has about 74% of its species with some reproductive mechanism fostering out-crossing (diclinous in some way, or dichogamous, etc.), and about 20% are wind pollinated. He developed an index of tendency toward outcrossing from zero (most autogamous) to five (most exogamous). He noted that in the Hawaiian flora values were generally high except for pioneer species, beach species, and a few genera from other habitats, which tended to be rated 0 or 1. The Galapagos flora also tended toward autogamy, and Carlquist considered it almost entirely a pioneer flora. The mean rating for the One-Tree flora was 1.8, 16 of the values being taken directly from Carlquist's listing and the rest constructed according to his general criteria. When only the established flora was considered, the mean rating dropped to 1.4. If one were to theorize about the optimum reproductive system for a species of plant that occupies far flung islands and archipelagos, such as some of the species occupying One-Tree Island, it might combine elements of both strategies. The species would need to colonize and establish easily, yet not suffer the long-term disadvantages of inbreeding. Allard (1965) has suggested that good colonizers are species whose 'genetic systems represent a compromise between the high recombinational potential of outbreeding species and the stability traditionally postulated for self-pollinated species'. An hermaphroditic self-compatible species, but which under appropriate conditions (conspecific plants available) would normally out-cross, would be such a species; an insect-pollinated rather than a wind-pollinated one would be indicated. Except that the extent of self-pollination is unknown, this description characterizes the common condition of the One-Tree flora. Thus, many of the One-Tree plants may optimize advantages for insular life by combining elements bestowing the advantages of out-crossing without sacrificing their success in establishing themselves on remote islands. Indeed, it has been suggested that two species of One-Tree plants follow just that strategy. Wilson (1977) interprets the floral morphology of *Ipomoea pes-caprae* as facilitating pollination by a variety of insects but providing an emergency selfing mechanism if cross-pollination has not been effective. He also said that '*Canavalia maritima* [= *C. rosea*] is apparently facultatively capable of self-pollination based on fertility of greenhouse plants' but that 'the presence of large bees at the flowers suggests outcrossing to be normal'.

Such a strategy would seem to be equally advantageous for ruderals and for strand species, the two groups which combined account for 52% of the total species on One-Tree Island and 64% of the established species (Table 5).

In conclusion, the One-Tree flora can be characterized as one which is largely sea-dispersed or bird-dispersed. Almost all species have bisexual flowers. Most can probably self-pollinate and be self-fertile but normally are cross-pollinated by insects; few have vegetative reproduction. There is a high proportion of (1) eurytopic species (often salt tolerant), (2) perennials, (3) chamaephytes and therophytes, and (4) small-leaved species. The general impression is of a flora which has wide ecological tolerances, is adapted to physiological dryness, and has properties favouring overwater dispersal.

These attributes can be related to Ehrendorfer's (1965) review of herbaceous species that are good colonizers. He pointed out that there were three main types that were successful colonizers, I – perennial polyploids, II – annual selfers, and III – annual dysploids. These can be looked upon as three different adaptive strategies in relation to colonizing ability. However, all three of these types had some features in common. They (1) were easily dispersed, (2) extended germination over a long period of time, (3) had a wide amplitude of plasticity and modifiability, (4) had relatively rapid individual development, (5) had rapid reproduction with a large production of progeny, and (6) possessed a genetic system permitting a balance between flexibility (high rate of variation output to be subjected to selection) and stability (possibilities for fixation and multiplication of successful biotypes). Not all of these characteristics are known for the species of the One-Tree flora. However, where data are available they tend to support Ehrendorfer's conclusions. For example, almost all the One-Tree plants had adaptations favouring wide dispersal, are eurytopic, and at least some of the species mature when very young (e.g. *Tribulus, Trachymene*) and/or produce large numbers of seeds (e.g. *Tribulus, Portulaca*); some have very extended germination times (e.g. *Portulaca*) (see Chapters 11 and 18). The genetic systems have already been discussed at length.

9. Plant biomass

At the beginning of the study the total biomass of living vegetation averaged slightly over 550 g dry wt/m^2 in the herb flat. This value was obtained in 1968 just before the drought. In the next sampling period (early 1969 after the drought) the plant biomass was reduced to less than half the pre-drought level (Figs. 6–3, 6–4, 25). Subsequently biomass gradually increased, but it was ten months (October 1968) before the original pre-drought levels were again reached. In 1970, biomass reached values even higher than the pre-drought level in the first part of the year but dropped to near the 1968 values in the latter part of the year, following the normal dry season. The wet period from the end of 1970 into early 1971 and the rain brought by 'Cyclone Fiona' was reflected in the high plant biomass in February–March 1971. The single value in April 1972 was approximately at the 1968 level and not unusually high despite 'Cyclone Emily'. Note, however, that that storm did not bring unusually high values of rainfall for the month of April.

Biomass in January 1976 was about at the lower end of the normal non-drought level after a year of nearly average rainfall for most months. In

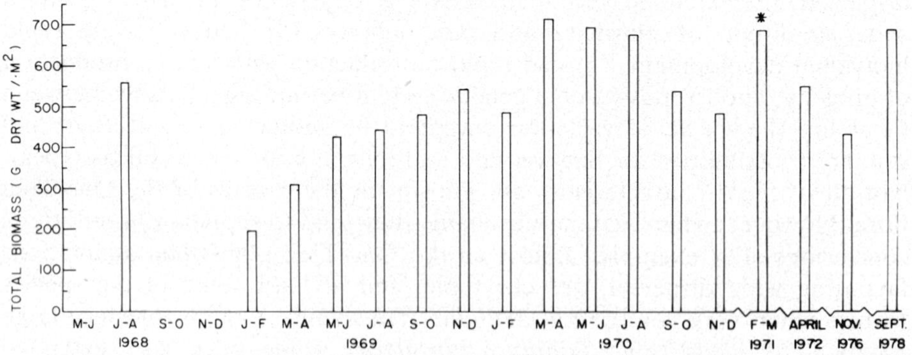

Fig. 25. Temporal change in total plant biomass (all species) in the herb flat. *Indicates that root biomass was not included. Letters are initial letters of names of months.

September 1978 biomass was high. There had been a drought eleven months previously and the early part of 1978 was slightly below the long-term average. However, for four months previous to the biomass determination, rainfall had been at or slightly above the long-term average and recovery from any effects of the drought was probably advanced.

It appears that unless unusual weather conditions occur, seasonal fluctuations in plant biomass are of low amplitude (from about 500 g dry wt/m^2 following the dry season to about 700 g dry wt/m^2 following the rainy one).

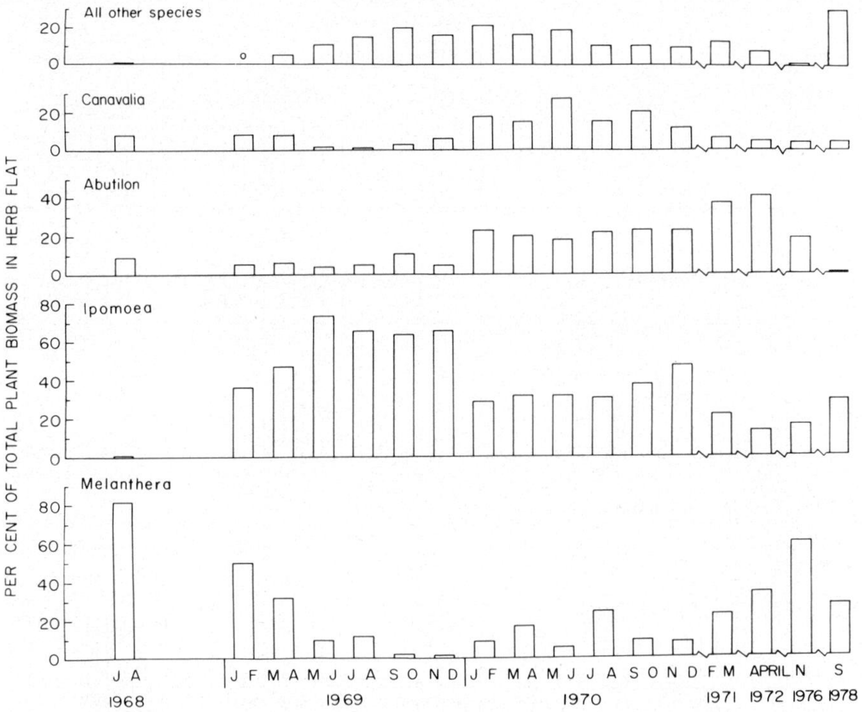

Fig. 26. Percentage of the total plant biomass of the herb flat contributed by the component species.

However, plant biomass on the herb flat seems to be particularly responsive to short-term rainfall patterns as evidenced by the facts that (1) it was reduced during the 1968 drought, (2) there was no reduction during the normal time for the dry season in 1969 when there were never two consecutive months of low rainfall, i.e. the dry season did not actually occur and (3) there was a rapid increase in biomass in February-March 1971, after some very wet months. This relationship between rainfall and plant biomass probably results from the limited water retaining capacity of the coral rubble substrate. The rubble is highly porous and permits rapid percolation of rain from the surface layers. Accumulated organic humus is scant and superficial and quickly dries to wilting point (see soils section). The plants are thus dependent upon evenly distributed rainfall for a continuous water supply. Even short droughts can have a significant effect on total plant biomass.

The pattern of temporal change in biomass shown by individual species varied and did not always conform to that of the total vegetation. *Melanthera biflora* was the dominant species in the herb flat prior to the drought (over 80% of the total plant biomass there; Fig. 26). After the drought its biomass dropped to less than a fourth of the pre-drought level (Fig. 27). It fluctuated

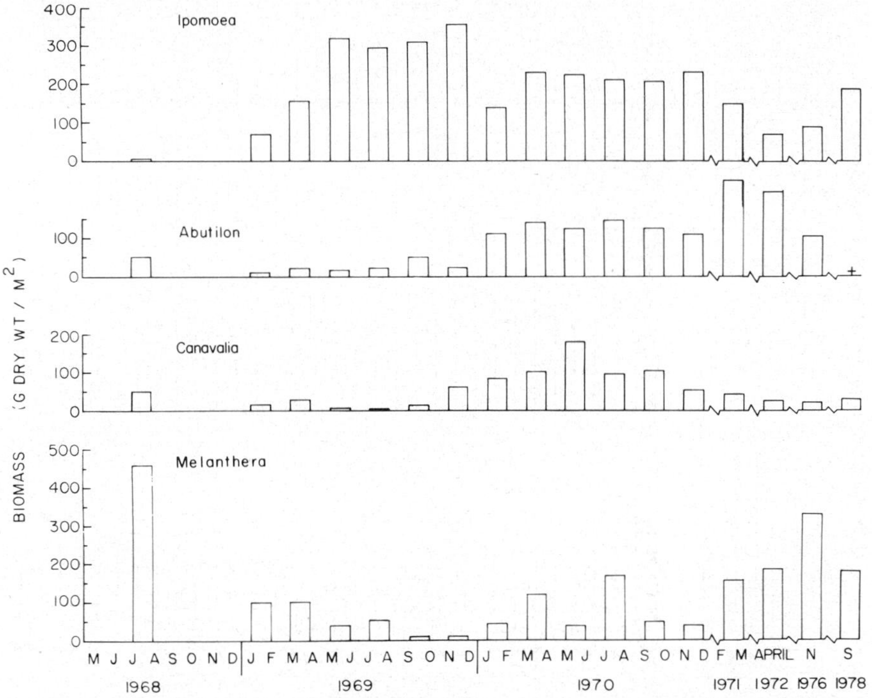

Fig. 27. Temporal change in the biomass of the four most important plant species in the herb flat.

somewhat throughout the remainder of the study but never recovered as much as half of the original biomass, even by April 1972. By 1976 it had returned to about 3/4 the original value but in 1978 was again low, probably because of the drought in 1977.

The pattern was quite different for *Ipomoea* which before the drought contributed a relatively insignificant proportion of the total herb flat biomass (Fig. 26). After the drought when *Melanthera* biomass was much reduced *Ipomoea* biomass increased rapidly until by May 1969 it was more than 60 times the pre-drought level. It remained at that general level (300–350 g dry wt/m^2) throughout the rest of that year but dropped in 1970 to a rather consistent level of about 2/3 the 1969 values.

Abutilon and *Canavalia* (Figs. 26, 27) had similar patterns of change in biomass during the first half of the study. Initially both had a biomass slightly over 50 g dry wt/m^2, which dropped by more than half following the drought. The biomass of both species remained at low levels until the end of 1969 when there was recovery to the original levels. In 1970 biomass rose to more than double the pre-drought value. *Abutilon* maintained this high level throughout the year and subsequently increased even more to the 1971 and 1972 values of nearly 200 g dry wt/m^2. It decreased again in 1976 and was at a very low level in 1978. *Canavalia* on the other hand peaked in June 1970 and subsequently declined; 1971 to 1978 values were of the same order of magnitude as before the drought.

Minor species of the herb flat collectively contributed small percentages of the total biomass and were poorly represented or absent in the samples of the initial phases of the study (Fig. 26). During the recovery phase after the drought both their absolute biomass and their proportional representation initially increased (Tables 6, 7; Fig. 26) and then declined (late 1970 and through the 1971, 1972 and 1976 sampling period; values were high again in 1978). Several individual species showed a pattern similar to that of all of the minor species collectively. They were *Achyranthes, Plumbago, Lepturus* and *Stenotaphrum*; in most cases the biomass of these species did not drop as low as the initial 1968 levels; the latter two did not show an increase in 1978. *Solanum* was a special case and is discussed in detail elsewhere (see Chapter 18).

Tetragonia was too rare in the samples to allow meaningful estimates of biomass. In general, when plant densities of minor species were low in the herb flat, species that were present but rare might not occur in the biomass samples merely by chance. This accounts for many of the zero values in Tables 6 and 7 (*Solanum* excepted). The larger area sampled by the releveés permits a better assessment of the minor species than does the biomass sampling. They will be treated in more detail under that section.

Because of the different patterns of temporal change in biomass in different species, there were shifts in relative contribution to total biomass among the four major species. Before the drought *Melanthera biflora* was undisputedly the dominant plant in the herb flat. This was clear just from casual inspection as the appearance of the vegetation was mainly attributable to this species (Fig. 6). Even though *Melanthera* suffered considerable mortality during the drought it still contributed half the biomass in early 1969 when total absolute levels were low. Subsequently it declined, and made up a relatively minor proportion of the biomass by late 1969. It increased again during subsequent years but although recovering somewhat it did not re-establish its dominance. By 1972 it accounted for about 34% of the total biomass (Fig. 26).

Table 6. Percent of total leaf biomass of the herb flat contributed by individual species.

	1968	1969						1970						1971	1972	1976	1978
	J–A	J–F	M–A	M–J	J–A	S–O	N–D	J–F	M–A	M–J	J–A	S–O	N–D	F–M	April	Nov.	Sept.
Species																	
Melanthera	81.5	54.4	6.8	5.6	19.6	6.6	3.2	7.4	15.3	6.5	14.6	12.1	8.3	24.2	48.7	69.8	47.1
Canavalia	7.2	7.8	9.4	2.3	2.7	9.5	25.8	22.1	20.4	30.2	44.0	37.1	23.2	7.4	4.3	13.1	10.6
Abutilon	8.2	2.9	17.7	5.3	11.4	22.8	8.0	17.8	23.0	24.3	28.8	29.3	19.9	25.4	18.9	7.5	3.2
Ipomoea	0	35.0	45.8	70.5	31.3	16.1	31.6	15.7	5.8	12.9	1.5	6.2	15.3	12.2	5.7	6.9	13.6
Minor species																	
Achyranthes	0	0	0	0.1	13.3	5.1	0	0.9	2.5	4.1	0.7	1.8	—	0	0.6	0.1	6.9
Plumbago	0	0	18.8	15.6	7.5	17.1	4.9	3.0	5.8	9.9	10.4	2.7	10.6	10.4	4.3	0.2	18.6
Tetragonia	0	0	0	0	0.1	0.4	0	0.2	0	0	0	0	0	0	0	0	0
Solanum	0	0	0	0	6.2	16.7	2.7	1.2	0	0	0	0	0	0	0	0	0.07
Lepturus	3.1	0	1.6	0.2	2.9	5.6	22.9	11.7	23.5	11.7	0	1.0	1.4	2.9	8.5	2.5	0
Stenotaphrum	0	0	0	0.3	5.0	0.1	0.8	20.0	3.7	0.4	0	9.8	21.4	17.6	9.0	0	+
Total for all minor species	3.1	0	20.4	16.2	35.0	45.0	31.3	37.0	35.5	26.1	11.1	15.3	33.4	30.9	22.4	2.8	25.8
Total	100.0	100.1	100.1	99.9	100.0	100.0	99.9	100.0	100.0	100.0	100.0	100.0	100.1	100.1	100.0	100.1	100.0

+ = present in very small quantities.

Table 7. Temporal changes in leaf biomass (g dry wt./m^2) in sample plots of the herb flat.

	1968	1969						1970						1971	1972	1976	1978
	J–A	J–F	M–A	M–J	J–A	S–O	N–D	J–F	M–A	M–J	J–A	S–O	N–D	F–M	April	Nov.	Sept.
Species																	
Melanthera	73.7	5.6	2.6	7.5	19.7	4.6	4.0	14.4	26.1	11.1	8.0	12.6	7.2[a]	52.9	71.9	60.5	6.6
Canavalia	6.5	0.8	3.6	3.1	2.7	6.7	32.2	43.2	34.7	51.7	24.1	38.7	20.2	16.1	6.3	11.4	1.5
Abutilon	7.4	0.3	6.8	7.1	11.5	16.0	10.0	34.7	39.2	41.6	15.8	30.6	17.3	55.5	27.9	6.5	0.4
Ipomoea	0	3.6	17.6	94.6	31.5	11.3	39.4	30.6	9.8	22.1	0.8	6.5	13.3	26.6	8.4	6.0	1.9
Minor species																	
Achyranthes	0	0	0	0.2	13.4	3.6	0	1.8	4.3	7.1	0.4	1.9	—	0	0.9	0.1	1.0
Plumbago	0	0	7.2	20.9	7.5	12.0	6.1	5.8	9.9	16.9	5.7	2.8	9.2	22.7	6.3	0.2	2.6
Tetragonia	0	0	0	0	0.1	0.3	0	0.3	0	0	0	0	0	0	0	0	0
Solanum	0	0	0	0	6.2	11.7	3.4	2.4	0	0	0	0	0	0	0	0	0.01
Lepturus stems & leaves	2.8	0	0.6	0.3	2.9	3.9	28.6	22.9	40.1	20.1	0	1.0	1.2	6.3	12.5[b]	2.1	0
Stenotaphrum stems & leaves	0	0	0	0.4	5.0	0.1	1.0	39.1	6.3	0.6	0	10.2	18.6	38.4	13.3[b]	0	—[c]
Total for all minor species	2.8	0	7.8	21.8	35.1	31.6	39.1	72.3	60.6	44.7	6.1	15.9	29.0	67.4	33.0	2.4	3.6
Total	90.4	10.3	38.4	134.1	100.5	70.2	124.7	195.2	170.4	171.2	54.8	104.3	87.0	218.5	147.5	86.8	19.0

[a] incomplete; [b] includes roots also; [c] indicates present in minute quantity.

The initial shift in dominance was from *Melanthera* to *Ipomoea*. Before the drought *Ipomoea* contributed so little toward the total plant biomass in the herb flat that it could have been considered one of the minor species (Fig. 26). However, immediately after the drought when total biomass was low *Ipomoea* accounted for a proportion (over 35%) second only to *Melanthera*. Subsequently, it became clearly dominant accounting for between 60 and 75% of the total biomass during the remainder of 1969. In 1970 although remaining the single most important contributor of biomass it accounted for a lower proportion than previously (about 30–50%). By 1971–72 it had again lost dominance and was overshadowed by both *Melanthera* and *Abutilon* in terms of biomass.

Abutilon and *Canavalia* each contributed less than 10% of the total biomass of the herb flat initially, and then between 9 and 10% during the post-drought recovery. During 1970 the proportion increased to 10–15% of the total and was consequently not far below the values for the dominant species at that time; both species generally contributed a greater proportion of the biomass than did the former dominant *Melanthera*. In 1971–72 *Abutilon* had become responsible for a greater proportion of the total biomass than any other species; *Canavalia* had declined to below the pre-drought representation. Although *Abutilon* was dominant in 1971–72, *Melanthera* approached it in biomass and may have been on its way toward re-establishment of dominancy. The 1977 drought seemed to have set the process back again.

Thus, it appears that there were two dominance shifts during the study, the first being from *Melanthera* to *Ipomoea* and the second being from *Ipomoea* to *Abutilon*. It is possible that the increase in proportional biomass of *Melanthera* in 1971–72 foreshadowed a further dominance shift from *Abutilon* to *Melanthera*, and a re-establishment of the original dominance structure. Although less frequent sampling occurred after 1971 the later data are consistent with the same kinds of shifts having occurred following the 1977 drought.

The reasons for these dominance shifts are not entirely clear. It does appear certain, however, that the drought was responsible for the *Melanthera* to *Ipomoea* shift. The high mortality of *Melanthera* was probably caused by the drought. *Ipomoea* is a drought-resistant strand species, often growing on sandy beaches or where fresh water is very limited (see Chapter 8). It survived the physical rigor of the drought, and increased in biomass when freed from competition with *Melanthera*. After the return of more consistently moist conditions the less drought-resistant species, perhaps with a competitive advantage over *Ipomoea* in other ways, increased in biomass relative to *Ipomoea*. The first stage in this process was the relative increase in biomass of *Canavalia* and *Abutilon* with the former becoming dominant, eventually being in turn approached by *Melanthera* in proportional biomass.

It appears that the drought may have set back the vegetation of the herb flat to a former successional stage by causing high mortality in the dominant plant species. The successive relative displacement of *Ipomoea* biomass by that of other species with the eventual establishment of *Abutilon* as a dominant may even have been stages in the succession through which the island originally passed. *Ipomoea* is often a colonizer of islands and is characteristic of early successional stages on islands (see Chapter 8). On an island such as One-Tree where there is a mixture of colonizer species as well as those characteristic of more mature communities, an environmental perturbation such as drought may renew the trend of succession, favouring the species characteristic of early stages. On return of more favourable conditions succession may then proceed, recapitulating in part the previous history of the island.

Indeed, the continued presence of *Ipomoea* in the central herb flat on One-Tree may require periodic environmental instability to reverse or retard succession and prevent the community from reaching the more mature stages in which it and other pioneer species would become completely eliminated. On many islands, the beaches appear to serve as a reservoir of pioneer species which invade the interior of the island following droughts or other disturbances.

Davis *et al.* (1938) have noted even greater short-term changes in the vegetation of a small island, than those described here; they did not identify the causative agents. Sauer (1962) found that on Mauritius recolonization of sites heavily damaged by storms was 'sparse, patchy and commonly displaced from the usual zones of the species involved, so that even without further disturbance many years may pass before microdistribution patterns are stabilized'. Some of the same genera, and in some cases the same species, that occur on One-Tree Island were involved. Before storm damage the 'outpost' species on very exposed coasts made up an outermost fringe of creeping grasses, prostrate plants (including *Sesuvium portulacastrum*) and long trailing vines (*Canavalia maritima* and *Ipomoea pes-caprae*), and an inner zone of shrubs including *Argusia argentea* and a species of *Scaevola*.

Some of these species, as well as others, were involved in the recolonization of other zones which were damaged. Areas of raw sand left by the storm waves were re-colonized mainly by *Stenotaphrum* vegetatively spreading from adjacent zones. Damaged beaches around the island were colonized by three main species, two of which were *I. pes-caprae* and *Canavalia*. Where waves had bitten deeply into the inner littoral vegetation, denuded areas were colonized by outpost species such as *I. pes-caprae, Canavalia* and *Scaevola*, as well as weedy or cultivated species mainly from further inland (*Portulaca oleracea* and a species of *Tetragonia* were among them). In general, various damaged areas were invaded by a mixture of species: some former occupants, some from inland and some from the coast.

Succession on One-Tree Island is discussed further in Chapter 23.

The changes on One-Tree Island discussed above were of total plant biomass including stems, roots, leaves, flowers and fruits. Assessment of the effect of the environment on the biomass of a particular species is incomplete unless it is known which plant parts are most affected. For example, a species in which drought kills the green parts leaving live stems or roots to sprout new vegetative growth upon the return of moisture plays quite a different role in a community than a species in which the reduction in biomass is via mortality of many entire plants. Consequently it is of considerable interest to investigate temporal changes in the relative distribution of biomass among stems, roots, leaves, and reproductive parts.

In the present study, roots, stems, leaves, fruits and flowers of the vegetation from the sample plots were weighed separately, and such an analysis could be made. In evaluating the results, the biomass values for roots are less reliable than those for the above-ground plant parts. This is because (1) small rootlets tend to break off and it is nearly impossible to get all root matter in the plot (consequently root biomass tends to be underestimated) and (2) species which have long trailing stems (e.g. *Canavalia, Ipomoea*) may have the shoots and leaves spreading long distances from the root. Thus a given sample may include the above-ground parts of a plant but miss the root(s) entirely. Larger numbers of samples than feasible on One-Tree Island would correct for this source of error.

During the drought when total *Melanthera* biomass dropped precipitously, most of the weight loss was through death of the leaves and stems, as live biomass of these two plant parts decreased very markedly whereas root biomass was initially unaffected (actually an apparent increase) (Table 8). In the absence of a large photosynthetic bulk in the plants, roots eventually died also and showed a drop in biomass about six months later than did the green parts (Table 8). This was also evident in the proportional contribution of different plant parts to the total biomass. After the drought, roots made up about 85–95% of the total live biomass of *Melanthera* whereas before the drought stems and leaves contributed the greatest proportion. After the decline of root biomass, and during the period when absolute biomass remained low, proportionate representation of different plant parts remained relatively constant with the above-ground parts contributing considerably more biomass than the roots (Table 9). The next significant change took place during the latter part of 1970 and especially into 1971 and 1972 when *Melanthera* appeared to be regaining its dominance. There was an overall increase in the biomass with the roots contributing a somewhat greater proportion than in the latter part of 1969. During recovery of the *Melanthera* there seemed to be a general increase in biomass of all plant parts including roots and not just a proliferation of the green parts.

Ipomoea was so poorly represented before the drought it was not sampled adequately. However, during both its phase of rapid increase (after the drought) and during its subsequent decline, the proportion of biomass in the various parts of the plants did not change greatly. It appears that whole plants

Table 8. Biomass (dry wt g/m^2) of different parts of *Melanthera* plants in the herb flat.

	1968	1969						1970						1971	1972	1976	1978
	J–A	J–F	M–A	M–J	J–A	S–O	N–D	J–F	M–A	M–J	J–A	S–O	N–D	F–M	April	Nov.	Sept.
Species																	
Roots	76.7	85.0	92.9	18.9	20.3	0.3	0.1	0.2	4.2	1.4	38.9	4.5	7.3	—[a]	24.7	14.5	0.9
Stems	306.6	8.3	3.2	14.7	13.4	5.7	5.7	28.7	89.2	24.5	118.9	30.1	22.5[b]	101.5	108.4	240.9	10.7
Leaves	73.7	5.6	2.6	7.5	19.7	4.6	4.0	14.4	26.1	11.1	8.0	12.6	7.2	52.9	71.9	60.5	6.6
Buds, fruits & flowers	0.2	0	0	0	0	0.4	0.1	0.1	0	0	0	0.03	0.3[b]	0	0	14.5	0.01
Total	457.2	98.9	98.7	41.1	53.4	11.0	9.9	43.4	119.5	37.0	165.8	47.2	37.3[b]	154.4	205.0	330.4	18.2

[a] No data; [b] Incomplete.

Table 9. Proportional contributions (%) of different plant parts to the total biomass of the four major plant species of the herb flat.

	1968	1969						1970						1971	1972	1976	1978
	J–A	J–F	M–A	M–J	J–A	S–O	N–D	J–F	M–A	M–J	J–A	S–O	N–D	F–M	April	Nov.	Sept.
Melanthera																	
Roots	16.8	86.0	94.1	46.0	38.0	2.7	1.0	0.5	3.5	3.8	23.5	9.5	19.6	—[b]	12.1	4.4	5.0
Stems	67.1	8.4	3.3	35.8	25.1	51.8	57.6	66.1	74.6	66.2	71.7	63.8	60.3	65.7	52.9	72.9	58.8
Leaves	16.1	5.7	2.6	18.3	36.9	41.8	40.4	33.2	21.8	30.0	4.8	26.7	19.3	34.3	35.1	18.3	36.2
Buds, fruits & flowers	0.04	0	0	0	0	3.6	1.0	0.2	0	0	0	0.1	0.8	—[b]	0	4.4	0.06
Total	100.0	100.1	100.0	100.1	100.0	99.9	100.0	100.0	99.9	100.0	100.0	100.1	100.0	100.0[b]	100.1	100.0	100,01
Ipomoea																	
Roots	0	38.4	26.5	10.3	16.5	13.4	14.9	—[a]	11.0	14.5	6.5	6.2	21.3	—[b]	14.4	0	0.3
Stems	100.0	56.6	61.3	59.2	72.0	82.5	73.9	78.3	84.6	75.6	93.1	90.7	73.0	82.3	74.2	93.2	89.7
Leaves	0	5.0	12.1	30.2	10.8	3.7	11.1	21.8	4.2	9.9	0.4	3.1	5.7	17.8	11.4	6.8	9.6
Buds, fruits & flowers	0	0	0.2	0.3	0.8	0.4	0.1	0	0.2	0.04	0	0	0	0	0	0	0.3
Total	100.0	100.0	100.1	100.0	100.1	100.0	100.0	100.1	100.0	100.0	100.0	100.0	100.0	100.1[b]	100.0	100.0	99.9
Canavalia																	
Roots	5.1	71.0	34.7	6.1	7.7	2.4	3.7	1.8	5.1	6.3	3.8	6.2	8.0	—[b]	13.4	0	8.0
Stems	66.5	24.1	50.6	24.5	23.1	44.0	41.6	44.9	58.2	58.0	66.8	57.2	52.8	62.0	63.2	48.1	41.6
Leaves	14.9	4.9	14.7	63.3	69.2	53.6	54.7	52.6	34.8	28.5	25.9	36.7	39.2	38.0	23.4	51.9	50.4
Buds, fruits & flowers	13.5	0	0	6.1	0	0	0	0.7	1.9	7.2	3.6	0	0	0	0	100.0	0
Total	100.0	100.0	100.0	100.0	100.0	100.0	100.0	100.0	100.0	100.0	100.1	100.1	100.0	100.0	100.0	—	100.0
Abutilon																	
Roots	9.6	96.2	17.3	12.6	2.5	15.5	14.3	11.4	10.5	8.2	12.9	9.6	14.6	—[b]	14.4	13.0	0.09
Stems	75.9	1.0	50.0	45.1	47.3	53.7	45.2	57.8	61.8	55.9	74.8	65.4	69.4	76.9	73.9	80.7	47.8
Leaves	14.5	2.9	32.7	40.6	48.5	30.7	38.6	30.2	27.6	33.9	11.0	24.2	15.5	22.2	11.7	6.4	47.8
Buds, fruits & flowers	0	0	0	1.7	1.7	0	1.9	0.6	0.1	2.0	1.3	0.9	0.5	0.9	0	0	0.07
Total	100.0	100.1	100.0	100.0	100.0	99.9	100.0	100.0	100.0	100.0	100.0	100.1	100.0	100.0	100.0	100.1	100.1

[a] Negligibly small amount; [b] Incomplete data.

were predominantly affected by the factors inducing biomass changes, not different plant parts differentially (Table 9).

Both *Canavalia* and *Abutilon* showed a brief increase in proportionate biomass attributable to roots after the drought, suggesting selective mortality of the above-ground shoots. Shoot recovery was rapid as indicated by the quick re-establishment of percentage contribution to near the original level. Subsequently, even in 1970–72 when total biomass was increasing, biomass of plant parts maintained a similar proportional relationship (Table 9).

Thus it seems that the drought killed the shoots of *Melanthera, Abutilon* and *Canavalia* and that the latter two recovered rather quickly and consequently did not suffer such long-term effects as *Melanthera*. In the latter, failure of shoot recovery over long periods eventually resulted in significant dying of the roots and consequent prolonged reduction in biomass and an extended loss of dominance.

Ipomoea by contrast was relatively unaffected by the drought and in the absence of a dense *Melanthera* population grew rapidly and took its place, the biomass increase involving proportionately similar increases in all plant parts.

The leaves are (in several senses) of more ecological importance than other parts of the plant. They are responsible for the greatest amount of photosynthesis, and hence production, and they serve as a food source for a major component of the herbivore fauna and hence of the trophic web dependent on those herbivores. Consequently, biomass changes in the leaves of the herb flat were analysed in more detail than those of other plant parts. Changes in leaf biomass were similar to but did not always parallel those of total biomass for a given species.

Before the drought the greatest leaf biomass was contributed by the dominant species, *Melanthera*. This also was true for a brief period after the drought (when absolute values for all species were low); *Ipomoea*, the new dominant, usually contributed the most leaf biomass from then until early 1970 when *Canavalia* and *Abutilon* superseded it (Table 6; absolute values in Table 7). Most departures from the pattern of total biomass were small; the most significant one, however, was the high values for leaf biomass from 1972 on when the contributions of *Melanthera* to total leaf biomass was the highest for any species, again suggesting that it may have been re-establishing its dominance.

10. Vegetation structure in the herb flat

Several features of the vegetation were studied, i.e., mean vegetation height, total percent cover, and relative cover (by species); these can be compared to the biomass values studied in the previous section. These habitat characteristics, although often interrelated with each other and with biomass, do have a degree of independence; they indicate different structural or ecological features, each of which plays its own distinctive role in the environment. Biomass indicates the amount of living plant material present, vegetation height its vertical distribution, and percent cover and relative cover its horizontal arrangement. Vegetation height might be expected to influence such factors as vertical stratification of arthropods (and possibly number of niches available), wind velocity near the ground, and availability of perches for some passerine birds. Cover is an important factor in ameliorating the thermal and hydric microclimates near the ground and in the upper layers of soil. Biomass is a measure of the potential food of various types present at any given time.

Vegetation height

The vegetation in the end plots of transects terminating at beaches was usually considerably shorter than that generally characteristic of the herb flat (Fig. 28, Table 10). This 'edge effect' persisted throughout the study. Vegetation height did not change as markedly in other ecotones. There was no gradual decrease in height before the sharp disjunction between the herb flat and the succulent mat, nor were there consistent changes in height at the boundaries between the herb flat and groves of *Pisonia* or *Pandanus* trees (Fig. 28, Table 10).

Within the herb flat, height varied considerably from plot to plot but the overall height characteristics of large areas were similar when edge and ecotonal quadrats were excluded (Table 10). The vegetation height of the area traversed by transect IV was consistently low in comparison to other

Table 10. Mean vegetation heights in releveés in the transects of the herb flat.

Transect No.	N	1969 M–J	1969 J–A	1969 S–O	1969 N–D	1970 J–F	1970 M–A	1970 M–J	1970 J–A	1970 S–O	1970 N–D	1971 F–M	1972 April	1978 Sept.
Herb flat														
I	26–38	36	33	35	38	49	48	48	40	54	41	63	41	58
II	24–31	36	36	31	37	44	45	44	38	48	40	63	51	68
III	6–16	46	26	38	40	54	44	39	42	48	33	57	—	—
IV	7–20	23	22	39	27	24	29	33	26	26	31	61	—	49
V	11–24	38	30	47	36	50	46	41	39	43	49	64	49	63
VI	8–15	31	33	37	42	50	50	44	45	53	38	56	—	48
VII	3–8	37	22	61	29	32	28	28	26	36	23	38	—	—
Grand mean of all transects except IV	82–127	37	32	37	38	48	46	43	39	48	40	61	46	58
Grand mean, dead height except transect IV	43–106	37	30	32	25	19	25	24	28	33	41	21	28	37
		(97%)	(95%)	(94%)	(67%)	(58%)	(54%)	(44%)	(70%)	(83%)	(78%)	(36%)	(78%)	
Destructive sample plots (live)	5–10	32	35	32	22	37	44	36	49	34	43	42	38	—
Destructive sample plots (dead)	2–10	32	29	29	12	12	21	14	25	21	31	20	21	—
Ectones with other vegetation types														
Pisonia	1–3	30	33	31	42	35	42	48	37	33	28	42	100	—
Pandanus	1–4	39	14	31	44	50	49	40	50	50	55	43	25	—
Sesuvium	2–3	45	25	31	23	33	33	37	42	43	35	48	38	—
Edges at beaches I–VII	6–11	15	21	21	29	25	26	31	27	29	22	40	28	—
Succulent mat														
II	9–12	11	11	13	15	16	17	11	14	18	16	18	18	15
III	5–8	17	15	19	20	20	17	17	13	16	17	25	—	—
Grand mean	15–19	13	13	15	17	18	17	13	14	17	17	20	—	—

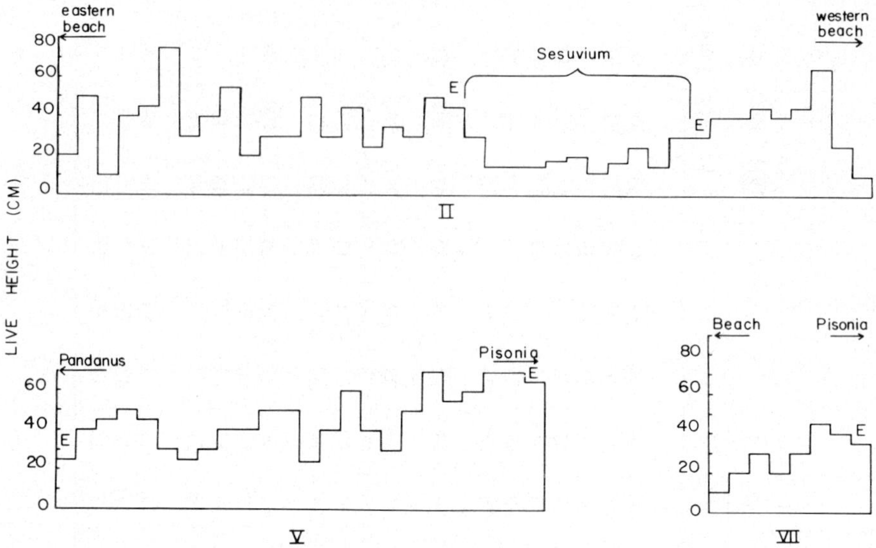

Fig. 28. Vegetation height in m^2 quadrats along transect lines II (Dec. 1970), V (April 1970) and VII (June 1970) in the herb flat. See Fig. 9 for locations of transects. Note changes in values at some ecotones.

parts of the herb flat, and this was visually obvious. The area was designated as the 'rubble flat' on the first visit to distinguish it from the rest of the herb flat. Also, the area traversed by transect VII tended to have relatively low vegetation at some times of year. However, ecotonal and other areas of somewhat shorter vegetation were small relative to the total area of the herb flat (approximately 18%). None of the biomass removal samples were taken in the rubble flat or in the vicinity of transect VII and consequently biomass estimates are representative for the greater part of the herb flat.

The mean height of vegetation in the destructively sampled plots, despite the fact there were only ten samples, were of the same general magnitude as the grand means for the much larger numbers of releveé plots (Table 10); the former can consequently be considered as providing a reasonable estimate of vegetation height during the part of the study prior to use of releveé transects.

The seasonal change in vegetation height of the herb flat was usually not great (Table 10). However, some changes did occur. In May, 1968, during the preliminary visit to the island, five measurements of vegetation height were made which averaged 56 cm. In July 1968, 22 such measurements averaged 67 cm, and the mean of the removal plots was 61 cm. Thus the pre-drought vegetation height in the herb flat was about 55 to 65 cm. After the drought, mean height (based on removal plots) had dropped to 25 cm (January–February, 1969) but had recovered to 39 cm by March–April, 1969. A height of 30–40 cm was maintained throughout the remainder of 1969, whereas in 1970 mean values were mostly between 40 and 50 cm. In the autumn of 1971, the height had reached the pre-drought level but it returned to the 1970 level by April 1972. Thus it appears that mean vegetation height in the herb flat varies from year to year from about 35 cm to 65 cm but that drought can reduce it as low as 25 cm.

The rubble flat, which had consistently lower vegetation, did not show as great a temporal change in live vegetation height as did the rest of the herb flat. Mean height in the rubble flat before the drought (July 1968) was 33 cm (N = 20). Measurements in January–February 1969 and March–April 1969 were 34 cm and 39 cm respectively (N = 10 in each case). Thereafter mean values ranged from 22 cm to 39 cm during the remainder of 1969 and 1970. Only in 1971 did the rubble flat develop a vegetation as high as the rest of the herb flat (Table 10). It was not measured in 1972.

The succulent mat showed the least temporal variability in thickness. Prior to the drought (July 1968) ten measurements averaged 14 cm. Following the drought, values of both transects combined ranged from 13 to 20 cm in the period from 1969 to 1972; the two transects were not greatly different from each other (Table 10).

The height of dead standing vegetation indicates the vertical distribution of material available to decomposers. In the herb flat, height of dead vegetation was measured only after the drought. In January–February and March–April 1969 it was 50 and 41 cm respectively (N = 10 in each case). Subsequently, it gradually decreased until early 1970 when height of dead standing material averaged 10–20 cm. There followed an increase until values of 30–40 cm were reached by the end of the year. The autumn values of 1971 and 1972 were of the same order of magnitude as those of 1970 (Table 10). However, the changes in mean height of dead standing material does not seem to have a regular seasonal pattern as values were high in the summer of 1970–71 but were low for the corresponding season of 1969–70.

Total cover

Total cover in the herb flat showed an edge effect similar to that of vegetation height, as percent cover generally decreased near beaches (Fig. 29, Table 11). Although this effect persisted throughout the study it semed to be less pronounced in some transects after the recovery from the drought.

In contrast to vegetation height, total cover showed a strong ecotonal effect, percent cover often being less in areas bordering *Pisonia* or *Pandanus* clumps although there was no decrease at the edge of the succulent mat (Fig. 29, Table 11); the difference in cover between ecotones and the rest of the herb flat was greatest just after drought. Later, the difference between the interior and the ecotones waned somewhat.

Figure 30 shows the seasonal changes in total cover values obtained from the releveés and compares them to the cover data from the destructively

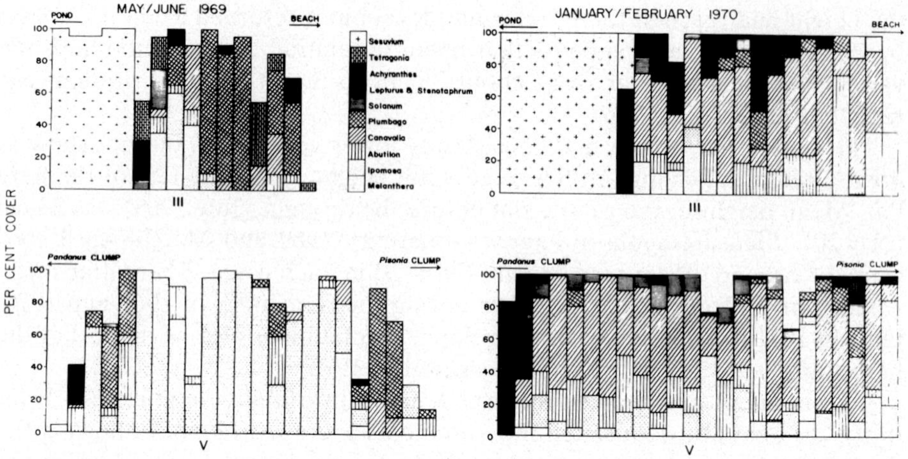

Fig. 29. Per cent cover along transects III and V (herb flat) at different times of year.

sampled plots. Prior to the drought, percent cover averaged more than 80% in the removal plots; afterwards it had dropped to less than 20% but had recovered to the pre-drought level by May–June 1969, when releveé sampling was begun. Thereafter, the releveé and removal plots presented very similar results both in pattern of change and in absolute levels; cover fluctuated but

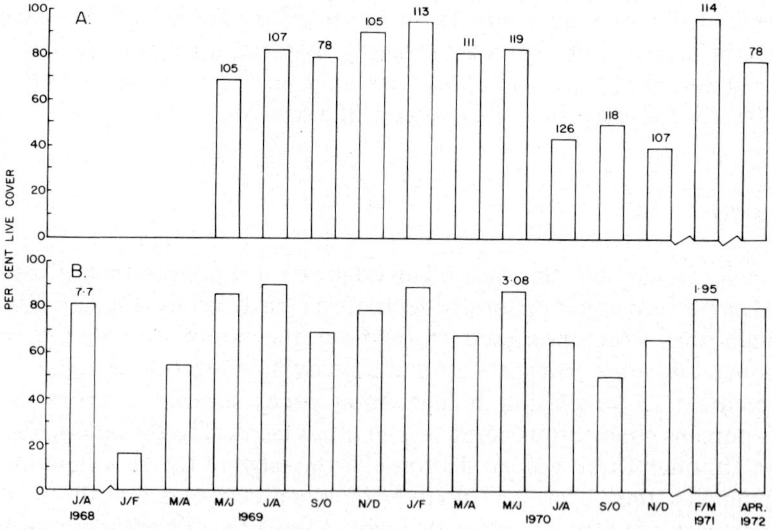

Fig. 30. Temporal change in per cent live cover of vegetation in the herb flat based on (A) releveés and on (B) destructively sampled plots. Area was 10 m^2 except for histograms in which sample area (m^2) appears at the top. Edge quadrats and ecotonal quadrats excluded. Data for (B) in 1978 was 84% (not shown on graph).

Table 11. Mean percent cover in relevеés along transects in herb flat. Dashes indicate absence of data. All data refer to live cover unless indicated otherwise. N = number of m^2 relevеés.

		1969						1970						1971	1972	1978
Transect No.	N	J–F	M–A	M–J	J–A	S–O	N–D	J–F	M–A	M–J	J–A	S–O	N–D	F–M	April	Sept.
I	26–38	—	—	60	86	85	90	94	81	82	34	45	40	98	81	84
II	22–28	—	—	73	74	79	91	98	87	82	49	50	40	94	74	74
III	6–16	—	—	86	86	82	97	99	81	89	42	43	39	97	—	—
IV	8–23	—	—	23	42	39	53	63	48	39	32	24	26	84	—	56
V	11–24	—	—	78	85	73	90	96	76	88	38	49	36	98	76	77
VI	8–15	—	—	63	82	67	87	93	80	88	69	65	47	97	—	70
VII	2–8	—	—	49	75	68	84	64	80	60	57	60	41	100	—	—
Grand mean of all plots except IV	78–126	—	—	69	82	79	90	95	81	83	44	50	40	97	78	78
Destructive plot samples	5–10	16	54	87	89	69	79	90	68	87	39	54	67	85	80	—
Destructive plot samples (dead)	10	20	10	11	—	7	—	—	—	1	26	0.3	7	—	1	—
% plots with dead cover	—	100	90	60	—	20	—	—	—	13	70	20	20	—	10	100
Ecotones with other vegetation types																
Pisonia	3	—	—	12	45	50	65	72	59	47	23	30	8	68	—	—
Pandanus	2	—	—	20	48	46	73	92	95	87	70	72	50	66	—	—
Sesuvium	2–3	—	—	59	91	58	86	95	74	77	50	52	56	100	70	—
Edges at beaches I–VII	1–13	—	—	18	48	50	62	48	44	59	30	17	17	68	59	25
Succulent Mat																
II	9–12	—	—	90	96	99	98	100	98	100	90	91	91	100	87	99
III	5–8	—	—	96	94	100	100	100	100	95	100	93	100	86	—	—
Grand mean	15–19	—	—	92	95	99	99	100	99	98	94	91	94	95	—	—

remained high until it dropped to levels near 50% in late 1970. The usual level near 80% was obtained both in the 1971, 1972 and 1978 samples (Fig. 30, Table 11).

The close correspondence between the results from the removal plots and the much larger number of samples from the releveés inspires confidence in the former as reliable indicators of cover during the period prior to releveé sampling.

The changes in cover paralleled those in biomass (compare Figs. 25 and 30). It would appear that change in biomass (either increase of decrease) has its greatest influence on the horizontal distribution of plant matter rather than on its vertical distribution; cover values showed a more marked change with change in leaf biomass (Fig. 31) than did vegetation height. Thus, the

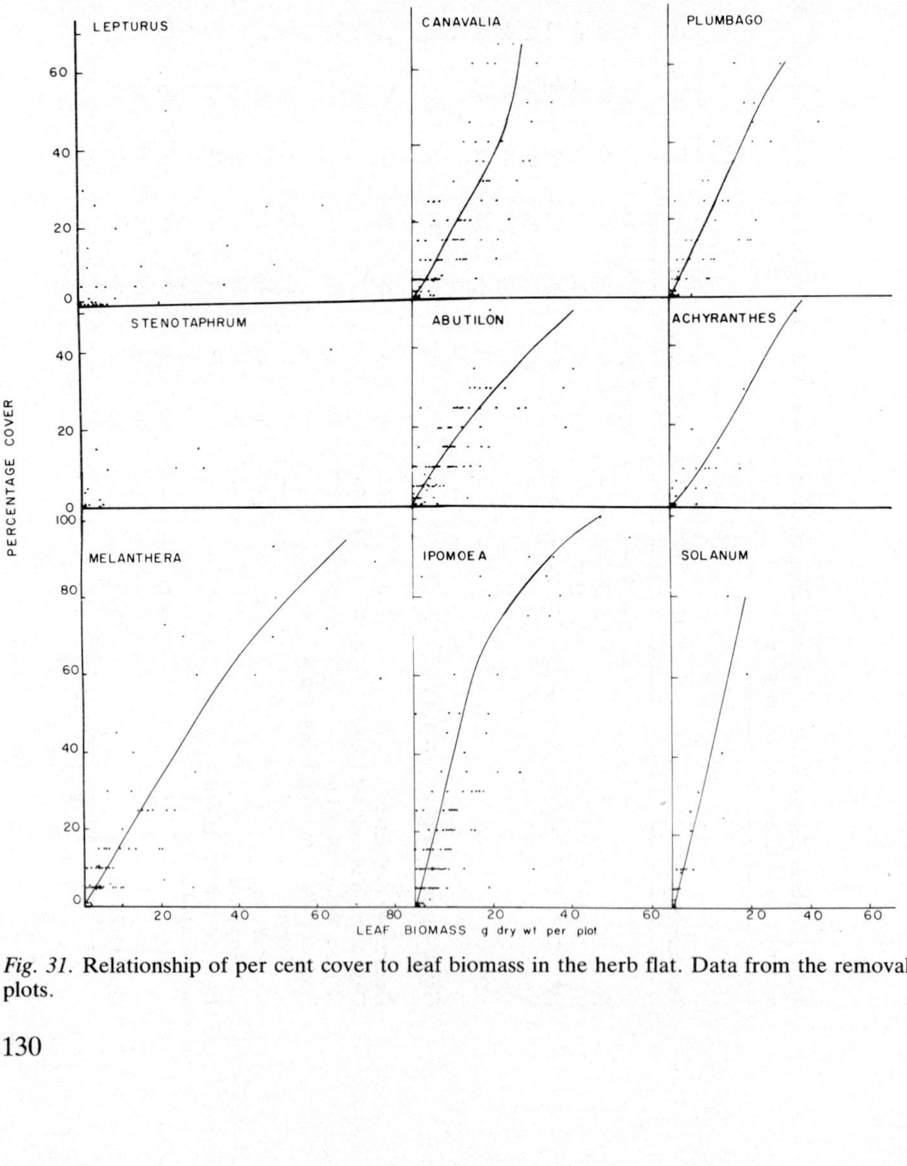

Fig. 31. Relationship of per cent cover to leaf biomass in the herb flat. Data from the removal plots.

ecological and microclimatic features affected by cover would be expected to be less stable than those influenced by height of vegetation.

Relative cover

The proportional representation of various species showed rather large differences even between adjacent plots. Superimposed upon this local variation, certain general areas of the herb flat tended to have a greater predominance of a particular species than did others so that regional differences could be detected. Thus, there could have been regional bias in the destructive sampling program, particular species' contributions being over- or underestimated depending on the specific region of the herb flat from which destructive samples were taken. The heterogeneity arising from local variation is not considered to be very important as a source of consistent bias as ten samples taken at 5 m intervals would encompass a number of different types of vegetation and there would be little probability of many of them falling on one particular variant. However, there would be a lot of variability in the results which would reduce the reliability of the data as precise estimates. The regional variation would be much more important as a consistent source of bias as it would be quite possible for a considerable proportion of the ten destructive plots (at 5 m intervals) to fall within one regional type. Regional heterogeneity in species composition as indicated by relative cover values, however, probably appears greater than would be the case if biomass estimates were used instead. An upright species of plant with small leaves clustered closely to the stem would show much less change in cover for each unit increase in biomass than would a creeping species with large, thin and non-overlapping leaves. Consequently, certain species are prone to wide changes in cover value with only slight change in biomass. The major species most likely to fit this category are *Ipomoea, Canavalia* and *Melanthera* as they have steep slopes of curves relating cover to leaf biomass (Fig. 31). These are the species that are most responsible for the regional differences in relative cover values. Consequently, the apparent regional heterogeneity, though certainly real, is not as important as might first seem to be the case. For this reason it is believed that the destructive sampling of the vegetation provides a reasonable estimate of temporal change in biomass.

The degree of regional heterogeneity changed with time. After recovery from the drought the spatial heterogeneity was rather great, thereafter decreasing. May 1969 was the time of greatest heterogeneity. For example, compare transects III and V in Fig. 29; in 1969 in transect III a high proportion of the herb flat cover was provided by *Plumbago* whereas in transect V most was provided by *Ipomoea*. In 1970, the two transects were much more similar to each other.

In comparing the profiles of relative cover for each date and transect, it

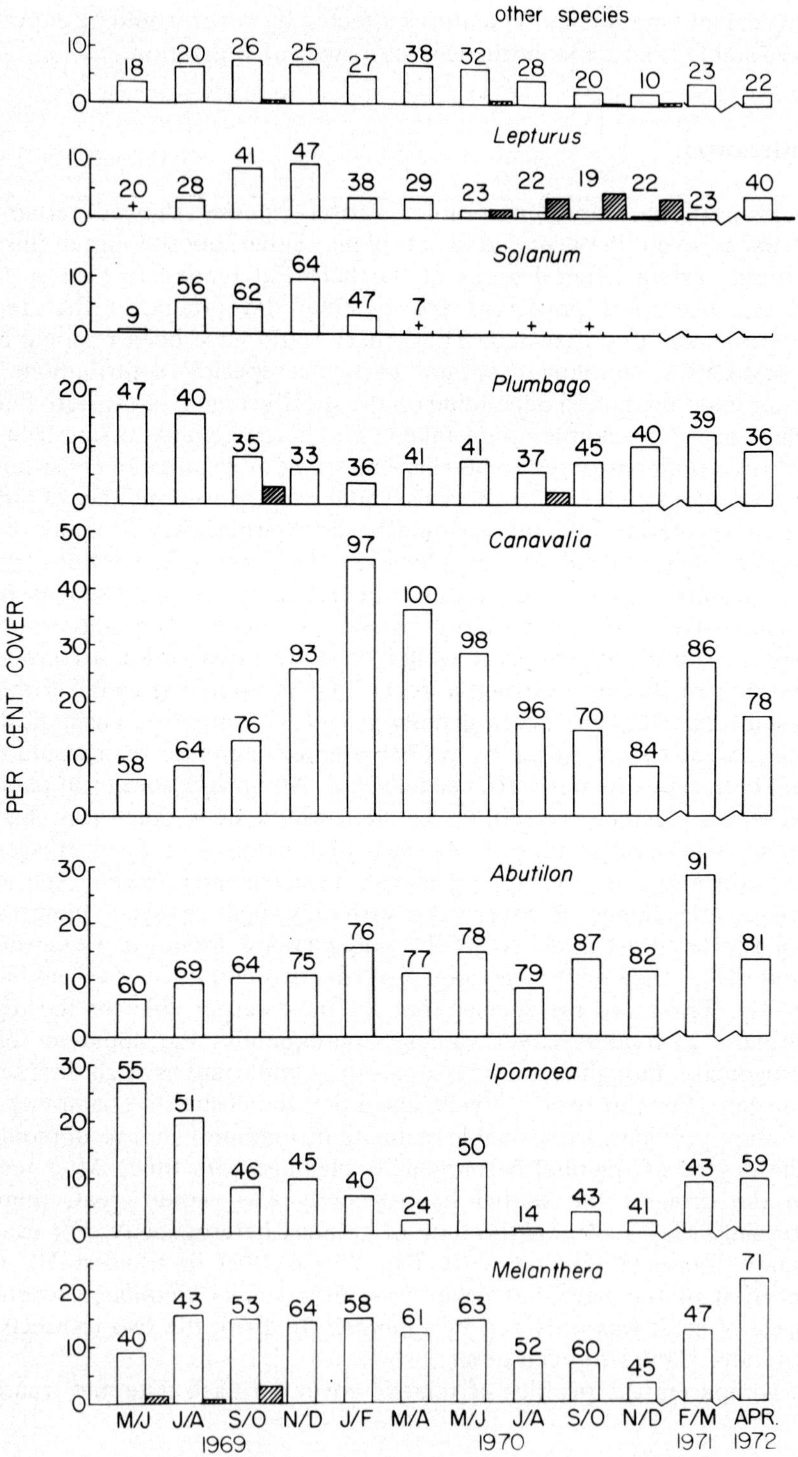

other species
Lepturus
Solanum
Plumbago
Canavalia
Abutilon
Ipomoea
Melanthera
PER CENT COVER
M/J
J/A
S/O
N/D
J/F
M/A
M/J
J/A
S/O
N/D
F/M
APR.
1969
1970
1971
1972

become clear that except immediately after recovery when heterogeneity was great, the herb flat was similar throughout[1] and that regional differences in relative cover largely arose from two specific areas which departed from the condition of the rest of the herb flat. These regions were (1) the area traversed by transect IV and (2) the eastern edge of transect I. None of the removal plots occurred in these two areas and hence they can be used to estimate the prevailing characteristics of the herb flat, but keeping in mind that there were two local areas which departed somewhat from that condition. The area of the two atypical regions was small compared to that of the entire herb flat.

That the cover values from the biomass plot samples parallel the seasonal changes in biomass (compare Figs. 25 and 30) more closely than do those from the releveés is to be expected. The fact that all three show similar trends lends confidence in accepting the conclusions drawn from the biomass study and attests to the adequacy of the sample sizes used.

Figure 32 presents the seasonal changes in percent cover of the various species of the herb flat, combining all transects but omitting edge and ecotonal quadrats. The pattern of seasonal change in cover for individual species correlated with the equivalent biomass values in a general way (compare Figs. 26 and 32). The pattern was generally similar for both measurements for *Melanthera, Ipomoea* and *Abutilon*. In *Canavalia* an increase in cover appeared in advance of the subsequent increase in total biomass and then declined. The former probably represents leaf production and the concomitant increase in cover; the later biomass increase is probably the result of photosynthetic activity and growth resulting from the functioning of the new leaves.

Dead cover in the removal plots was relatively high during the drought-recovery phase (reflecting mortality during the drought) but subsequently declined except for one high value in July–August 1970 (Table 11). This was paralleled by the decrease (from 100%) in proportion of plots with standing dead plants at the beginning of the recovery phase to levels of 10–20% except for the inexplicably high value of July–August 1970.

In general, the releveés showed similar trends and tended to support the data gained from the plot sampling except where there were ready explanations for divergences. The former were designed to provide a check upon the latter because of the small numbers of removal plots which could be employed. Convergence of the two separate methods inspires confidence in the estimates.

[1] See also Chapter 23 for a discussion of heterogeneity induced by the sampling methods.

Fig. 32. Temporal changes in average per cent cover of the various species in the releveés in the herb flat, omitting edge and ecotonal quadrats and those from transect IV (see text). The numbers above the histograms indicate the percentage of the quadrats in which the particular species was represented. Cross-hatched histograms are cover values of dead standing plants. + indicates values too small to be plotted easily. 'Other species' includes *Achyranthes, Stenotaphrum, Portulaca* and *Tetragonia*.

11. Phenology, flowering and fruiting

The seasonal extent of flowering and fruiting were obtained from counts of buds, flowers, and fruits in the releveés and from qualitative observations; data from *Argusia* were obtained by direct counting of all flower heads on the island.

Herb flat

Melanthera had at least a few flowers or fruits in evidence in the herb flat at all times of year. In 1969 they were sufficiently abundant to first appear in small numbers in the releveés in May, at which time both buds and flowers were present; fruits first appeared in the releveés during the July sampling period. Flowering reached its peak during the September–October sampling period and fruiting by November–December. Flowering had practically ceased by the following April (Fig. 33). Thus the principal reproductive period for *Melanthera* in 1969 was from July to late March, i.e., late winter, spring and summer. This seems to be an annual cycle as a similar pattern was initiated in 1971 although at a somewhat later date; values in February–March 1971 and April 1972 were very low and corresponded to the near absence of reproduction during those periods in other years.

Plumbago also had a well defined reproductive period. When regular sampling was initiated in May 1969, flowering was well underway; the fruiting peak was reached by late July and was over by November (Fig. 33). In 1970, reproduction occurred at approximately the same time of year with flowering being well advanced by May and fruiting by July-August with some activity continuing into November. However, the number of fruits produced was only a small fraction of the 1969 yield. The late summer and early autumn periods were times of low reproductive activity in all years.

Ipomoea tuba showed a flowering peak in July–August of 1969, with reproduction continuing at least into December; only occasional flowers and

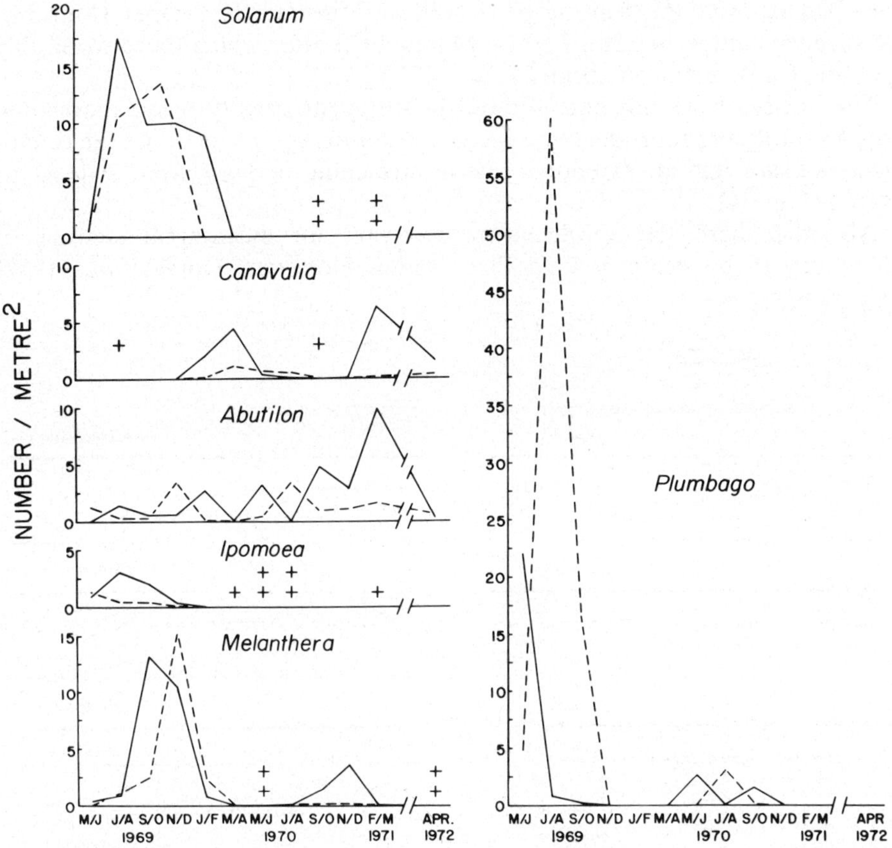

Fig. 33. Temporal changes in the numbers of flowers and fruits of the principal plants in the herb flat. Solid line indicates buds plus flowers. Dashed line indicates fruits, upper +'s indicate presence of a very small number of fruits present, lower +'s indicate presence of a very small number of buds and flowers.

fruits were observed in 1970. In no year was flower density high, never reaching more than 1.5 per m^2 (Fig. 33).

Abutilon showed almost continuous flowering and fruiting throughout the year with minor oscillations perhaps resulting from sampling error. Density of flowers and fruits was always low. There are 3 seeds per locule and on the average 20.8 locules per capsule (S.E. = 0.48; range = 18–24; N = 15) for a mean of 64.2 seeds per capsule. The mean dry weight of seeds is 18.9 mg (N = 72).

Canavalia flowered and fruited over a wide seasonal spread but without well defined periods of reproduction. At any given time flowers and fruits were produced in low numbers.

Solanum began flowering in May–June 1969 and reached a peak in

July–August with the fruiting peak following in late September (Fig. 33). Most reproduction had finished by March 1970 after which there were only sporadic fruits or flowers found.

The grasses were not encountered in sufficient density to do more than indicate probable reproductive seasons. *Lepturus* seems to fruit year round. *Stenotaphrum* has an extended season stretching at least from August to March (Fig. 34).

All other herb flat species were too rare for meaningful analysis of phenology to be made at One-Tree Island. However, Holm *et al.* (1977)

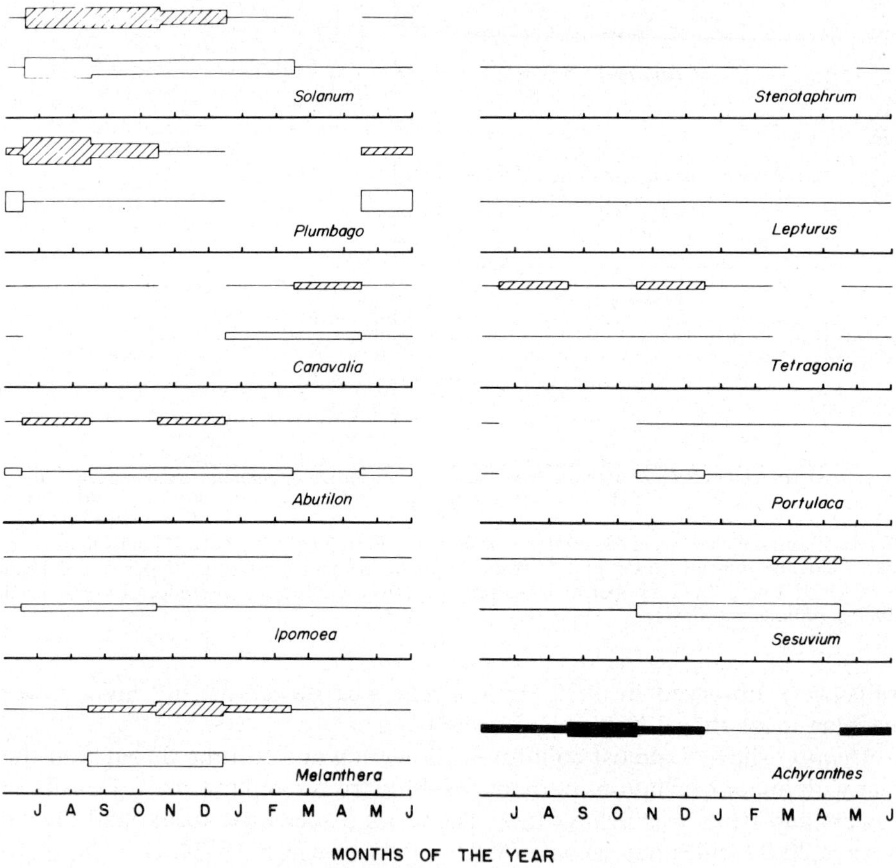

Fig. 34. Flowering and fruiting of the herb flat flora. Cross-hatched figures indicate fruits, open figures buds and flowers. Black figures indicate inflorescences (buds, flowers and fruits) of *Achyranthes*. Data from all years combined. The wider the figure the larger the number of fruits or flowers:

indicate that *Tribulus cistoides* grows rapidly, and produces fruit when only a few cm tall; it flowers all year round. Individual plants can produce up to 10,000 seeds.

In the succulent mat the flowering of *Sesuvium* was followed by counting the flowers in the releveés on each visit. *Sesuvium* flowers nearly throughout the year. Although there were months in which no flowers were reported for particular years, over the entire study flowering was observed in all months. Flowering is slight during the winter and most prolific in the summer months. There are considerable year to year differences in numbers of flowers produced. Flowering was much heavier in 1970 and 1971 than in 1969.

In July 1968, there were on the average 2.5 flowers per m^2 of habitat and 0.3 fruits per m^2. In 1969 values for flowers never exceeded 1 per m^2 from March through October and sometimes no flowers were present. Flowering was heavier from November 1969 to June 1970 with numbers of flowers averaging 4.7, 6.9, 5.6 and 1.4 for the 4 consecutive sampling periods. In July–August no flowers were present and in September–October the mean crop was only 0.5 flowers per m^2. In November–December 1970 and in February–March 1971 *Sesuvium* flowered prolifically (30.8 and 21.3 flowers per m^2 respectively) and many fruits were present.

Portulaca was not abundant enough for quantitative treatment of its phenology. We observed it in flower in all months from September through April inclusive but not in the months May, June, July or August. Holm *et al.* (1977) have noted that fruiting of *Portulaca oleracea* may be prolific; as many as 10,000 seeds per plant are sometimes produced.

Several important features emerge from the above data. First of all, different species have different reproductive periods. Although there are broad seasonal overlaps in flowering and fruiting, each species tends to peak at a different time of year, and at least some flowering is occurring in each season (Figs. 33, 34).

This has implications for the nectar feeding insects as generalized nectarivores would have a food source year round. Had all species flowered simultaneously, periods of food over-abundance and food scarcity would probably have resulted. Asynchronous flowering is also beneficial to the plant as maintenance of a generalized nectarivore population year round insures that pollinators will be available for any insect-pollinated species at the time it flowers.

A second important feature was the reduced fruit production of *Plumbago, Ipomoea* and *Solanum* in 1970 compared to 1969. The causes were of two kinds. In September–October, 1969, when flowering in *Solanum* was at its peak this species accounted for 17% of the total leaf biomass (Table 6), whereas in 1970 it did not occur in releveés at all. Similarly, in 1970 the biomass of *Ipomoea* was usually much less than it was in 1969 (Table 7). Thus for these two species, the reduction in fruit production was at least in part a consequence of fewer plants rather than merely failure of individual plants to set seed. In the case of *Plumbago*, however, the leaf biomass in the peak season of fruiting (July–August) was about the same in 1970 as the same months in 1969 (Table 7) and consequently the lower seed production in 1970 must have resulted from a lower seed production per plant.

On the available data it is perhaps premature to draw conclusions concerning the influence of climatic or weather variables on fruiting as one cannot easily separate distal and proximal factors. It can be noted merely that both *Melanthera* and *Ipomoea* fruited at about the same time on two consecutive years even though amount and pattern of rainfall varied considerably between the two years. This suggests that the timing of the flowering season is not dependent on the rainfall pattern, at least within certain limits.

Argusia argentea

Perhaps the most conspicuous and abundant flowering on the island is that of the shrub *Argusia argentea* (Fig. 23). Although flowering occurred all times of year, there were definite peaks in 1969 and 1970. These peaks were seasonally constant, despite markedly different rainfall patterns between the two years. Buds and flowers began to appear in abundance in May and flowering reached a peak in October–November (Fig. 35). Fruiting reached a peak in November–December. The similarity of pattern in years of very different

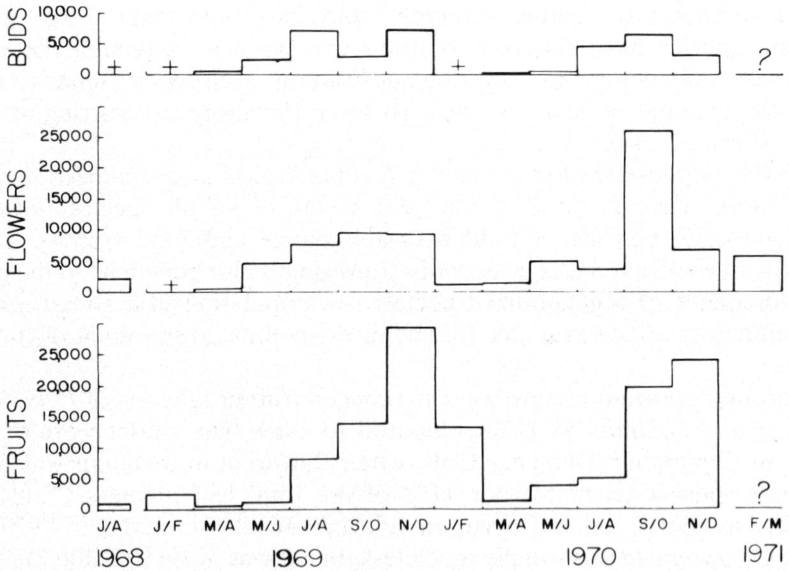

Fig. 35. Temporal changes in number of cymes of buds, flowers and fruits of *Argusia* as censused for the entire island. +'s indicate numbers too small to be plotted on the scale of the figure.

rainfall suggests that flowering and fruiting in *Argusia* is independent of short-term weather conditions, at least within certain ranges. The abundance of *Argusia* flowers between September and December would greatly increase the food supply for nectarivores above that generally available at other times of year. November and December would be expected to be especially rich in food for nectarivores, as at that time there is a coincidence of heavy flowering by several of the most abundant species (*Argusia, Melanthera, Sesuvium, Abutilon, Pisonia* and when present in high biomass, *Solanum*) (compare Figs. 33, 34 and 35).

Ten inflorescences with fruits in various stages of maturity were collected. If it is assumed that all the fruits would eventually have reached maturity, immature and mature fruits both can be included in counts. The average number of fruits produced per inflorescence was 864 (s.e. = 189; range = 556–1,575). On the other hand, if one assumes that the very small fruits were abortive, and only the well-developed ones are included in the counts, the values become 230 (s.e. = 43; range = 48–568). Fruit production per inflorescence is quite variable among inflorescences. However, the real value probably falls somewhere between the above two means.

Twenty-five ripe fruits were dissected and all had either 2 or 3 seeds (mean =2.12). Thus, the seed production per *Argusia* inflorescence probably lies somewhere between 488 and 1,832. If one considers the inflorescences at peak flowering period as near the total annual production (an underestimate) the total seed production for *Argusia* for the entire island is estimated to lie between about 22.5 million and 84 million seeds for 1969 and between about 25.5 million and 95.5 million seeds for 1970. Although this estimate is not precise, it does indicate that a very large number of seeds are produced, i.e., certainly not less than 20–25 million and probably considerably more. During the course of the study less than five saplings or seedlings of this species were observed on the island. *Argusia* is primarily sea-dispersed (see Chapter 8) and it is likely that few of the seeds that are produced ever reach a suitable germination site. The high seed production must be interpreted in this context.

Pisonia grandis

Canopy samples indicated that fruits and/or flowers of *Pisonia* occurred in small numbers in widely scattered periods (e.g. January–February, May–June, and July; Table 12). However, the flowering and fruiting peaks seem to be in October to December. In February 1969, well after the fruiting peak for the season we counted 415 inflorescences in fruit over the whole island. In October of that year we again censused the *Pisonia* inflorescences. There were 3,371 mostly in the bud stage. This number was believed to be near the yearly peak. By December few inflorescences were found and most of these

Table 12. Characteristics of the canopy of *Pisonia* groves on One-Tree Island.

Date	Mean canopy thickness (cm)[a] (Range in parentheses)		Leaf biomass (g/m^2 of canopy)		Flower biomass (g/m^2 of canopy)		Fruit biomass (g/m^2 of canopy)	
	Top	Side	Top	Side	Top	Side	Top	Side
July 1968	65 (49–99)	89 (68–119)	337	166	0.2	0.6	0	0
Jan.–Feb. 1969	78 (50–100)	— ——	133	—	—	—	2.3	—
Mar.–Apr. 1969	78 (60–100)	81 (45–100)	114	174	0	0	0	0
Sept.–Oct. 1969	65 (47–100)	57 (45–70)	124	—	0	0	0	0
May–June 1970	55 (40–100)	45 (30–60)	308	261	0	0	0.3	0
Nov.–Dec. 1970	87 (70–105)	75 (52–97)	157	—	0	0	—	—

[a] Mean canopy thickness (side) measured during sampling of leaf fauna, Sept.–Oct. 1969 to Feb.–March 1971 was 71 cm. S.E. = 2.33; Range = 15–135; N = 85.

had mature fruit. In February 1970 there were a number of buds and flowers (not counted) but no fruits. In October 1970 there were numerous inflorescences in the bud and flower stage, but not in fruit. Some of the buds opened into flowers during our 12-day stay. By December they were mostly in flower and fruit and a direct census indicated 1,824 for the entire island, again believed to be near the annual peak.

The inflorescences produced are much fewer than for *Argusia*. The peak value for *Pisonia* for 1969 and 1970 was only about 7% and 4% respectively that for *Argusia*. The low flowering and fruiting incidence of *Pisonia* on One-Tree Island may be related to the exposed, windy conditions prevailing there. The inflorescences are not produced throughout the canopy; they were most abundant on the sides of the groves (often near the ground) on the lee side or where sheltered but not heavily shaded. There were only a few on the tops or windward side of these wind-pruned groves. It is our impression that fruiting of this species is much more prolific on Heron Island.

Ten ripe inflorescences were collected and number of seeds per inflorescence counted. The mean was 123.5 (s.e. = 14.6; range = 48–270), giving an estimated total production of over 416,000 and 225,000 seeds for 1969 and 1970, respectively. There are no young plants on the island and no seedlings were observed throughout the study. It appears that the *Pisonia* now covers all areas to which it is edaphically suited except for that occupied by *Pandanus*. Its seed production would seem to be significant primarily in terms of export to other islands via seabirds (see Chapter 8).

Pandanus sp.

Fruiting in *Pandanus* occurred continuously throughout the study. Fruits were censused for the entire island; numbers ranged from 62–121 at different visits except for one period (Sept–Dec 1969) when it dropped to very low values (17–20 fruits) before recovering toward the original values during the next sampling period (49 in Jan–Feb 1970) (Fig. 36).

Fruits were not initially tallied by stage of maturity. However, during the decline in number in late 1969 the number of green fruits dropped rapidly until in Nov–Dec, none were present at all (Fig. 36). Throughout 1970 almost all the fruits were green suggesting high production and/or low rate of maturation. During late 1969 when many fruits were ripening, large numbers of old fruit stalks (left when ripened seeds fell) were in evidence. The scarcity of old fruit stalks in 1970 suggested that little maturation of fruit was occurring. However, it is evident that on at least some of the censuses, old stalks and/or fruits were missed as the number of green fruits plus old fruit stalks of July–Aug 1970 did not add up to the total number of green fruits for the preceding sampling period (Fig. 36). Flowers were apparently quite ephemeral as very few were observed during the study.

Neither the fruiting season nor standing crop of fruits showed any relation to weather factors, nor did they occur according to a regular seasonal pattern. Fruiting in this species appears to be complex and the factors affecting it remain unexplained.

The average number of seeds per fruit averaged 57.7 (R = 52–67; N = 3) and consequently standing crop of seeds ranged from about 3,580 to 6,980 for

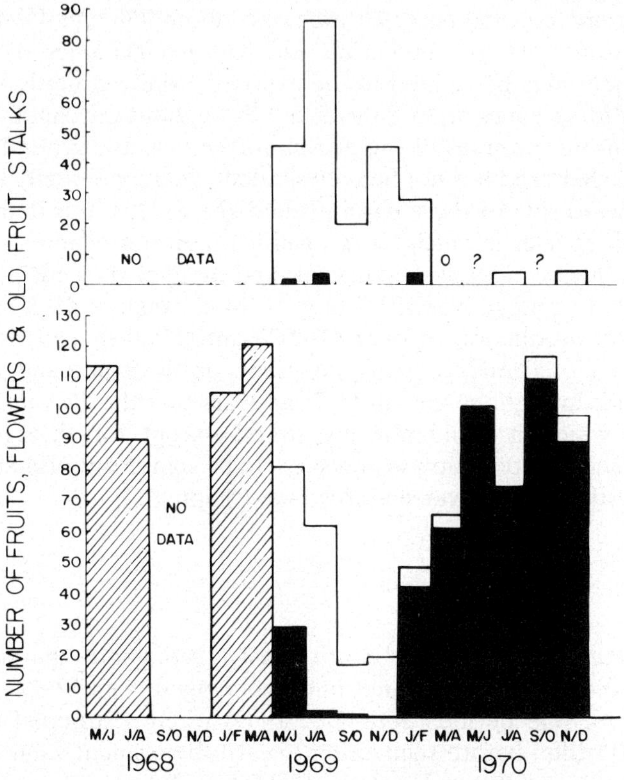

Fig. 36. Temporal changes in number of *Pandanus* fruits and flowers present on the island. Upper: old fruit stalks and flowers. Lower: fruits. Cross-hatched bars indicate total number of fruits before separate counts of green and ripe fruits were made. Open parts of bars indicate ripe fruits and black parts of bars indicate green fruits.

the different visits except for the period of low fruiting and recovery when the values were 980 to 2.830. Mean dry weight of ripe seeds was 11.3 g (R = 8.7–14.1; N = 10). Standing crop biomass was consequently 11.1 kg to 32.0 kg during the low fruiting period but reached 40.5 kg to 78.9 kg at other times of year.

Pandanus fruits require a long time to mature. For example, a green fruit marked in February 1969 had all but six of the seeds ripe by the end of September, indicating a maturation time of seven months for that fruit. A second green fruit 17 cm long marked by a tag around its stalk in July 1969 had ripened and the orange seeds had fallen by September of that year. There was thus much variation in the length of time it took for individual fruits to mature. Another indication of long maturation time, at least in 1970, is the persistence of numerous green fruits for many months with the appearance of very few ripe ones or of old fruit stalks that year.

Production of seeds was not precisely measured. However, a rough estimate for 1969 can be made by adding the number of fruits in July–August (two months before the number of fruits dropped to very low values) to the number of old fruit stalks still present at the same time. The number is 148 fruits, or 8,540 seeds for a seed biomass of 96.5 kg.

12. Production and growth

Leaf production by shrubs and trees

The relative uniformity of thickness and density of foliage in the *Pisonia* groves (Table 12) made sampling of leaf biomass simpler than for other tree and shrub species. Before the drought, standing crop biomass was high in the *Pisonia* canopy but lower on the sides of the groves. After the drought, values had dropped such that top and sides had similar and lower biomass per unit area of canopy; these low values were maintained until May 1970, whereupon they increased markedly and remained relatively high throughout the rest of that year (Fig. 37).

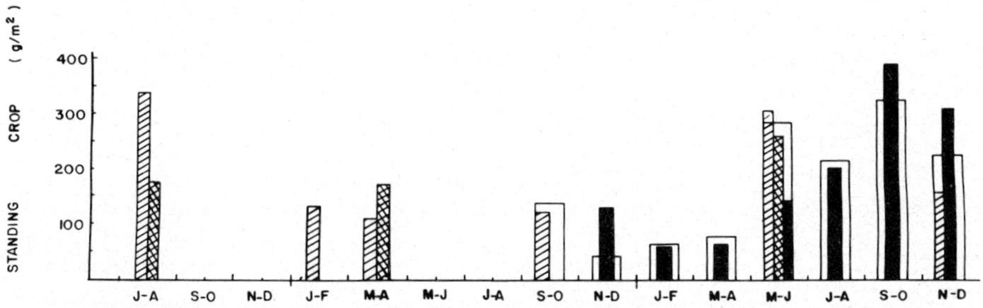

Fig. 37. Temporal change in leaf biomass (dry weight) of the *Pisonia* canopy. Open histograms are estimates based on sampling the number of twigs per m^2 in foliage quadrats. The values are mean number of twigs per m^2 times the mean number of leaves per twig times the mean leaf weight of a sample of leaves. The solid histograms are based on marked twigs. Values are previous biomass plus net production in intervening period. Other histograms are based on m^2 removal sampling of the top (hatched) and sides (cross-hatched) of the grove.

The standing crop, although indicating the amount of food available at a given time for higher trophic levels, does not by itself give a very good picture of leaf production since it represents an equilibrium between the production of new leaves and the abscission of old ones. Therefore, both of these processes were monitored during the latter part of the study (see Chapter 7).

Production of new *Pisonia* leaves by 2-month periods is shown in Fig. 38. There was an increase in leaf production in late 1969 followed by a progressive decline into August 1970. In late 1970 there was a period of very high leaf production followed by a subsequent decline. Leaf fall showed a

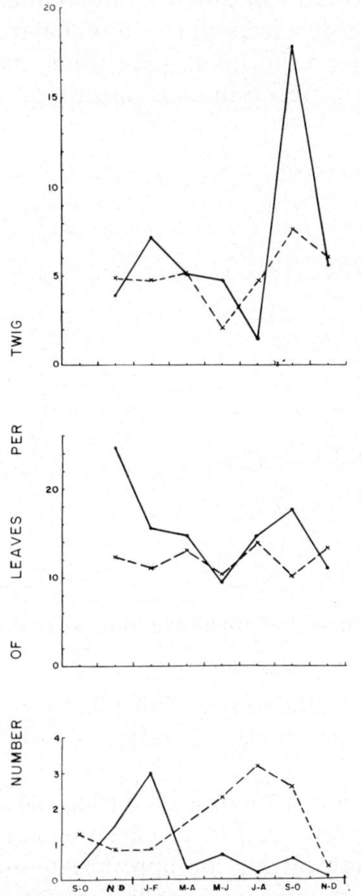

Fig. 38. Leaf fall and leaf production in three species of shrubs and trees on One-Tree Island. Solid dots and unbroken lines indicate production of new leaves and X's and broken lines indicate leaf fall. Upper: *Pisonia.* Centre: *Argusia.* Lower: *Pandanus.*

generally similar trend to the extent that there was a low point in mid 1970 and a high point in late 1970. However, leaf fall was much more constant and subject to less variation than was leaf production; consequently most of the temporal change in net leaf production was due to fluctuation in leaf production rather than seasonal variation in rate of leaf fall.

Net leaf production[1] was calculated from data on leaf production and leaf

[1] A positive production is a net gain of leaves; negative production figures indicate a net loss of leaves for the period.

fall (Fig. 39). Between September–October and November–December, 1969, there was a net loss of leaves and a resulting drop in the standing crop; the next period showed a net gain of leaves and the standing crop increased. Throughout the subsequent sampling periods net production of leaves was always reflected in increases in estimates of standing crop, and net losses in decreases in standing crop except once. In that case, estimated standing crop increased slightly between the January–February and March–April sampling periods whereas net production was estimated to be zero. The small

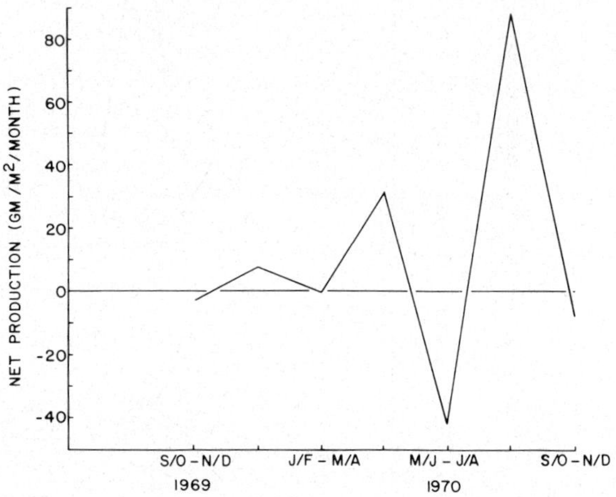

Fig. 39. Temporal changes in net leaf production (or loss) by *Pisonia* trees at One-Tree Island.

discrepancy can be attributed to sampling error, especially since standing crop and net production were measured independently (i.e. one not calculated from the other).

The net leaf production (gain or loss) during the time intervening between two sampling periods was used to predict the standing crop at the second period by adding or subtracting the appropriate amount from the previous estimate based on direct biomass sampling. These calculated estimates were then compared to those arising directly from quadrat sampling of canopy foliage (Fig. 37). With several exceptions the estimates were close and can be considered reasonable estimates of leaf standing crop for the *Pisonia* groves. It should be noted that this latter method is probably less reliable than the other two as it can compound sampling errors.

The early 1969 decrease in leaf biomass (Fig. 37) probably resulted from the drought. However, the much later increase in production and standing crop did not correspond to any marked changes in weather. Minor fluctuations in standing crop or production in the post-drought period did not correlate with temperature or rainfall patterns. It would appear that, at least during periods of normal rainfall, any effect of rainfall upon net production of leaves in *Pisonia* at One-Tree is delayed sufficiently and/or modified by other factors to such an extent that any such relationship is obscured. The last few months of the year in both 1969 and 1970 were characterized by low net production; however, with data from only two seasons, one cannot ascertain whether this is a regular seasonal phenomenon or not. Thus, the factors influencing net leaf production are complex and have not been identified by the present study; it is probable that the factors influencing new leaf production are different from those affecting leaf fall, and that the interaction

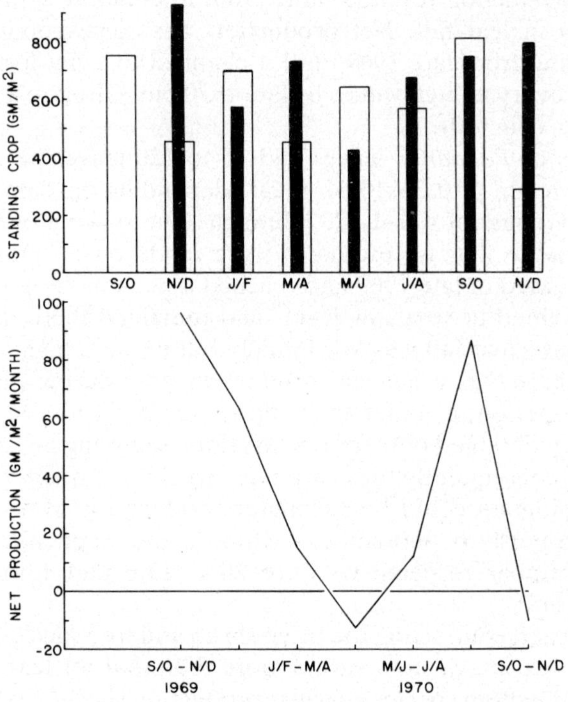

Fig. 40. Standing crop of leaves (upper) and net production (or loss) of leaves (lower) in *Argusia* at One-Tree Island. Symbols as in Fig. 37.

of both sets of factors is responsible for the complex pattern of net production.

The thickness of the canopy of *Argusia* (measured during the sampling for leaf fauna) averaged 23.5 cm (s.e. = 0.71; R = 12–45; N = 100). Thickness was much less than that of the *Pisonia* canopy which averaged 71 cm over the same period of time (Table 12).

The estimates of standing crop of *Argusia* foliage varied considerably among the various sampling periods but seemed to vary around a nearly constant mean rather than show consistent temporal trends. Furthermore, the estimation of standing crop from actual net production data did not correspond closely with the standing crop estimates based on foliage quadrats (Fig. 40). These findings suggest that sampling error was large in this species. Consequently, standing crop estimates probably reflect order of magnitude only.

The temporal pattern in *Argusia* was a marked decline in leaf fall in late 1969 and 1970, reaching a low point in May–June; as in *Pisonia* it then rose to a peak production in September–October 1970 and then declined again. Leaf fall showed relatively little seasonal fluctuation, and as in *Pisonia*, temporal change in net production resulted more from fluctuation in production than from variability in leaf fall. Net production was correspondingly high but rapidly decreasing from late 1969 until it changed to a net loss in mid 1970. There was a recovery to high values by late 1970 but a drop to a net loss by the end of that year (Fig. 40).

The branches of *Pandanus* leaves had 45 to 120 leaves per branch and a mean leaf dry weight of 9.36–15.61 g/leaf (depending on time of year) for a standing crop biomass of 421–1,873 g/branch. The spacing of clumps was so diffuse and variable that no estimates were made on an areal basis. Leaf production increased in late 1969 and reached a peak in early 1970 (Fig. 38); thereafter it declined to very low levels and remained there throughout the study. By contrast, leaf fall was low initially but became high later (Fig. 38). Consequently there was a net leaf production only during part of the late 1969-early 1970 period; at other times there was a net loss of leaves. Mean leaf weights were obtained only for the last three sampling periods of 1970 for this species and consequently these are the only times a quantitative estimate of net loss in biomass could be calculated. Values for May–June to July–August, July–August to September–October, and September–October to November–December respectively were 20.4, 15.6 and 1.4 g dry wt. per clump per month.

Summarizing and contrasting the three shrub and tree species, the temporal pattern of new leaf production varied more than that for leaf fall in *Pisonia* and *Argusia* and in both species new leaf production showed a decline in early 1970 and a peak production in late 1970. This pattern was reflected in net production and net loss. *Pandanus* showed a net loss of foliage during much of the period measurements were made. In none of the three species was the pattern of change a regular seasonal phenomenon.

Leaf production by herbaceous species

Leaf production of the herbaceous species was studied only for two sampling periods in late 1970. The results are summarized in Table 13, for the four species studied. In all species, leaf production was higher in the period July–August to September–October than from September–October to

Table 13. Net production or loss of leaves in four species from the herb flat of One-Tree Island in the latter part of 1970.

Species	Period	Mean no. new leaves/shoot	Mean no. leaves died/shoot	Mean leaf wt (g dry wt)	Net production (g dry wt/shoot/month)
Abutilon	J–A to S–O	12.9	6.0	0.0454	0.156
indicum	S–O to N–D	5.4	6.0	0.0675	−0.021
Melanthera	J–A to S–O	41.4	5.7	0.0753	1.347
biflora	S–O to N–D	18.4	13.1	0.0841	0.223
Canavalia	J–A to S–O	9.2	5.4	0.0836	0.159
rosea	S–O to N–D	6.3	15.7	0.1361	−0.635
Ipomoea	J–A to S–O	15.0	9.9	0.0563	0.144
macrantha	S–O to N–D	8.3	5.3	0.0739	0.111

November–December. In two species (*Abutilon* and *Canavalia*) the latter period showed a net loss of leaves. Thus, the trends in the herbaceous species were similar to that of *Pisonia* and *Argusia* during the same period. In *Canavalia*, net production between July–August and September–October corresponded to an estimated increase in standing crop biomass and the subsequent net loss to a decrease in estimated biomass (Table 7, Fig. 27). Similarly, the two types of estimates corresponded in *Ipomoea macrantha* in that the small net production occurred at times when biomass was estimated to have increased slightly or stayed about the same. However, the two types of estimates gave opposite trends for both sampling periods in *Melanthera* and for one of them in *Abutilon*.

Leaf longevity of *Pandanus*

Pandanus leaves are produced at the terminus of the branch in a spike-like bud. As the branch grows, the spike opens and the leaves project horizontally. As new leaves are formed and the branch extends, the older leaves become located progressively further from the apex. Those most distant from the apex eventually die, turn brown, hang on the stem for a while and eventually drop off. The leaves marked in the present study provide some indication of longevity.

At the time of initial marking of the leaves on the ten trees only two parastichies had all the leaves live and green, seven had one leaf of the parastichy dead (usually the bottom one, but sometimes the one next to the bottom), and one had four dead leaves in the parastichy. Since there are three parastichies on each branch, the number per branch would be approximately tripled. The leaves were examined again four months later, and by that time the average number that had died in the meantime was 1 (range 0–3) per parastichy. All the dead leaves were those which had been fourth or less from the bottom, i.e., the oldest leaves of the parastichy. By the next observation period an average of 0.6 additional leaves per parastichy had died (range 0–2), after which no further leaves were noted to have died by the end of the series of observations. In some cases, the oldest live leaf of the parastichy at the beginning of the study was still green at the end, about a year later. In no case did a new leaf produced during the period of observation die and none younger than the eighth oldest one died during the study. There was an average of 18.5 leaves per parastichy and on the average there were 2.2 new leaves per parastichy (6.7 per branch) produced during the year of observation. With an average of 1.6 leaves dying per parastichy in the same time, a rough estimate of leaf turnover would be about 2 per year. In even a relatively small parastichy of about 10 (branch of 30) leaves it would take about 5 years for a new leaf to become the oldest on the parastichy and die.

Even after the leaves die they may persist for long periods. Although most dry leaves had dropped within several months after dying, some that were dead at the beginning of the study were still on the stem about a year later. Thus, leaf death and leaf fall are not closely correlated and it may be some time before dead leaves become available to the decomposer organisms in the litter. Leaf fall may be more closely related to factors such as wind strength or storms, than it is to numbers of leaves dying (see Fig. 38 for values of leaf fall, as opposed to leaf death).

In summary, it can be concluded that *Pandanus* leaves are extremely long-lived, certainly several years, and that they may persist on the stem long after death. These characteristics, plus the relatively large number of leaves on a branch at any one time means that leaf production and leaf fall may be relatively unrelated over any particular short period of time.

The leaf production figures provide a means of ageing the One-Tree *Pandanus* trees, assuming the observed rate of leaf production to have been the same in the past as it was during the study.

Based on leaf production rates and the pattern of distribution of leaf scars around the stem we estimated that if one counts leaf scars in a straight line up one side of the stem, every four scars represents about one year's production. Thus, if the numbers of rows of scars are divided by four, the age of the stem can be ascertained. According to this method a small *Pandanus* tree, 2.5 m high at the edge of the clump was estimated to be 60 years old. Unfortunately, in the older ones the leaf scars become obscured with age and can't be counted on the lower parts of the stem. However, on One-Tree, the scars were visible after about 2.2 m above the ground. If it is assumed that that section was about 50 years old (based on the age of the 2.5 m tree) then counting the scars above that height gave an age of 191 years. Another one was estimated to be 120 years old. The age of the largest tree could not be estimated but would exceed the above values considerably. Such ages are reasonable in view of the fact that the eagle's nest, on the ground but supported by the bases of several *Pandanus* trees, had been there for at least 127 years as of 1970 (Chapter 17), the time when the leaf scars were counted.

Twig growth

Increase in length during the months of June through September 1969 was measured on the twigs marked for leaf production studies in *Argusia* shrubs and *Pisonia* trees. Both species had twigs of similar diameters, but *Argusia* increased in length at a much more rapid rate than did *Pisonia*. During the four months growth was measured, the mean increase in twig length of *Argusia* was 58 mm (S.E. = 8.9; range = 7–102 mm; N= 10) whereas it was only 5.5 mm (S.E. = 1.8; R = 0–18; N = 10) for *Pisonia*. The mean number of twigs per square metre of canopy also differed in the two species, being 49.2 (S.E. = 1.8; R = 12–134; N = 100) for *Argusia* and 27.7 (S.E. = 1.41;

R = 8–68; N = 84) for *Pisonia*. Taking both these measurements into account, the total length of stem added during four months per m^2 of canopy was 2.85 m, or 8.55 m per m^2 per year for *Argusia* and 0.15 m (0.45 m per m^2 per year) for *Pisonia*. Thus, *Argusia* produces about 19 times as much terminal twig growth per m^2 of canopy as does *Pisonia*.

The estimated total area of canopy (note: actual canopy area, not ground area covered by trees) was 3,181 m^2 for *Argusia* and 4,190 m^2 for *Pisonia*. Consequently, total length of new twigs added per year for the whole island was 27,198 m and 1,886 m for *Argusia* and *Pisonia*, respectively. On an island-wide basis then, slightly more than 14 times as much new twig length is produced by *Argusia* as by *Pisonia*. Both species are rather brittle and lose a considerable amount of dead twig wood to the litter (see Chaper 13).

The above values refer to lengths of twigs produced terminally but do not indicate biomass. Total wood production would need to include increase in girth of trunks, branches and twigs. These were not measured. However, a rough biomass estimate for terminal growth is available for *Argusia*. The terminal 5 cm of 20 twigs were cut into 1 cm lengths and the individual lengths then oven-dried and weighed. The mean dry weight for the terminal cm was 0.080 g with respective values for sections 2 to 5 of 0.127 g, 0.144 g, 0.161 g and 0.174 g. Thus, a centimetre length of twig increased in dry weight (increase in girth) by 0.047 g (59%) between the time it represented the terminal cm and when an additional cm of twig had grown. The proportional increase in dry weight decreased with further distance from the tip until that between the fourth and fifth cm length was only 0.013 g (8%). Considering radial growth to be negligible proximal to the terminal 5 cm of twig, the mean biomass produced was 2.475 g dry weight per twig per year or 121.8 g dry weight per m^2 of canopy. This represents 387 kg dry weight of *Argusia* twigs produced each year for One-Tree Island as a whole.

Koranda (personal communication) has provided some overall values of biomass for *Argusia argentea* from the Marshall Islands. A 66.7 kg (dry weight) shrub had its above-ground biomass partitioned as follows: leaves 6.6%, wood 0–2.5 cm in diameter 18.0%, wood 2.5–7.5 cm in diameter 29.4%, and wood greater than 7.5 cm in diameter 46.0%. A 139.5 kg (dry weight) plant had corresponding values of 7.7%, 8.7%, 10.2% and 73.4%. Dead wood was 6.1 and 13.9 kg (dry weight) for these two trees or 9.2% and 10.0% as heavy as total live materials. Koranda *et al.* (1979) report on a 223 kg plant that was not broken down into as many categories of biomass; it was 7.6% leaves, 12% young wood, 58% branches and 14% trunk, the rest being presumably dead wood. Total tree biomass increased linearly with basal area up to very large basal areas, and relative leaf biomass decreased with increasing total biomass and basal area.

13. Litter and dead standing stems

The dead remains of plants have significance to several areas of community metabolism; (1) their breakdown in the soil by decay micro-organisms releases nutrients which are then available for recycling by plants and (2) they provide food for some arthropods and hence indirectly for higher trophic levels. Thus litter accumulation is important to both plants and animals in the community, and the amount and rate of supply of such material should be assessed.

The removal plots were used to sample standing crop of dead standing stems and ground litter in the herb flat. The biomass of dead standing stems in

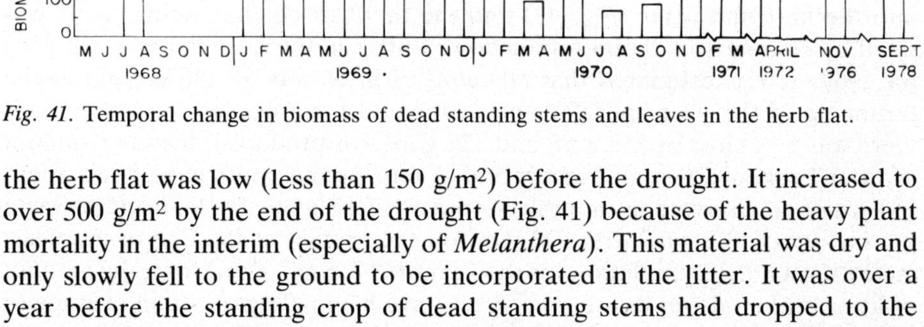

Fig. 41. Temporal change in biomass of dead standing stems and leaves in the herb flat.

the herb flat was low (less than 150 g/m^2) before the drought. It increased to over 500 g/m^2 by the end of the drought (Fig. 41) because of the heavy plant mortality in the interim (especially of *Melanthera*). This material was dry and only slowly fell to the ground to be incorporated in the litter. It was over a year before the standing crop of dead standing stems had dropped to the pre-drought level. It remained at that level for about eight months; then there was an increase to a higher level which persisted throughout the remainder of the study.

The slow incorporation of the dead standing stems into the litter meant that despite a marked increase in amount of dead organic matter after the drought, it did not become availabe to decay micro-organisms, detritivores and plants as a sudden pulse; rather there was a lag and its energetic input into the higher trophic levels was gradual.

The incorporation of stems into the litter was also gauged by periodic sampling by litter trays. The total weight of stems estimated to have fallen from May–June 1969 to September–October 1970 by means of litter trays was 173 g/m^2, a figure lower than the independently estimated total decrease in biomass of dead standing stems for the same period (265 g/m^2). Despite the temporal fluctuations in amount of organic matter dying and its rate of

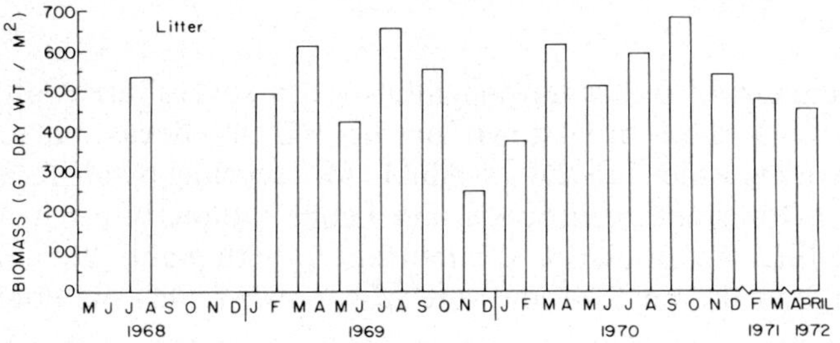

Fig. 42. Temporal change in biomass of litter in the herb flat. Data from litter traps.

incorporation into the litter, caused by the drought, the amount of litter remained remarkably constant throughout the study and pre-drought and post-drought values were quite similar (Fig. 42).

The standing crop of litter is dependent upon interaction of two opposing processes, (1) the rate of fall of dead material to the ground and (2) its subsequent decay and breakdown into humus. The gradual incorporation of the dead stems resulting from the drought, did not markedly elevate the litter standing crop. Leaves accounted for the greatest proportion of organic input into the herb flat litter (Fig. 43) and the input of dead standing stems was small by comparison. Total production of litter in the herb flat was 525 g/m^2 for 1969. It was estimated that 508 g/m^2 of litter was on the ground at the beginning of the year and 250 g/m^2 at the end for a net loss of 258 g/m^2. If there was a net loss of 258 g/m^2 and 525 g/m^2 was produced, then the amount of litter broken down and incorporated into humus must have been about 783 g/m^2, or an average monthly turnover of 65 g/m^2. In 1970 there was a production of 502 g/m^2 with 543 g/m^2 on the ground at the end of the year, indicating a net accumulation for the year of 293 g/m^2. Therefore, the amount of litter broken down into humus must have been 209 g/m^2 or an average of 17 g/m^2/month. In September 1978 the standing crop of litter in the herb flat was 492 g/m^2.

Thus, although total litter production was very similar in the two years, they differed markedly in other features; 1969 was characterized by extensive

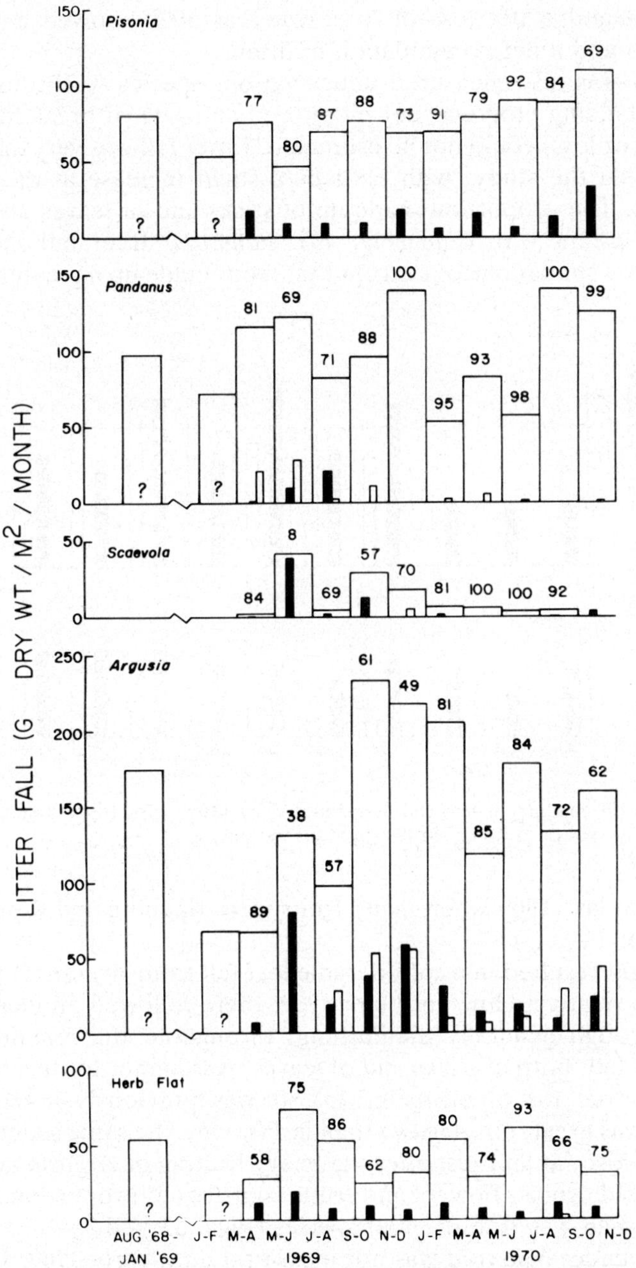

Fig. 43. Litter fall in the herb flat and under various shrubs and trees. Large open histograms indicate total litter fall, small open ones that part consisting of fallen seeds, fruits, and flowers, and old cymes, and the small dark histograms that part consisting of fallen twigs and stems. Number above histograms indicate the percentage of the total weight of litter fall contributed by leaves. A question mark indicates that in the early part of the study total litter was not separated into its component parts and only daily total litter weights are available. Data from litter traps.

humification and a net loss of litter whereas 1970 showed a low rate of humification and a net accumulation of litter.

Litter fall was also measured under various species of shrubs and trees. However, standing crop was not monitored and rate of breakdown and net production (or loss) could not be estimated. Litter fall was very iniform under *Pisonia* during the study, with perhaps a slight increase at the end of the second year. The proportion made up of sticks and of leaves also remained relatively constant with time (Fig. 43). *Pandanus* litter fall showed more variation but a similar pattern except that seeds made up a greater proportion

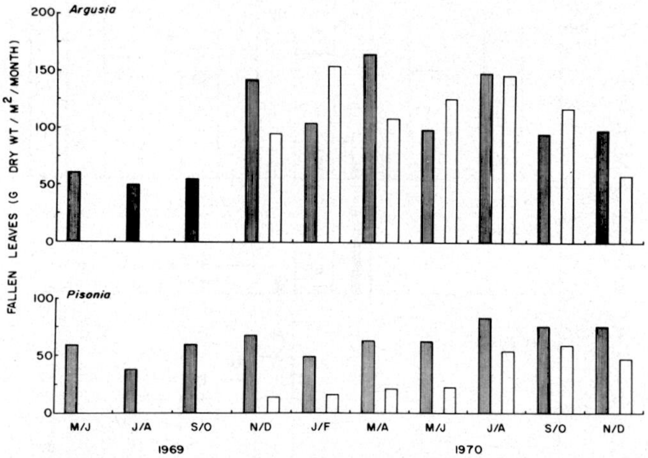

Fig. 44. Temporal changes in leaf fall as estimated by litter traps (dark histograms) and by marking of live leaves on trees or shrubs (light histograms).

during mid to late 1969 when many fruits were ripening and dropping seeds (see Fig. 36).

The drought resulted in a considerable leaf fall from *Argusia* (Fig. 43) after which it was relatively low until late 1969; there followed an increase which persisted, though generally diminishing, throughout the remainder of the study. Litter fall, both of sticks and of leaves, was higher for this species than was usually true for others (Fig. 43). It was previously noted that twig production was greater in *Argusia* than in *Pisonia*. The same seems to be true of twig fall. Also, at and just after the heavy fruiting of *Argusia* at the end of both years, old cymes, flowers and fruits contributed proportionately higher to the litter than was true of other species (Figs. 35, 43).

Litter fall under *Scaevola* was not measured until early 1969. In late 1969 there was a slight increase in litter fall (Fig. 43), although the higher values were low by other species' standards and most of the year *Scaevola* contributed almost negligible amounts of litter.

Combining all species of trees and shrubs, the mean monthly production of litter for the island as a whole was 808 kg dry weight/hectare (Table 14). During the recovery from drought, values were low but rose in late 1969 and remained high throughout 1970.

There were two species (*Argusia* and *Pisonia*) for which leaf fall was estimated by two independent methods, leaf fall traps and rate of disappearance of leaves from marked twigs (see Chapters 7 and 12). In *Pisonia*, leaf fall as measured by litter traps was rather constant throughout the year. The leaf longevity method always provided a lower estimate, especially early in the study (Fig. 44). For *Argusia*, the two methods gave more comparable values that were usually of the same order of magnitude and the pattern of temporal change was similar in both cases. These results lend confidence to the estimates obtained for this species. Regardless of which method is selected, *Argusia* was responsible for a greater amount of leaf fall on a unit area basis than was *Pisonia* (Fig. 44). This was not always true on an island-wide basis because of the greater area covered by *Pisonia*. Sometimes *Argusia* produced the higher amount of litter, and sometimes *Pisonia* did. Over the entire study period their mean percentage contributions were very similar. These two species together usually produced more than 80% of the litter from trees and shrubs (Table 14).

The suggestion of Domm and Recher (1973) that there is a seasonal leaf fall in most One-Tree species (except *Pandanus*) between October and December was not supported by the present data. Certainly such deciduousness is not a regular phenomenon; it may, however, have occurred in a particular year or series of years during their observations which were mostly after the intensive part of the present study was carried out. The most constant species in terms of rate of leaf fall during the study was *Pisonia*. It is known that after prolonged dry spells this species may undergo virtual defoliation (Cribb 1969) and severe storms were observed to cause heavy loss of leaves during the present study. It must be pointed out, however, that rate of leaf fall and the extent of defoliation are not necessarily related. A species with heavy leaf fall occurring at the same time as heavy production of new leaves will not appear bare whereas if leaf production is low, defoliation will eventually occur even at a constant, moderate leaf fall. It is possible that drought affects leaf production in *Pisonia* more than it affects abscission. Unfortunately collection of data on leaf turnover had not begun yet during the drought at the beginning of this study.

Litter fall under all three major types of trees and shrubs on One-Tree Island produced more litter per year than either rainforest or wet sclerophyll forest near Brisbane (Table 15); the methods used in the latter vegetation types were comparable to those employed at One-Tree (Plowman 1979) and both studies were conducted in 1969 and 1970. Only the *Scaevola* clump had low litter production on One-Tree.

The herb flat litter production (502–525 g dry weight/year) was more than half that of rainforest and wet sclerophyll forest but only about a third that

Table 14. Total litter production by trees and shrubs. Except where noted otherwise, values are in grams dry weight per month for the entire island; percentages are in parentheses.

	1969				1970						
	M–J	J–A	S–O	N–D	J–F	M–A	M–J	J–A			Mean
Species											
Argusia	144,535	119,689	132,112	342,105	252,517	397,530	239,139	357,633	230,300	236,989	245,255
	(34.4)	(38.3)	(34.5)	(53.4)	(47.6)	(60.6)	(46.5)	(51.8)	(38.5)	(39.7)	(44.5)
Scaevola	45	95	92	454	342	140	148	101	101		152
	(0.02)	(0.03)	(0.02)	(0.07)	(0.07)	(0.02)	(0.03)	(0.02)	(0.02)	?[a]	(0.03)
Pisonia	205,306	129,703	207,733	234,784	172,706	220,218	217,097	290,272	262,528	266,689	220,704
	(48.9)	(41.5)	(54.2)	(36.6)	(32.6)	(33.6)	(42.3)	(42.1)	(43.9)	(44.7)	(42.0)
Pandanus	69,975	63,150	43,500	63,750	104,925	37,650	57,450	42,075	105,075	93,300	68,085
	(16.7)	(20.2)	(11.4)	(9.9)	(19.8)	(5.7)	(11.2)	(6.1)	(17.6)	(15.6)	(13.4)
Total	419,861	312,637	383,437	641,093	530,490	655,538	513,834	690,081	598,004	596,978	534,195
Total Kg/hectare of area covered by shrubs and trees	636	474	581	971	804	993	763	1,046	906	905	808

[a] No data for this category. It is usually so small as to be negligible and was considered 0 for the purposes of calculating percentages.

Table 15. Summary of total annual litter fall (g dry wt/m^2) under the various types of arborescent vegetation on One-Tree Island. All data based on litter traps; those from rainforest and wet sclerophyll forest from Plowman (1979).

	1969	1970	$\bar{X}$
Species			
Pisonia	846	1063	955
Pandanus	1069	1183	1126
Argusia	1572	1786	1679
Scaevola (mid 1969 to mid 1970)	—	—	200
Rainforest (Sept. 1969–Sept. 1970)	—	—	913
Wet Sclerophyll Forest (Sept. 1969–Sept. 1970)	—	—	887

under *Argusia*. It is possible that litter fall in Plowman's (1979) study refers only to *leaf* litter. The mean proportions of the total litter fall on One-Tree made up by leaves were 82%, 89%, 68%, and 73% for *Pisonia, Pandanus, Argusia* and *Scaevola* respectively. Adjusting the mean annual total litter production to mean annual leaf litter production gives values of 783, 1002, 1141, and 146 g dry weight/m^2 respectively for these same species. This adjustment brings the mean annual *Pisonia* production below that of the two mainland forests but still leaves that of *Pandanus* and *Argusia* higher.

14. Epigeal arthropods[1]

The arthropods (exclusive of the soil microarthropods which are treated separately) are referred to here as epigeal arthropods, even though a few of them such as silverfish, earwigs and centipedes may be cryptozoic or inhabit litter or crevices in the soil.

There were 396 species of epigeal arthropods recorded from One-Tree Island during the study. Those which have been identified to genus or species are listed in appendix II. There were 17 orders of insects, three groups of arachnids (pseudoscorpions, spiders and mites) and centipedes and isopods (Table 16). The Hymenoptera made up 24.5% of the total species, the Diptera accounted for 22.7%, the Lepidoptera for 14.4%, the Hemiptera for 9.9% and the Coleoptera for 9.1%. These five most common insect orders account for 319 (80.6%) of the species known from the island with all other groups of epigeal arthropods accounting for only 77 species or 19.4%.

The One-Tree fauna is imbalanced in comparison with that of the Australian mainland (Table 17). This imbalance arises from a disproportionately larger number of species of Hymenoptera and Diptera, and to a lesser extent of Hemiptera, and unusually small numbers of species of Lepidoptera and Coleoptera on One-Tree compared to the mainland.

On the mainland, beetles are by far the richest order, whereas on One-Tree they are the least numerous of the five orders. The differences in the number of species of the various groups observed on One-Tree Island compared to the expected values based on proportional representation on the mainland is highly significant ($P < 0.001$; see Table 17). Such a situation could arise either if beetles are less easily dispersed to islands then the over-represented groups such as Hymenoptera and Diptera, or if the latter could become more easily established and/or persist longer.

[1] For treatment of the arthropods of the Bunker–Capricorn Group generally, see Musgrave (1925–27b), Chadwick (1962), Fletcher (1973), Smithers (1975) and Great Barrier Reef Committee (1978).

Species richness in the herb flat

For a technique to adequately sample the number of species present, a sufficient area must be included in the sample that the likelihood of species being missed is low. Adequacy of the area sampled by the cage method was assessed by plotting species-area curves for the ten successive plots at each sampling period (Fig. 45). Species numbers initially increased rapidly with the addition of each cage plot but in most cases eventually approached an asymptote. Thus the sample size of ten cage plots generally seems adequate for sampling the number of species of arthropod groups susceptible to capture by this method. There were several sampling periods, however, when it appeared as though the asymptote had not yet been reached, e.g., March–April 1969, 1970; November–December 1970 and especially February–March 1971 when only five plots were cleared. In these periods, numbers of species undoubtedly were underestimated. However, some of the missed species may have been picked up in the sweeps and when the two methods are combined to estimate total species present in the herb flat the error would be reduced.

Similarly, curves of cumulative species obtained with successive sweeping were plotted (Fig. 45). In all sampling periods, the asymptote was approached before the area had been swept ten times and the method would seem to be adequate for estimating numbers of species amenable to capture by sweeping.

Of the species of arthropods sampled in the herb flat, 41% occurred at some time in both sweeps and plots, 38% were taken only in sweeps, and 21% were found only in caged plots. Thus, (1) had only plots been used, more than 1/3 of the species would have been overlooked, and (2) had only the sweep method been employed, about 1/5 of the species would have been missed. Clearly, as an estimator of species richness the sweep method was superior to cage sampling but neither method alone approached the effectiveness of the two in combination; both are required.[2] For either method, most of the species that were missed were rare or ephemeral as indicated by the fact that the great majority caught by a single method were captured only during one sampling period (Fig. 46). No doubt some rare ephemerals were missed altogether.

The species that were captured by only one method and yet were taken on more than one sampling period probably represent the truly exclusive species and warrant closer examination. There were 13 species captured exclusively by the plot method but in more than one period. Of these, four were spiders, two were mites, two were moths (including caterpillars) and there was one species each of centipede, thysanuran, phasmid, cricket (*Teleogryllus oceanicus*), and small wasp. Thus 69% (9 out of 13) were either normally ground dwellers (and usually out of reach of sweep sampling) or were flightless and might be expected to adopt the escape strategy of letisimulation and dropping

[2] Hespenheide (1979) has briefly reviewed sampling biases in arthropod sampling; his information confirms our view that one method alone is inefficient.

to the ground when the vegetation on which they rest is moved. By contrast the 26 species exclusive to the sweep method and captured in more than one sampling period, included: nine flying hymenoptera, eight flies, two moths, two thrips, and one species each of grasshopper, damselfly, cockroach, aphid

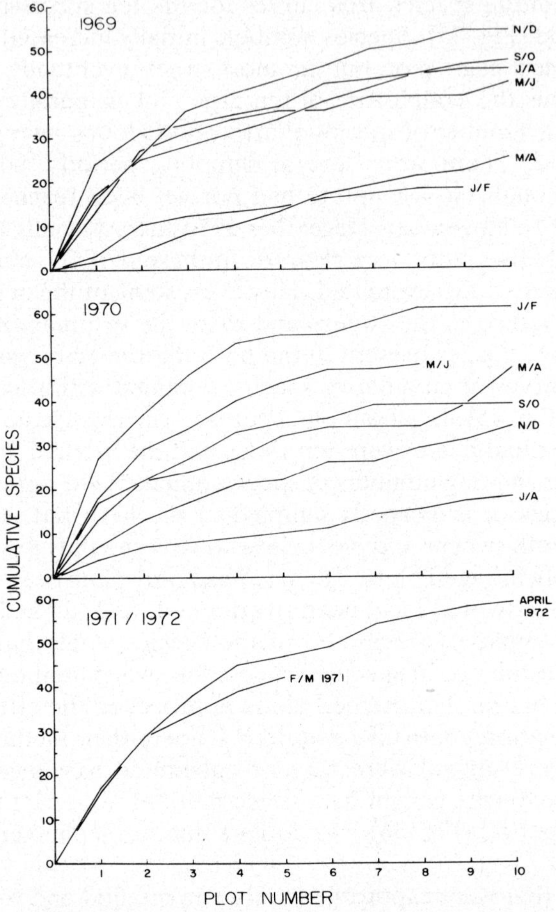

Fig. 45. Cumulative number of arthropod species obtained in successive cage plots (left) and with successive sweeping of a 100 m^2 area (right).

and lacewing. Seventy-two per cent (21 of 29 species) can be considered visually oriented, rather strong fliers. These would be likely to have the escape strategy of flying away from approaching objects and to be less likely to be trapped by a cage than by the more rapid sweep net. It would appear that each method is effective in sampling different subsets of the arthropod fauna but that both are required for an adequate assessment of the total number of species present.

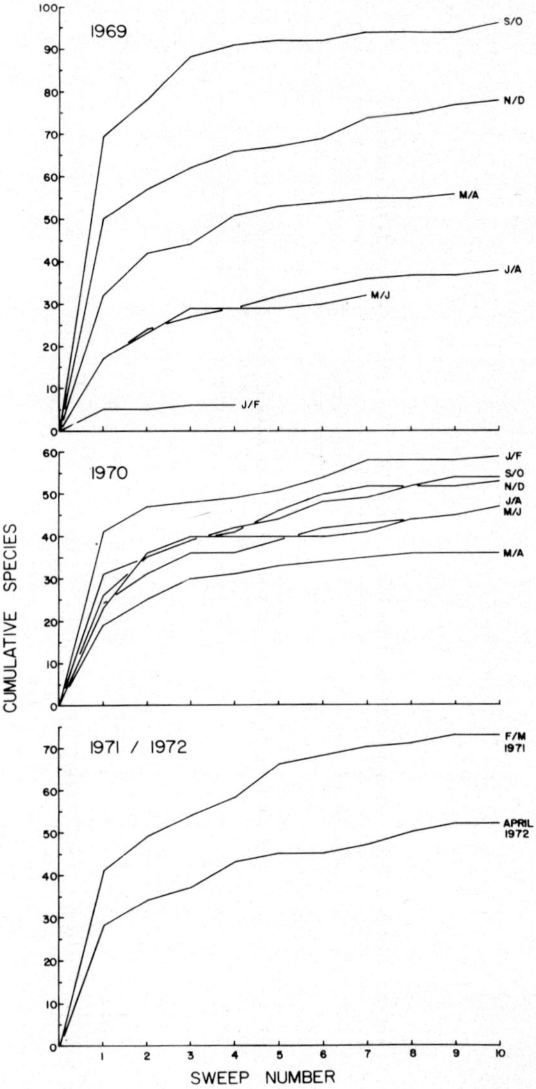

Some epigeal species may not be adequately sampled by either method, e.g., extremely rapid fliers such as dragonflies and some butterflies that could avoid both the cage and the sweep net. However, these are conspicuous insects that can be observed directly; they were noted to be infrequent and in low abundance. For example, on the rare occasions that dragonflies were present there were usually only a few individuals around the pond; occasionally they would fly over the herb flat. Unsampled species of butterflies were observed but were represented on the island by only a few individuals.

Table 16. Numbers and percentages of various taxa in different habitats on One-Tree Island, 1968–1972. Soil microarthropods are excluded.

Taxon		One habitat only					Shared by two habitats only					Shared by three habitats only		Shared by five habitats	Total No. of species	% of total species
		HF	AR	Pi	B	OP	HF & AR	HF & Pi	HF & OP	AR & Pi	AR & OP	HF & AR & Pi	HF & AR & OP	HF & AR & Pi & B & OP		
Isopoda	No.					1									1	0.3
	%					(100.0)										
Pseudoscor-pionida	No.					1									1*	0.3
	%					(100.0)										
Acarina	No.	5			4										9	2.3
	%	(55.6)			(44.4)											
Araneida	No.	13	2	4	1	3		1		1		7			32	8.1
	%	(40.6)	(6.3)	(12.5)	(3.1)	(9.4)		(3.1)		(3.1)						
Chilopoda	No.	1							2			21.9			3	0.8
	%	(33.3)							(66.7)							
Collembola	No.	1				2									3	0.8
	%	(33.3)				(66.7)										
Thysanura	No.	1													1	0.3
	%	(100.0)														
Odonata	No.	1				2									3	0.8
	%	(33.3)				(66.7)										
Blattodea	No.	1				1					1				3	0.8
	%	(33.3)				(33.3)					(33.3)					
Isoptera	No.	1													1	0.3
	%	(100.0)														
Mantodea	No.	1													1	0.3
	%	(100.0)														

Dermaptera	No.				1										1	0.3
	%				(100.0)											
Orthoptera	No.	3							2			3	1	1	10	2.5
	%	(30.0)							(20.0)			(30.0)	(10.0)	(10.0)		
Psocoptera	No.						1					1			2	0.5
	%						(50.0)					(50.0)				
Phthiroptera	No.					1									1	0.3
	%					(100.0)										
Hemiptera	No.	18	5	1		1	4		1	3		6			39	9.9
	%	(46.2)	(12.8)	(2.6)		(2.6)	(10.3)		(2.6)	(7.7)		(15.4)				
Thysanoptera	No.								2						2	0.5
	%								(100.0)							
Neuroptera	No.	3													3	0.8
	%	(100.0)														
Coleoptera	No.	17	2			12			3			2			36	9.1
	%	(47.2)	(5.6)			(33.3)			(8.3)			(5.6)				
Diptera	No.	34	15	1		14	7	1			1	17			90	22.7
	%	(37.8)	(16.7)	(1.1)		(15.6)	(7.8)	(1.1)			(1.1)	(18.9)				
Lepidoptera	No.	26	11	1		13			2	2		2			57	14.4
	%	(45.6)	(19.3)	(1.8)		(22.8)			(3.5)	(3.5)		(3.5)				
Hymenoptera (non-ants)	No.	55	12	8		1	4	2		3			4		89	22.5
	%	(61.8)	(13.5)	(9.0)		(1.1)	(4.5)	(2.3)		(3.4)			(4.5)			
Hymenoptera (ants)	No.	5						1				2			8	2.0
	%	(62.5)						(12.5)				(25.0)				
Total No.		186	47	15	6	52	16	5	12	9	2	40	5	1	396	100.6

HF = Herb flat; AR = *Argusia* shrubs; Pi = *Pisonia* forest; B = beach; OP = opportunistic collections.
* Muchmore reports the collection to contain 2 species rather than one.

Table 17. Comparison of the relative number of species of the 5 most common orders of insects on One-Tree Island with that of the same orders for Australia generally. Data on total Australian species from Riek *et al.* (1970), Colless and McAlpine (1970), Common (1970), Britton (1970) and Woodward *et al.* (1970).

	Hymenoptera	Diptera	Lepidoptera	Coleoptera	Hemiptera	Total
No. of described Australian species (as of 1970)	8,834	6,256	11,221	19,219	3,661	49,191
% of total species for the 5 orders	18.0	12.7	22.8	39.1	7.4	100.0
Expected no. of species for One-Tree Island	57	41	73	125	24	320
Observed no. of species on One-Tree Island	97	90	57	36	39	319

Chi-square test $\chi^2 = 163$; d.f. $= 4$; $P < 0.001$.

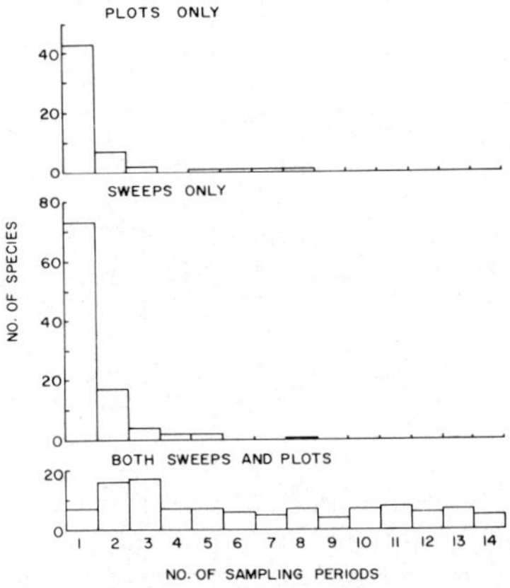

Fig. 46. Number of sampling periods species were caught exclusively in plots, exclusively in sweeps and by both methods.

Population densities in the herb flat

With one important exception, it was possible to obtain population estimates of all herb-flat arthropod groups. The exceptional taxon was the ants which occurred above-ground for only part of the diel period and although they were captured in sufficient numbers that the number of species was known, their population densities would have been greatly underestimated by both the cage and sweep methods. Furthermore, their extreme clumping prevented their proper sampling by methods employed for the soil fauna. Ants are discussed separately (see Chapter 18).

For other taxa, the cage and sweep methods differed in their effectiveness in estimating population densities. The sweep method almost certainly underestimated population densities of some species of arthropods, especially those non-fliers that drop to the ground when disturbed. Conversely the cage method probably underestimated visually oriented, fast fliers that sometimes escape the cage. Since in both cases the result is an underestimate of individuals present, for any species the best estimate of population density is obtained by the method giving the highest value. Fig. 47 compares the total

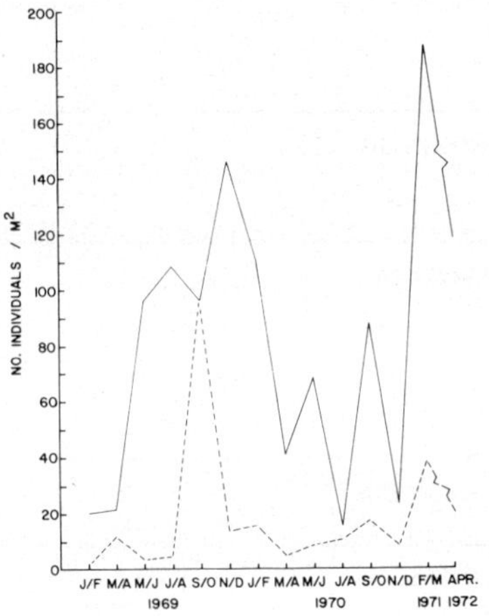

Fig. 47. Numbers of individuals of epigeal arthropods in the herb flat as estimated from cage plots (solid line) and sweep sampling (broken line).

population densities of arthropods in the herb flat as estimated by the two methods. The plot method consistently provided a higher value than sweeps. However, the plot method produced the more erratic results largely because of the smaller area sampled. The sweep samples involved 100 m^2 at each sampling period whereas the cage plots involved only 3.85 m^2. Consequently differences of only a few individuals in cage plots had a larger proportional effect on the estimate.

The two methods gave similar overall patterns of temporal change except for an unusually high estimate obtained from sweeping in September–October 1969 which was at variance with the results from cage sampling. Of the 312 times individuals of the same species were captured by both methods, the cage method gave the highest estimate of population density 296 times (94.9%) and sweeps the highest 16 times (5.1%). In contrast to estimation of species richness, estimation of population densities is best achieved by the cage method rather than the sweep method. The discrepancy probably occurred because most of the rare or ephemeral species are those fast fliers that are easily dispersed and hence more amenable to capture by sweeping, whereas the bulk of the individuals present belong to the slower flying or non-flying taxa.

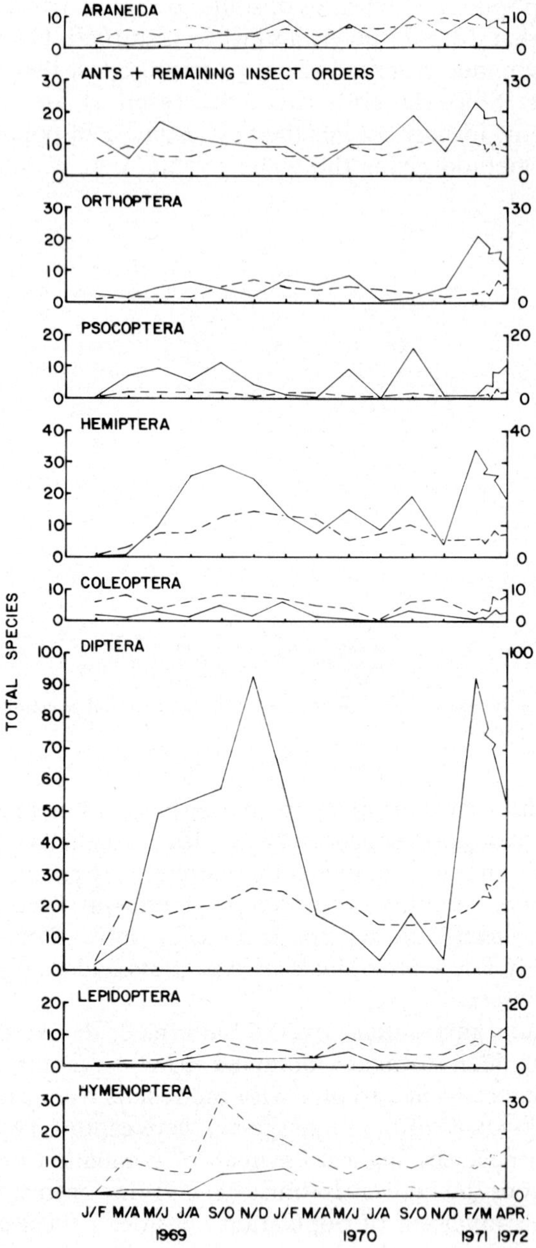

Fig. 48. Numbers of species (broken line) and population density (solid line) of various groups of epigeal arthropods from the herb flat estimated by a combination of plot and sweep methods. The category 'Hymenoptera' in the bottom graph is exclusive of ants.

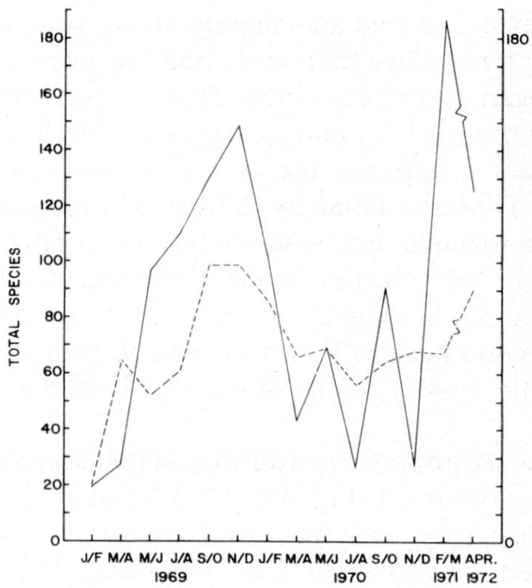

Fig. 49. Numbers of species (broken line) and population density (solid line) of all epigeal arthropods in the herb flat as estimated by a combination of plot and sweep methods.

For each occasion, the method giving the highest estimate at a particular time was selected and individual species densities were summed for various taxa (Figs. 48, 49). Unfortunately, quantitative studies on arthropods were not initiated until after the drought and the effect of the drought cannot be directly assessed except for a few species specially investigated. However, indirect evidence suggests that population densities were reduced by the drought. In almost all groups, population densities were very low at the first sampling period after the drought (January–February 1969). Subsequently they rose in most species, the length of time until an increase occurred varying greatly among taxa. In most groups, the level of population densities then fluctuated with different taxa showing different temporal patterns of change (Fig. 48). The spiders showed minor fluctuations around a general tendency toward increasing density. The beetles and psocoptera were characterized by slightly fluctuating but consistently low densities. Lepidoptera showed a similar pattern but with a tendency to increase in population density at the very end of the study.

Other groups displayed temporal patterns of change that warrant more detailed treatment. The Orthoptera had consistently low density throughout the period of quantitative sampling except for an increase in February–March 1971. However, qualitative data suggest that there was a massive population crash (either by death or emigration or both) during the drought (see Chapter 18).

No hemipterans were present in samples just after the drought (Fig. 48). However, populations subsequently built up and reached a peak in September–October 1969. After this time they declined to lower, but fluctuating, levels and again rose to a second and slightly higher peak in February–March 1971. The Hymenoptera (other than ants) had low population densities after the drought but began to increase after July–August 1969; they reached a peak by January–February 1970 and subsequently declined.

The most abundant group and the one with the most marked temporal fluctuation was the Diptera. Total fly density was extremely low (less than $2/m^2$) just after the drought but rose steeply to a peak of over $90/m^2$ by November–December 1969. It then dropped sharply and fluctuated from less than $4/m^2$ to slightly more than $25/m^2$ between May and December 1970. There followed a second peak in February–March 1971 of approximately the same magnitude as the first one (Fig. 48); a year later the density was slightly over $50/m^2$.

When all herb flat arthropods are considered collectively, two major peaks in population density occurred (Fig. 49). One (of almost $150/m^2$) occurred in November–December 1970 and the other (of over $185/m^2$) in February–March 1971. Between the two peaks density fluctuated between about 25 and 90 arthropods/m^2. The density just after the drought was less than $20/m^2$.

The causative factors of the temporal change in density of herb flat arthropod populations are difficult to determine. On the one hand different species undoubtedly respond to different environmental factors in different ways and when an entire order, or all arthropods collectively, are considered, one can get only an average response pattern. Treatment of all species individually is beyond the scope of this book (but see Chapter 18). A second difficulty is that apparent changes in population density may arise from different levels of mortality, natality, persistence of resistant stages, and immigration at different times.

(1) Specific, immediate weather conditions may influence natality, mortality, or both and thereby result in changes in population density. Such density changes should correlate with weather and be as regular, or as unpredictable as the weather conditions on which they depend.

Frith (1949) indicated that rainfall controlled insect population density on Aldabra Atoll. A probable example from the present study is the generally low densities following the drought, presumably resulting from massive mortality. However, the effect on a species' continuity of such fluctuations may be less severe than superficially appears, if the species involved survive unfavourable conditions in resistant stages. For example, the rapid build-up of arthropod numbers after the drought may have resulted in part from the hatching of drought-resistant eggs deposited prior to the drought by the much larger population that probably existed at that time. In other cases, like the grasshoppers (see Chapter 18), the crash during the drought was real.

(2) There may be regular, seasonal periodicity in life history phenomena controlled either by an internal rhythm or some regularly fluctuating environmental factor. Again, fluctuations may appear greater than is actually the case because of difficulty in sampling some life history stages, e.g. eggs.

(3) Certain weather conditions may result in increased immigration of animals to the island and inflate the numbers of both species and individuals occurring there.[3]

Methods of assessing these types of population changes are available. If immigration is a significant factor one would expect that during times when densities are peaking, there would be a corresponding increase in numbers of species as well. This was true in some cases (Fig. 48). The September–October 1969 increase in density of hymenopterans was accompanied by a

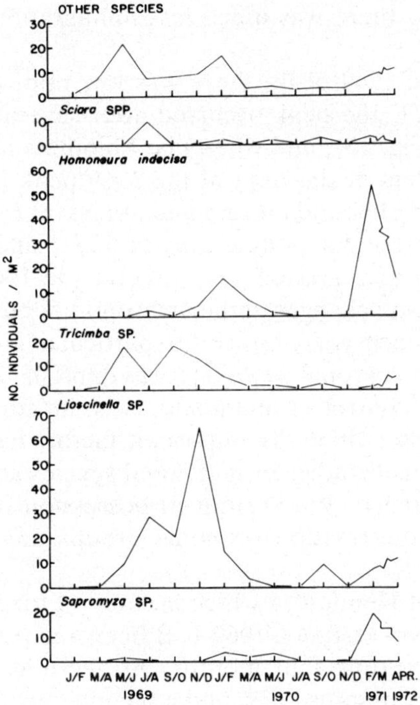

Fig. 50. Temporal changes in the population densities of some of the most abundant flies in the herb flat, estimated by a combination of cage and sweep methods.

marked rise in species number. The first peak for Hemiptera (1969) also occurred at a time of high species number for that group whereas there was no increase in species number during the second (1971) peak of bug population density.

In the Diptera, there were no marked changes in number of species associated with either population peak. The changes in population density were largely the result of changes in a few abundant species (Fig. 50). In all

[3] Certainly large numbers of conspecific individuals may migrate from mainlands to islands under appropriate conditions (Fletcher 1973, Hulsman 1974b).

other groups species numbers remained relatively constant throughout (Fig. 48). However, when all species are lumped, the cumulative effect was a marked elevation in species number coinciding with overall peak population density in 1969 but not in 1971 (Fig. 49). This suggests that immigration may have played a significant role in the elevation of individual numbers in 1969 but natural increase of resident populations was chiefly involved in the 1971 peak. Fletcher (1973) indicates that a northwesterly air stream favours massive immigration of insects from the mainland to Heron Island (19 km from One-Tree). Although favourable winds for aerial migration to One-Tree occurred both years, there was much less immigration in 1970 than in 1969 (see Chapter 21).

At the time of the 1969 peak, there was low rainfall following months of variable rain. In 1971, the peak occurred after several months of heavy and increasing rainfall (Fig. 13), and when plant biomass in the herb flat was at a much higher level than at the time of the 1969 peak (Fig. 25). It is possible that the combination of abundant rain and abundant food was responsible for natural increase in resident populations in 1971, and that special weather conditions favouring immigration were involved in 1969.[4]

It is clear that density changes in the arthropod populations at One-Tree do not recur regularly each year, either for particular groups or for all groups collectively. Neither seasonal biological rhythms in life histories not direct temperature-related control of abundance account for the observed fluctuations. Although some potentially important factors have been noted above, only detailed ecological studies of individual species and groups can quantify their relative importance. We are not in possession of such complete data. However available information on some groups are informative (see also Chapter 18).

The five species of Hemiptera which accounted for 95% of the population increase in September–October 1969 had been recorded in the plots in low densities for the preceding four months. All were also present two months later, one absent four months later, and a second one absent six months later. Thus, the population increase of bugs in September–October was caused by increase in species already present and/or arrivals of more individuals of the same species.

The one species that accounted for 70% of the 1969 hymenopteran population explosion was not recorded in the preceding four months, either in the herb flat or elsewhere. It was present in very low densities in the sweeps two months later but absent at the next two-monthly visits. All of the six species which made up an additional 20% of the peak numbers were absent at least four months before and after the higher population levels. It would

[4] An additional explanation for 1969 is that species, too rare to be included in the samples before the peak, increased in population densities and then became susceptible to sampling, thereby elevating number of species and of individuals. The two explanations are not mutually exclusive.

appear that the high densities resulted either from (1) immigration of large numbers of adults, or of a few adults which reproduced rapidly, or from (2) numerous hatchings from eggs of individuals which had previously died or been reduced to very low densities.

The species of fly (*Lioscinella* sp.) that accounted for a great proportion of the peak fly density in November–December 1969 had been well established for the preceding seven months and persisted several months afterwards; it was at very low densities the equivalent season the following year. A second species (*Tricimba* sp.) showed a similar pattern but at lower densities. A third species of fly (*Homoneura indecisa*) had been absent from the herb flat since about June of 1970 but suddenly rose to high population densities in early 1971. However, it had been present in low densities at *Argusia* flowers during that period and some of the reappearance in the herb flat may have resulted from movement from the shrubs. Other factors were also involved, however, as their density in the shrubs could not have accounted entirely for the peak values in the herb flat.

Sweeps of the succulent mat and arborescent vegetation

It is clear from data presented above that sweeping alone is insufficient to characterize adequately the epigeal arthropod fauna of an area. However, not all habitats were amenable to all the other techniques employed, arborescent vegetation for mechanical reasons, and the succulent mat because of its small size. Consequently we swept the vegetation in these areas on a few occasions in order to provide at least some information on the fauna of the succulent mat and to supplement other data from the arborescent vegetation.

Sweeps were carried out in the succulent mat in July, October and December 1970. On the first occasion 100 sweeps yielded 15 individuals of 5 species (0.15 individuals per sweep). In October, 150 sweeps yielded only 16 individual arthropods representing 8 species (0.11 individuals per sweep). The arthropods from the first 100 sweeps were analyzed separately from those captured by the following 50 sweeps. These two subsets of the sweeps yielded 11 and 5 individuals respectively; there was only one species common to them both. In December, 100 sweeps yielded 30 individuals of 10 species (0.3 individuals per sweep). Only 2 species were common to all 3 periods of sweeping, and only 3 species common to both the October and December sweeps. The total captured fauna contained 7 species of flies, 6 of flying hymenopterans (these two orders account for 77% of the total species), 1 species of bug, 1 species of ant, 1 species of cricket and 1 species of spider. The facts that (1) the catch was exceptionally poor both in numbers of individuals and numbers of species, (2) multiple sweeping showed little overlap in species composition, (3) an exceptionally high proportion of the captured animals are from strong-flying taxa (see below) and (4) all of them were well represented in the herb flat at the time, suggests that the arthropod fauna of the succulent mat is a depauperate one composed mainly of temporary strays from the herb flat.

A thorough search of the succulent mat in May 1968 yielded only 23 species, over half of which were Diptera and flying Hymenoptera. All occasions except July 1970 on which the arthropods of the succulents were examined were times when numbers of species and individuals were generally high on the island.

On the same day in October 1970 that the 100 sweeps of the succulent mat were carried out, the foliage of *Argusia* shrubs, *Pisonia* trees and *Pandanus* trees were also swept 100 times each. The total numbers of individuals captured were 69, 57 and 39 for the three habitats respectively. The *Pisonia* and *Argusia* thus had considerably higher numbers of arthropods than did *Pandanus. Pandanus* was nearly as poor in numbers of individuals as was the succulent mat.

The same pattern was reflected in numbers of species represented in the sweeps, i.e., 25, 26, and 10 for *Argusia, Pisonia* and *Pandanus* respectively. Most of the species captured on *Argusia* and *Pisonia* were either Diptera or flying Hymenoptera (48% and 42% respectively). *Pandanus* differed in that no flying hymenopterans were swept from it; however, flies acounted for 50% of the total species. *Argusia* and *Pisonia* shared 13 species, *Argusia* and *Pandanus* 6, and *Pisonia* and *Pandanus* 7.

In order to gain a further insight into the nature of the fauna of these habitats, a white sheet was placed on the ground below a tree or shrub and the foliage beaten with a stick. The arthropods that fell on the sheet were collected and examined. Few individuals and few species were obtained and none of the species were different from those obtained by sweeping.

It can be concluded that the succulent mat and *Pandanus* have a very depauperate fauna in comparison to the *Pisonia* groves or *Argusia* shrubs. The latter two are examined in a more quantitative way below.

Species richness of taxa associated with *Argusia* flowers

The number of arthropods attending *Argusia* flowers varied considerably with time (Fig. 51). The number of species was low before the drought in 1968. At that time the abundance of *Argusia* flowers was also low (Fig. 35). After the drought the number of species was somewhat higher and subsequently fluctuated but usually remained at a relatively high level throughout most of the rest of the study. The highest values were found in July–August and November–December of 1970. Since it is not possible to overestimate species numbers by either of the methods used, the one giving the highest estimate should be the most accurate. When there was any extensive disagreement between the methods, the activity-peak method usually gave the highest estimate and was consequently considered to provide the best data on species numbers.

Throughout the study, the total number of species associated with *Argusia* flowers ranged from 10 to 35. Of those most were nectarivores. On one occasion all the species observed at the flowers were nectarivores, and the lowest percentage was about 71%; on the average about 89% of the species were nectarivores with the other 11% being divided among predators, leaf eaters and other trophic levels (Table 18). The most common non-nectarivores observed on flowers were caterpillars (*Utetheisa pulchelloides*),

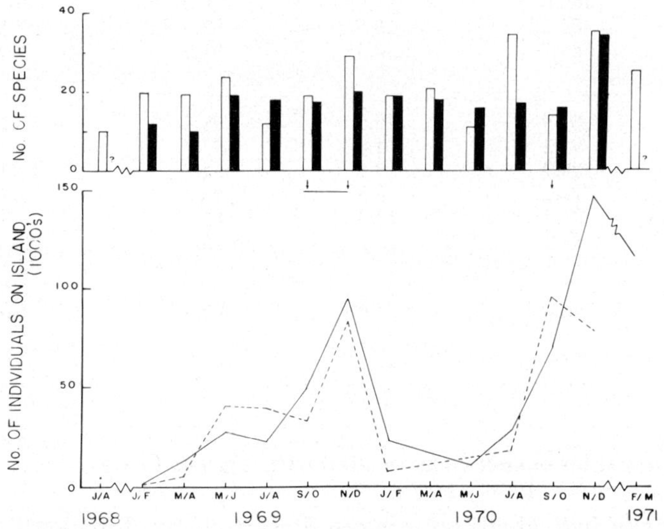

Fig. 51. Total number of species and individuals associated with *Argusia* flowers. Open histograms and solid line indicate estimates obtained from peak activity periods. Dark histograms and broken line indicate estimates obtained from the Hanson (1968) removal method. A question mark indicates that no appropriate data were collected for that time period. Arrows indicate peak flowering periods of *Argusia*.

spiders, beetles, crickets, and grasshoppers. Some were probably accidentals, i.e., just resting on the flower and neither attracted to nectar nor to the animals feeding on the nectar.

The three most abundant groups in terms of individuals (see below) were the ants, flies and bugs. Their patterns of change in species numbers were very different. The ants were always represented by two or three species only. The species richness of bugs and flies varied inversely (Fig. 52).

Spiders were the most common predators associated with the inflorescences. None were found just after the drought, but as species numbers of other groups built up more species of spiders appeared. They never reached a high species-richness, however. The largest spider fauna (6 species) occurred in November–December of 1970 (Fig. 52).

Table 18. Fauna at *Argusia* flowers. Values are percent of total. Based on activity-peak method.

	Species		Individuals	
Date	Nectarivores	Other	Nectarivores	Other
1968				
July	80.0	20.0	82.6	17.4
1969				
J–F	80.0	20.0	93.4	6.6
M–A	89.5	10.5	98.5	1.5
M–J	70.8	29.2	83.8	16.2
J–A	100.0	0	100.0	0
S–O	89.5	10.5	92.3	7.7
N–D	89.7	10.3	91.0	9.0
1970				
J–F	89.5	10.5	98.8	1.2
M–A	95.2	4.8	97.5	2.5
M–J	90.9	9.1	91.7	8.3
J–A	91.2	8.8	96.3	3.7
S–O	85.7	14.3	85.2	14.8
N–D	97.1	2.9	96.7	3.3
1971				
F–M	96.0	4.0	99.0	1.0
Mean	88.9	11.1	93.3	6.7

Population densities of species associated with *Argusia* flowers

The number of individuals on *Argusia* flowers changed markedly with time (Fig. 51). Numbers were low before and after the drought at times when flower density was low. Numbers increased during 1969 reaching a peak in November–December. The numbers then declined until mid-1970 when they rose toward a peak late in the year. The late 1970 peak was higher than that of late 1969. Both coincided with, or immediately followed the peak flowering of *Argusia*. The two methods agreed well in estimating the number of individuals except for the late 1970 peak in which the Hanson method gave a considerably lower estimate than did the activity-peak method.

The estimates of total numbers of individuals by both methods relies on the flower census, as estimates of the number at a given number of inflorescences are adjusted in accordance with the total number of inflorescences to give the total number of arthropods on *Argusia* flowers for the whole island. Consequently the information on population levels of arthropods as presented in Fig. 51 does not assess the intensity of use of the flowers by insects. The number of individuals per inflorescence, as shown in Fig. 52 is a better indication of that. In March–April 1969 the numbers of arthropods per flower head was high but then dropped to a low level. There were other high peaks

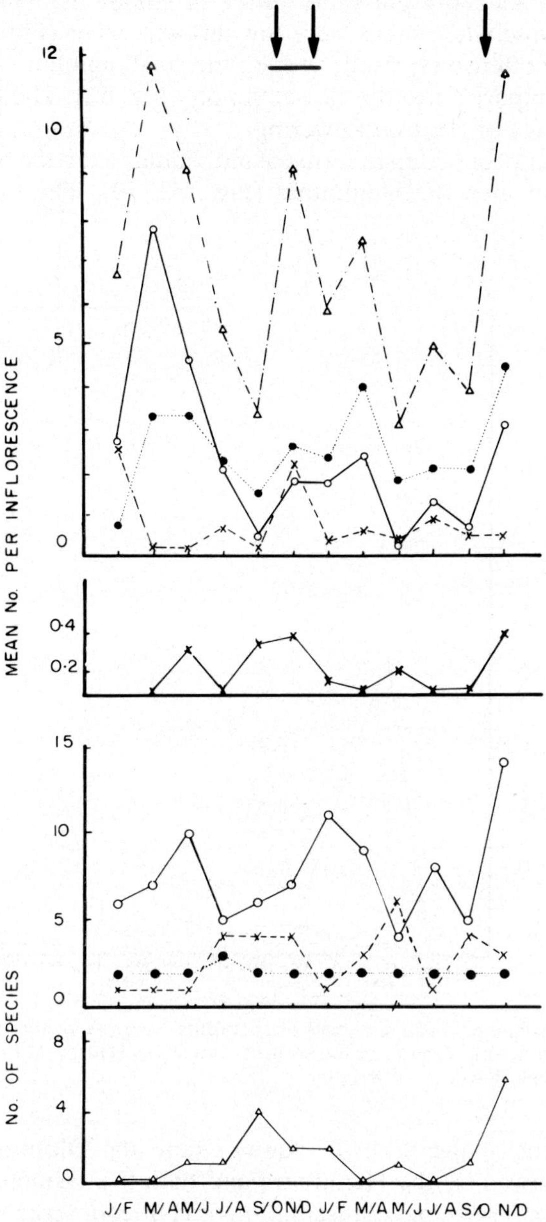

Fig. 52. Number of arthropods per inflorescence (upper two graphs) and numbers of species (lower two graphs) of arthropods associated with *Argusia* inflorescences in 1969 and 1970. Arrows indicate peak flowering of *Argusia.* All data based on the Hanson (1968) removal method. Upper graph and second graph from bottom: triangles = all species. Circles = flies. Dots = ants. x's = bugs. Other two graphs are of spiders.

in the winter to spring of 1969–70 and again in November–December 1970. The various peaks represent quite different situations. That in early 1969 occurred at a time when there were few inflorescences (Fig. 35). Yet, those few were being intensively used, as even the small number of insects present was large in comparison to the flower supply (Fig. 52). The other two peaks occurred at peaks of *Argusia* flowering.

The most abundant group in terms of individuals were the ants, followed by the Diptera and then the Hemiptera (Figs. 52, 53). The flower fauna thus

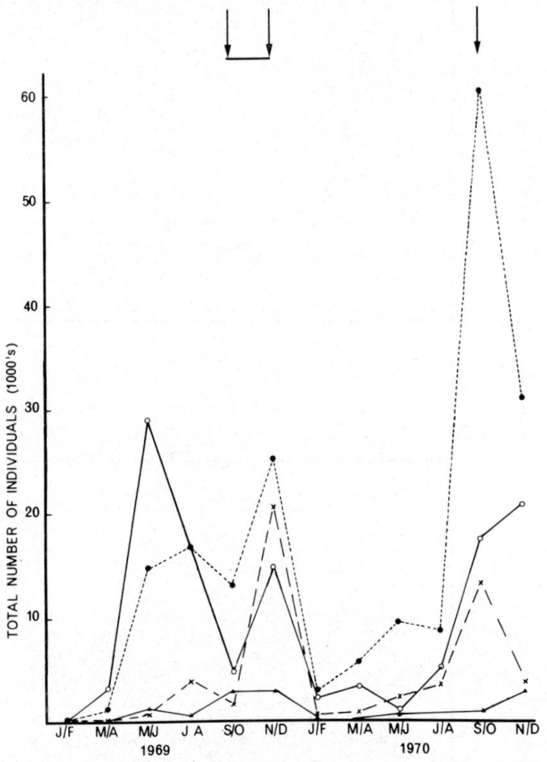

Fig. 53. Changes in numbers of individuals of ants (dots), bugs (x's), flies (circles) and spiders (triangles) associated with *Argusia* inflorescences. Based on Hanson (1968) removal method. Arrows indicate peak flowering of *Argusia*.

differs from that of the herb flat fauna where the Diptera were the most numerous, followed by the Hemiptera and then the Hymenoptera exclusive of ants (Fig. 48). The poor representation of Hymenoptera other than ants on *Argusia* is surprising as many species of that group are nectarivorous.

The ants are a special case as their representation in samples is not a good indication of their population density. They live in nests which are located in protected places, and under unfavourable conditions they remain in the nests and are not subject to sampling. Of even more significance is the fact that they are omnivorous, and their appearance at flowers is partly determined by the extent to which they are consuming alternative foods.

In early 1969 there were few ants on the sparse *Argusia* inflorescences and ants used the flowers at low intensities (Figs. 52, 53). As the year progressed and the number of flowers increased, so did the numbers of ants. They used the flowers more intensively initially, but the later, lower intensities did not alter greatly thereafter (Fig. 52) and did not correlate consistently with peaks and valleys in the flowering cycle of *Argusia*; however, at the flowering peak of 1970 the greatest intensity of usage by ants occurred.

It seemed that the number of ants visiting inflorescences was determined by the number of inflorescences available. The total numbers of ants at flowers (Fig. 53) followed a very similar curve to that of the numbers of *Argusia* inflorescences (Fig. 35).

Hemiptera differed from the ants in that they showed an intensive use of inflorescences during the initial low densities of flowers and insects. As numbers of flowers increased bug numbers and intensity of usage remained low until the late 1969 flowering peak when their population density increased markedly, as did the number of individuals per inflorescence (Figs. 52, 53). Both measures dropped to low values again for most of 1970. At the late 1970 peak of *Argusia* flowering absolute numbers of bugs increased but the number per inflorescence did not.

The Diptera showed an altogether different pattern. After the initial low numbers following the drought, density of flies soon built up to quite high levels, in fact the highest ones observed in the entire study (Fig. 53). As inflorescences were not abundant at that time, the number of flies per inflorescence was high (Fig. 52). Both values decreased and during the flowering peak of late 1969, absolute numbers of flies increased but not out of proportion to the number of inflorescences, as the intensity of usage did not alter greatly. However, at the 1970 flowering peak, both fly density per inflorescence and absolute number of individuals increased.

The spiders maintained low absolute densities throughout but did show fluctuations related to the densities of potential prey. The number of spiders per inflorescence reached a peak during the initial peak in the fly population, and again at each of the overall peaks in insect numbers correlated with flowering peaks (Fig. 52).

It is clear that the relationship between the flower fauna and *Argusia* flowering is complex. Different arthropod groups respond differently, and a given group may respond differently in different years. In some groups, e.g., ants, the response may be largely a behavioural one; a species may alter its foraging pattern and shift from other resources to *Argusia* flowers in proportion to the number of flowers available. Other responses may be demographic in that population size may actually increase or decrease as influenced by food supply. This would be expected to be most important in species whose feeding is restricted to nectar, especially to the nectar of one particular species.

A third response may be a migratory one. Generalized nectarivores may move from the herb flat or elsewhere to *Argusia* when it flowers heavily, or may leave it during periods of low flowering. In view of this possibility, it is important to compare flowering in the herb flat with the *Argusia* cycle, and to relate insect abundances, especially nectarivores, in the two areas.

The total numbers of flowers of all species in the herb flat and the inflorescences of *Argusia* both peaked at the same time of year in late 1969 and both then declined in early 1970. By contrast, the late 1970 peak of flowering of *Argusia* was not accompanied by a flowering peak in the herb flat. At all times the total numbers of flowers in the herb flat far exceeded the number of *Argusia* inflorescences (Figs. 35, 54, note different scales for the

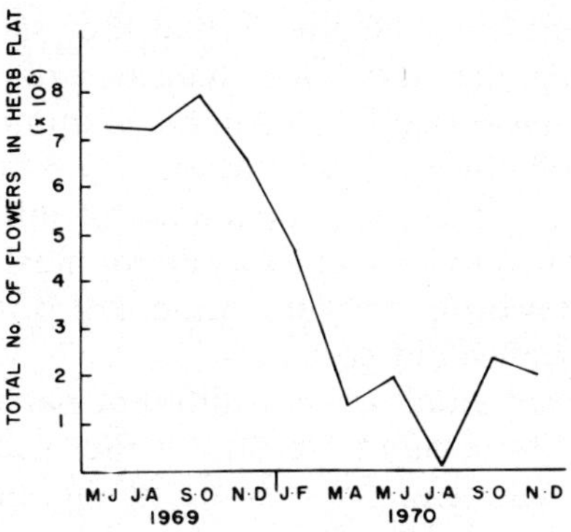

Fig. 54. Temporal changes in total numbers of flowers of all species in the herb flat.

different graphs); the *Argusia* inflorescences, however, were bigger than any flowers in the herb flat and were more concentrated in space. It is possible that such a concentrated source of nectar would attract nectarivores from a source of widely scattered small units, such as were present in the herb flat. The nectarivore population in the herb flat was usually much higher than that associated with *Argusia* (compare Figs. 51 and 55) and thus constituted a large reservoir of individuals which could move to the *Argusia* flowers as they became available. The nectarivore population in the herb flat averaged 14.3 times as high as that of the *Argusia* shrubs (range 0.4–43.9). The only occasion in which there were greater numbers of nectarivores at *Argusia* than in the herb flat was in November–December 1970, the population peak for *Argusia* nectarivores and just after the flowering peak of that shrub. At that time flowers were relatively few (slightly over 200,000) in the herb flat (Fig. 54), and it is possible that some generalized nectarivores had moved from the herb flat to the *Argusia* shrubs.

Nectarivores were not studied as intensively on other species. However, some observations were made on the fauna visiting 5 *Pisonia* inflorescences on 4 October 1969. At 0900 hrs one ant (*Iridomyrmex glaber*) was feeding on nectar. At 1200 hrs there was a total of 1 ant of the same species plus a fly. At 1300 hrs and 1400 hrs there were no arthropod visitors. At 1600 hrs there was a total of 2 flies (each of a different species) and 1 moth. At 1700 hrs there was one fly. At 1900 hrs there were 7 small black beetles. During the same observation periods no arthropods were observed at all on 5 nearby inflorescences that were in bud instead of in flower.

In comparison to *Argusia* the number of inflorescences produced by *Pisonia* is rather small (see Chapter 11). The combination of its low number of inflorescences and the paucity of insects that visit them indicates that the

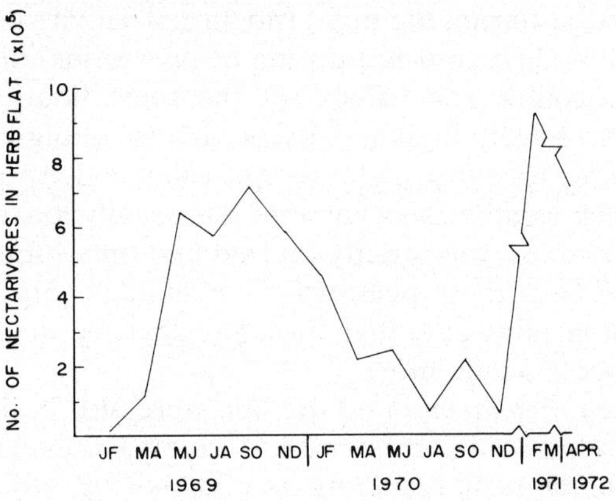

Fig. 55. Temporal change in total numbers of individual nectarivores in the herb flat.

nectarivores of the *Pisonia* do not contribute greatly to the totals for the island as a whole.

In May 1968, flowers of several herb flat and succulent mat species were examined for insects associated specifically with them. Twenty-seven *Sesuvium* flowers were carefully dissected. They contained 58 adults and 11 immatures of a species of thrip but no other animals (total of 69 thrips of all stages or 2.6 per flower). These thrips were present on the island throughout the study. Forty-two flowers of *Canavalia* contained 35 thrips (0.8 per flower; 34 adults and 1 immature) plus 6 ants (*Iridomyrmex glaber*). No insects were found in flowers of either *Abutilon* or *Plumbago*.

In December 1970, 25 *Melanthera* flowers were dissected. There were 8 pupae of one species of fly and 11 pupae of another, plus 1 caterpillar, 1 small parasitic hymenopteran and 3 ants. Some of the developing seeds were damaged.

Daily activity cycles of some nectarivores associated with *Argusia*

Utilization of limited resources by potentially competing species may be facilitated by separation of their activity cycles. For this reason, the four most consistently abundant species of nectarivores at *Argusia* flowers were subjected to an analysis of their activity patterns. In order of decreasing relative abundance these species were the ant *Iridomyrmex glaber* (Mayr), the flies *Tricimba* sp. A (Chloropidae) and *Lamprolonchaea metatarsata* (Kert.) (Lonchaeidae) and an unidentified bee.

Iridomyrmex glaber were active at all hours of the day and night at which observations were made (0400–2400 hours). However, they were not uniformly active throughout the diel period; the time of activity peaks depended on weather conditions. On fine, sunny days, activity was low during most of the day but increased during the night (nocturnal activity peak). By contrast, on overcast or rainy days (but not during heavy storms) diurnal activity was often more than double the values for the same hours during fine days (Fig. 56), with excessively high activity occurring during early morning and during late evening to early night (crepuscular activity peaks); nocturnal activity during such weather was variable but usually low or non-existent.

By contrast, *Tricimba* was strictly diurnal and only foraged between 0700 and 1900 hours. The activity peak was near midday (Fig. 57). It was more active on overcast or rainy days than on sunny days, by an order or magnitude of 2–4 fold and occasionally more.

Lamprolonchaea metatarsata and the bee were also both entirely diurnal, the former with a bimodal pattern with morning and evening peaks and the latter with a unimodal peak centering on midday (Fig. 58). *Lamprolonchaea* tended to be more active during overcast, rainy weather than during sunny days. There was no consistent relation between activity and weather for the bee.

Although there are differences in activity patterns among the predominant nectarivores associated with *Argusia* flowers, they are not sharply separated in time and there is broad overlap among species. Of the four, only the ant feeds at night. During the day all species forage in all kinds of weather. The tendency for several species to be more active during overcast or rainy weather, especially in the morning and evening, means that the nectar resources are most heavily utilized at such times. Measurements of nectar production were not undertaken, unfortunately, and it is not known to what extent daily nectar resources were depleted.

In general, the fauna associated with *Argusia* inflorescences tended toward diurnality; on the average only 13% were nocturnal, another 13% were active both during the day and night and 74% were predominantly diurnal (Table 19). This situation did not change greatly over time. However, during the early part of the recovery from the drought, when absolute numbers were

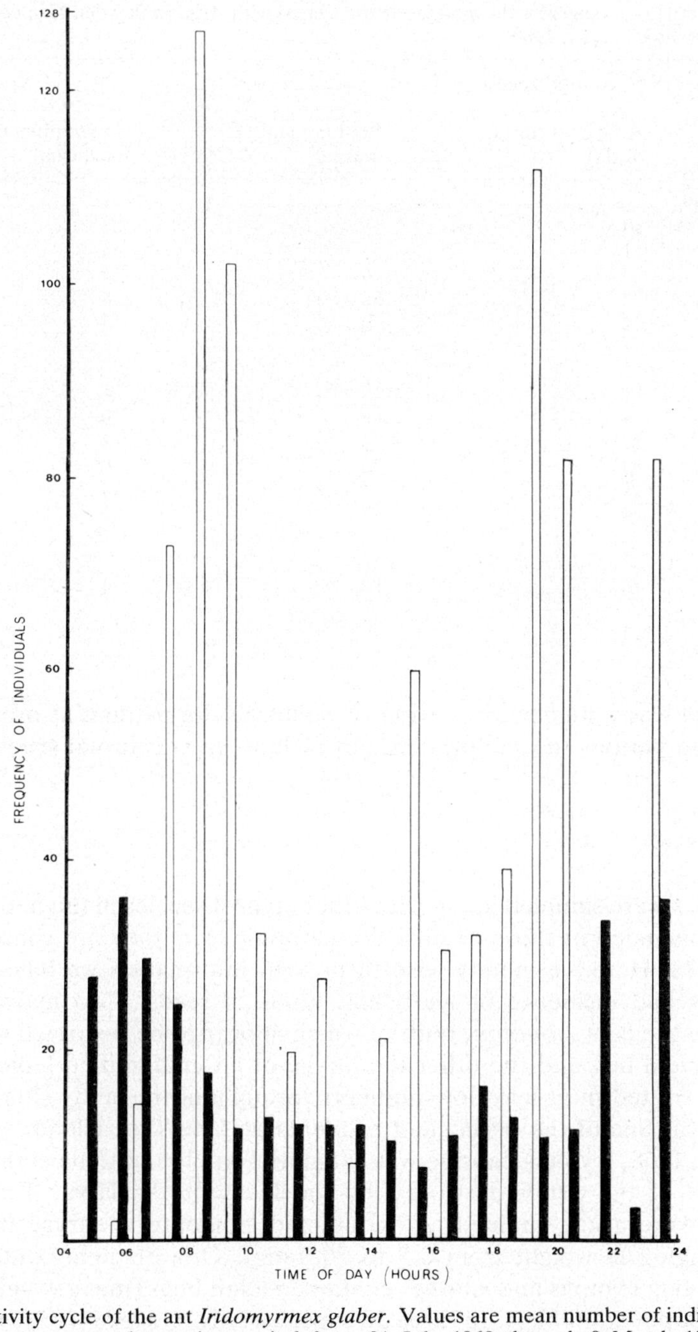

Fig. 56. Activity cycle of the ant *Iridomyrmex glaber*. Values are mean number of individuals on ten inflorescences per observation period from 21 July 1968 through 9 March 1971. Open histograms indicate values obtained during overcast or rainy weather, dark histograms are those obtained on fine days.

Table 19. Activity of epigeal arthropod fauna associated with *Argusia* flowers. No nocturnal data available for July–August 1969.

Date	% of total species		
	Active all times of day	Predominantly diurnal	Predominantly nocturnal
1969			
J–F	10	55	35
M–A	11	33	56
M–J	7	79	14
S–O	9	86	5
N–D	21	72	7
1970			
J–F	17	78	6
M–A	5	86	9
M–J	17	75	8
J–A	10	90	0
S–O	14	71	14
N–D	24	77	0
1971			
F–M	16	80	4
Mean	13	74	13

low, there was a higher proportion of nocturnal forms than at other times; during two periods the samples did not include any nocturnal species.

Granivores

Granivores were sampled along with other epigeal species in the herb flat and they are included in those results. As a trophic level they are considered in Chapter 22. Here we merely wish to present information we have on food specificity and incidence on particular kinds of seeds. Two major species attack the seeds of *Abutilon*, both of which appear to be restricted to it. One was a lygaeid bug and the other the larvae of an unidentified lepidopteran. *Abutilon* fruited in at least low numbers throughout the study (Fig. 34) and the bug was one of the permanent residents on One-Tree Island.

In May 1968, 1 young capsule was examined and found to have the ovaries and some of the petals destroyed by the lepidopteran larva. Ten mature capsules were also examined. All contained lepidopteran larvae of various sizes (ranging in weight from 0.5 to 26.0 mg). One of them contained, in addition, bug nymphs and another contained adult bugs (mean weight 5.6 mg dry weight per adult bug). In these 10 capsules only 63 seeds (10%) remained undamaged out of a total potential production of 639 seeds (19–23 locules per capsule, 3 seeds per locule). Some capsules had no seeds remaining and the

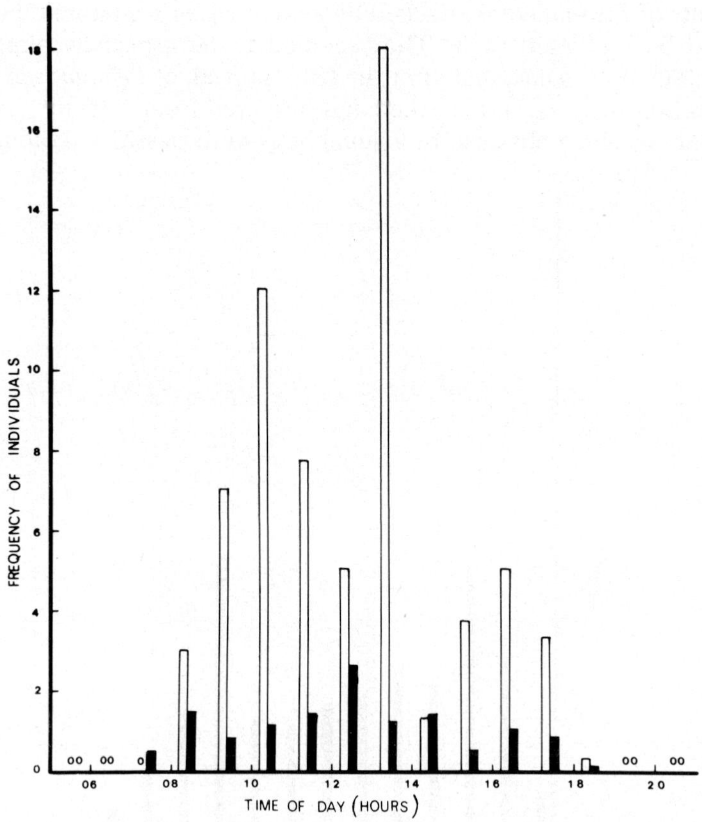

Fig. 57. Activity cycle of the fly *Tricimba* sp. A (Chloropidae). Symbols as in Fig. 56. Note, however, the different scale.

maximum number in any capsule was 23 (37% of the potential for that capsule), with an average of 6.3 seeds per capsule produced, or a mean seed biomass of 18.9 mg per capsule. Thus, seed production at that time was severely curtailed by granivores especially the lepidopteran larvae. Later, consumption was much less severe.

In May 1969, 20 dry capsules contained 25 adult bugs (mean = 1.25 bugs/capsule). In 1970 there was only 1 bug in 11 dry capsules (0.09 per capsule) in March, but there were 3 caterpillars and 1 pupa (0.4 per capsule) of the lepidopteran. In August, 38 dry capsules were examined. In 7, the seeds were completely destroyed, 20 had some of the seeds destroyed and 11 were completely healthy. Three lepidopteran larvae and 1 ant were found in these capsules. In December, 39 dry capsules contained 7 lygaeids (0.18 per capsule), 2 caterpillars and 1 ant. In February 1971, 99 green capsules were examined and found to contain 68 bugs and 4 caterpillars (0.69 and 0.04 per capsule respectively).

The fruits of *Pisonia* were attacked by a large green pentatomid bug which seemed to be restricted to it. They seemed to be especially abundant in February 1969, i.e., somewhat after the fruiting peak of *Pisonia* (see Chapter 11). Ten inflorescences were marked and observed from 0900 to 1700 hrs, at hourly intervals on 5 February. Individual bugs were classified as adults, early

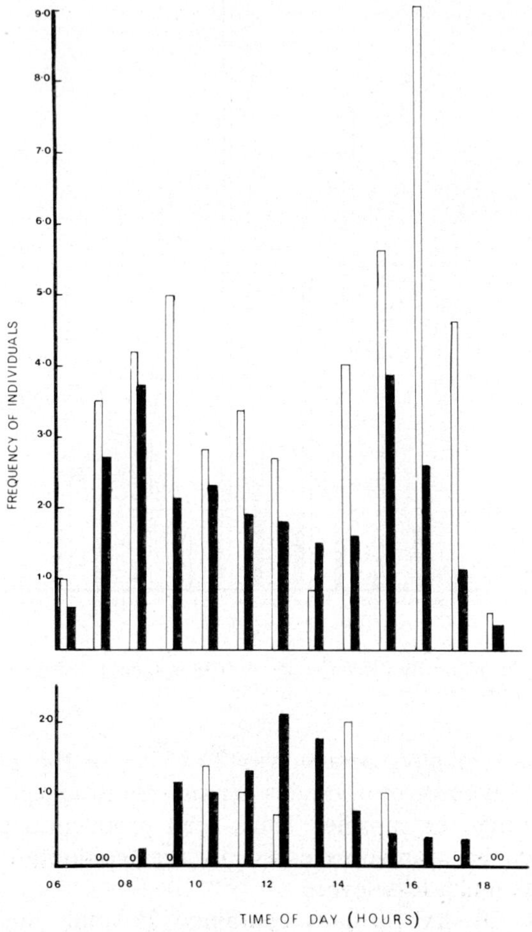

Fig. 58. Activity cycle of the fly *Lamprolonchaea metatarsata* (Kert.) (Lonchaeidae) (upper) and an unidentified bee (lower). Symbols as in Fig. 56.

nymphs or late nymphs and the numbers of each counted at each observation period. There was no apparent trend in activity level during the day. Numbers of adults ranged from 38 to 45 for the 10 inflorescences, those of late nymphs from 23 to 38 and those of early nymphs from 32 to 42. The mean number per inflorescence for the three age groups and over all of the observation periods respectively were 4.2, 3.0, and 3.7, or a total of 10.9 bugs of all stages per inflorescence. If one takes the maximum number of each group as indicative of the numbers present when all animals were active on the fruits, then the total is 12.5. The total number of fruiting inflorescences on the island at that time was 415, giving an estimated number of bugs of this species of 5,188 for the entire island. The 10 inflorescences had 21 clutches of bug eggs deposited upon them for an estimated 8,715 clutches for the whole island.

There were only 3 other species observed to be associated with the *Pisonia* fruits. They were 1 species each of spider, weevil and ant with mean numbers per inflorescence of 0.1, 0.03 and 0.01 respectively. The weevils were probably feeding on the *Pisonia* and the ant may have been taking sticky secretions from the fruits. The spiders probably preyed on granivores.

Pandanus seeds were attacked by an insect. Emergence of the granivore, or its hymenopterous parasite, left characteristic holes in the seeds which were periodically counted. The incidence of infestation was very high. From May 1968 to February 1970, and December 1970 the average infestation rate was 61% (range 30%–84%) based on examination of a sample of 17 to 121 freshly fallen ripe seeds per sampling period, depending on availability. Multiple emergence holes were characteristic, indicating that rearing of several granivores and/or parasites in the same seed was successful. Up to 5 emergence holes were observed in one seed. The average number of emergence holes per seed was 1.5 (range 0.69 to 2.88 on different sampling periods) and the corresponding value per infested seed was about 2.5.

The sampling period in March 1971 revealed a very different situation. No fresh emergence holes were found even though 340 seeds were examined. It is possible that the granivore had become extinct.

Very few other animals were associated with *Pandanus* fruits. One which had loose seeds was found on the ground. It was sealed in a plastic bag and later examined for insects. It contained 1 fly, 1 fly larva, 1 cockroach and 1 weevil. Some of these may have merely sought shelter in the fruit. Only occasionally were seeds observed that seemed to have been chewed from outside.

The fauna on the foliage of trees and shrubs

The number of species sampled from the foliage for given periods of time was 8–29 for *Pisonia* and 10–31 for *Argusia*. The temporal pattern was similar for both types of vegetation (Fig. 59). There were peaks of species numbers in both types of foliage in November–December 1969, May–June 1970 and September–October 1970 and species number was rising again at the end of the intensive study.

There were 67 species found on *Argusia* foliage and 68 species on *Pisonia* foliage. Of those on *Argusia*, 24 (35.8%) were found only on that species and 43 (64.2%) were shared with *Pisonia*. Similarly 25 (36.8%) of those on

Pisonia were found only on that species and the 43 species shared with *Argusia* represented 63.2% of the total number on *Pisonia*. Of the combined foliage fauna of 92 species, 26.1% were on *Argusia* alone, 27.2% were on *Pisonia* alone and 46.7% were shared. A treatment of the similarity of the

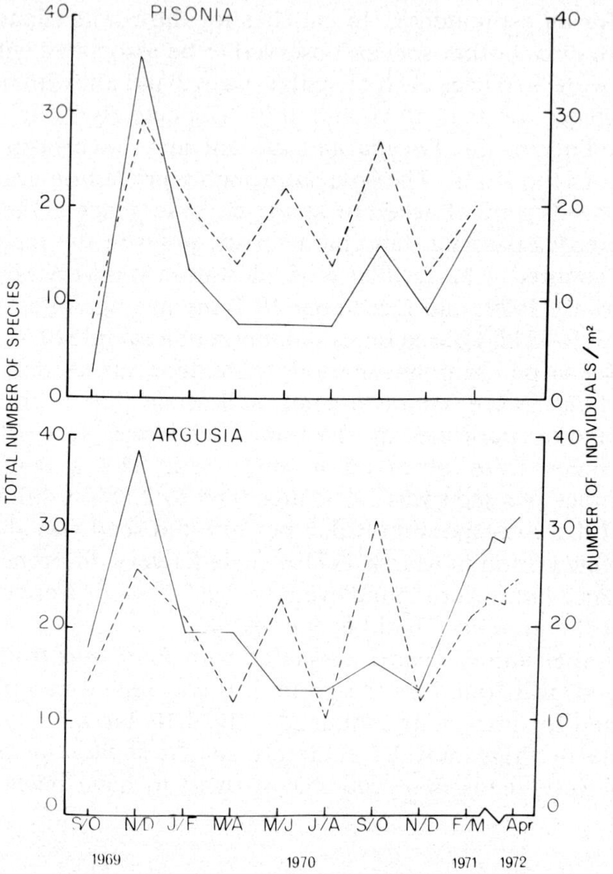

Fig. 59. Numbers of species (broken line) and individuals (solid line) of arthropods from the foliage of *Pisonia* (upper) and *Argusia* (lower).

total fauna (including the foliage fauna, nectarivores, etc.) is presented in the following section.

The numbers of individuals of arthropods were remarkably similar on both types of foliage (Fig. 59). In both cases densities were low at the initial sampling (September–October 1969), then rose to a very high level by November–December of that year. Values then dropped and for the rest of the study stayed at a relatively constant low level until November–December 1970 when densities again began to rise. There was one exception to this pattern. There was a small peak in number of individuals on *Pisonia* foliage in

September–October 1970. However, that peak was almost entirely because of intensive foraging activity there by the ant *Iridomyrmex glaber* (Fig. 60) and did not represent a rise in numbers of other species of the foliage fauna. During that period, 52.7% of the total number of individuals on *Pisonia* foliage were ants of this species. Excluding this date the mean proportion of ants in the total number of individuals on *Pisonia* foliage was 21.4% (range 8.3–35.6%). On *Argusia* foliage the mean proportion of individual arthropods which were ants was 13.9% (range 3.4–21.5%).

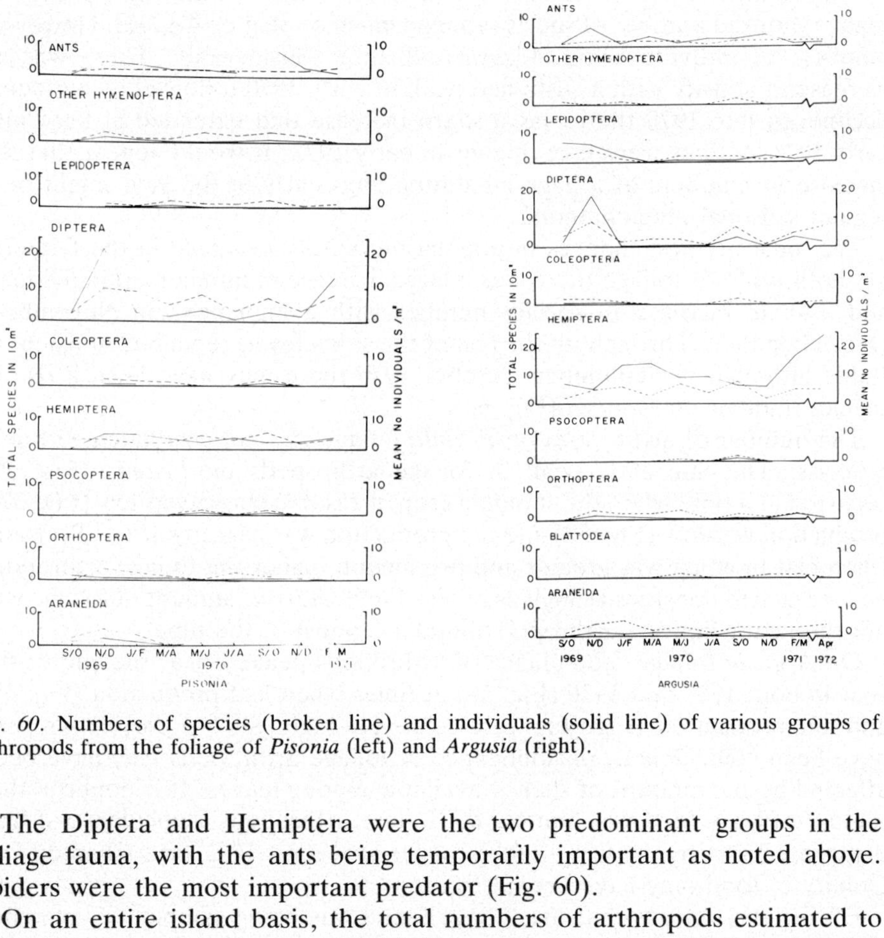

Fig. 60. Numbers of species (broken line) and individuals (solid line) of various groups of arthropods from the foliage of *Pisonia* (left) and *Argusia* (right).

The Diptera and Hemiptera were the two predominant groups in the foliage fauna, with the ants being temporarily important as noted above. Spiders were the most important predator (Fig. 60).

On an entire island basis, the total numbers of arthropods estimated to inhabit foliage ranged from 67,617 (September–October 1969) to 268,180 (November–December 1969) with a mean value for the study of 116,564. Excluding ants, the mean becomes 95,278 (minimum 54,320 November–December 1970; maximum 231,899 November–December 1969).

Many of the taxa represented in the foliage fauna were relatively constant in numbers of species and numbers of individuals throughout the study and in both types of foliage, e.g. Hymenoptera exclusive of ants, Orthoptera, Araneida. Others were in low numbers during some periods and did not appear in others (Psocoptera, Blattodea, Coleoptera). The rest of the groups showed considerable temporal fluctuation, at least on one type of foliage (Fig. 60). The Lepidoptera were abundant early in the study on *Argusia* foliage (see Chapter 18 for detailed treatment of *Utetheisa*, the most important species), but later declined to lower, rather constant levels. On *Pisonia* numbers of lepidopterans were constantly low.

The Hemiptera did not fluctuate in species or individuals much on *Pisonia* foliage, nor did number of species change much on that of *Argusia*. However, numbers of individuals on *Argusia* did alter considerably. There was an increase in density with a sustained peak in early 1970 followed by a gradual decline; in late 1970 there was a sharp increase that extended at least into early 1971. Values were even higher in early 1972. It would appear that the increase in numbers of foliage-inhabiting bugs early in the year might be a regular seasonal phenomenon.

The most marked changes in population density occurred in the Diptera. On both types of foliage there was a large increase in numbers of individuals and also an increase in species number with a high peak in November–December 1969. Throughout the rest of the study levels remained much more stable although in September–October 1970 there may have been a rise in species number on *Argusia* (Fig. 60).

The number of arthropods on *Pisonia* foliage was not correlated with leaf biomass. The late 1969 peak in foliage arthropods on *Pisonia* (Fig. 59) occurred at a time when the standing crop of *Pisonia* leaves was low (Fig. 37), production was low (Fig. 38) and net production was negative (Fig. 39). Later when leaf biomass was greater and production higher the foliage arthropods never reached densities as high as in late 1969. Clearly, amount of foliage was not the factor limiting arthropod abundance most of the time.

On *Argusia* foliage, abundance of arthropods peaked near the end of the year in both 1969 and 1970 (Fig. 59) at times when leaf production (Fig. 38) and net production (Fig. 40) were relatively high. Such a correlation may have been coincidental, or abundance of foliage arthropods may have been affected by the amount of shelter available among leaves. It is doubtful that amount of leafy food was limiting the foliage arthropods as incidence of leaf damage caused by chewing insects never exceeded 15% (see Chapter 15). Quality of food may have been a factor.

During the periods that the foliage fauna was being sampled (September 1969 through April 1972) the total number of individuals of chewing herbivores in the herb flat averaged 307 times the total number on *Argusia* and *Pisonia* foliage combined (range 25–1,860; omitting the upper extreme value the mean was 135 and the range was 25–360). Thus, a relatively large number of chewing herbivores were available from the herb flat at all times. The extent of movement between the two areas is unknown. Grasshoppers were known to feed on foliage in both areas; other chewing herbivores like *Utetheisa* caterpillars were restricted to one host plant (*Argusia*).

The relative abundance of different taxonomic groups changed temporally (Fig. 61). As indicated previously the Hemiptera, Diptera, and ants were the most numerically abundant on the foliage of *Pisonia* and *Argusia*. In late 1969 the *Diptera* were the predominant group on *Pisonia* foliage. In 1970 their relative importance was reduced and they never accounted for more than 30% of the individuals of foliage arthropods. In early 1971 flies had again become numerically dominant and accounted for over 70% of the insects on

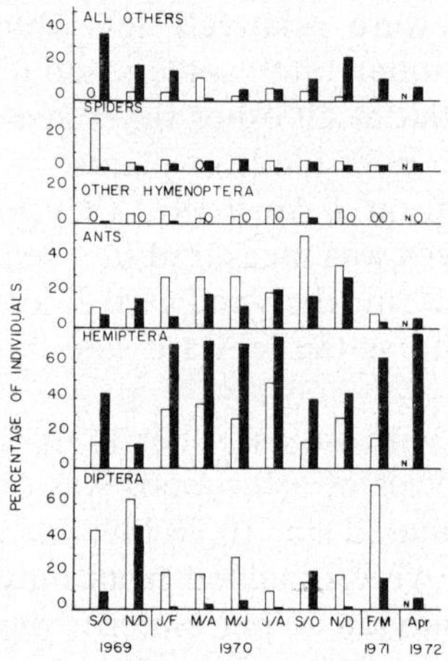

Fig. 61. Relative abundances of different taxonomic groups in the foliage fauna of *Pisonia* trees (open histograms) and *Argusia* shrubs (solid histograms). N = no data available.

Pisonia foliage. Flies did not make up such a great proportion of the total foliage fauna on *Argusia* and in all but one case values were lower than those for *Pisonia*. There were more flies relative to other groups in November–December 1969 than at any other time; at that time almost half the individuals on the *Argusia* foliage were flies. At all other periods, values were always less than 22%.

Bugs showed the reverse pattern. On *Pisonia* they made up a smaller percentage of the total numbers of individuals than they did on *Argusia* except for one period (November–December 1969). On *Pisonia* their abundance relative to other groups was 15% or less in late 1969 but rose to higher values for much of 1970 with a peak of almost 50% in July–August followed by a decline. On *Argusia*, proportional representation of bugs dropped from about 43% to about 14% in late 1969, then rose to values exceeding 70% for most of 1970 before dropping to about 40% by the end of the year. Values were high in both of the later sampling periods (1971 and 1972).

The percentage of the total number of individuals represented by ants depended in part on the abundance of ants and partly on their foraging activities. Unlike the other groups they did not remain among the foliage when they were not foraging. In *Pisonia*, their representation among the foliage fauna was low in late 1969, increased to somewhat higher values in 1970 and at one time (September–October 1970), ants accounted for half of the arthropods among *Pisonia* foliage. In 1971 values were again low.

Ants accounted for a lower percentage of the total number of arthropods in *Argusia* than they did in *Pisonia*, and changes in proportional representation were not great; at most sampling periods they accounted for less than 20% of the individuals in the foliage and values never reached 30%.

Other groups usually were relatively less abundant (Fig. 61). On one occasion (September–October 1969) spiders had a rather high representation (about 30%) in *Pisonia* but at all other times accounted for less than 7% of the foliage arthropods on both kinds of plants.

The effect of the storm on 2 April 1972 ('Cyclone Emily') on the foliage fauna of the *Pisonia* groves was measured in a semi-quantitative way. It was observed that after the storm, the windward side of the groves had suffered extensive defoliation whereas the leeward side did not appear to have been damaged.

Both sides were swept with an insect net 11 days after the cyclone. A total of 11 individuals of 5 species of arthropods was obtained in 40 sweeps from the damaged, partly defoliated side (0.28 individuals per sweep), whereas 35 individuals of 14 species were obtained from only 20 sweeps on the intact, undamaged side (1.75 individuals per sweep); the cyclone-damaged foliage thus contained only about 16% as many arthropods as the undamaged foliage.

Habitat specificity

The various habitats on One-Tree Island each had their unique epigeal arthropods as well as a number of species that overlapped two or more habitats either regularly, or at particular times. For purposes of analysis, five major 'habitats' were considered: (1) the herb flat, (2) the *Pisonia* forests, (3) the *Argusia* shrubs, (4) the beach and (5) opportunistic collections[5].

Over 3/4 of the total number of species were found only in one habitat and only 89 species (22.5%) were shared between 2 and 3 habitats collectively. The number of species shared by more than 3 habitats was only 1 (Table 20).

The herb flat had the greatest number of species. About 2/3 of the total

[5] The succulent mat and *Pandanus* grove were not included separately in this analysis as they were sampled semiquantitatively and only infrequently (see above). The category 'opportunistic collections' includes specimens collected from the *Pandanus* grove (including a search through leaf litter), the succulent mat, under *Pisonia* bark, in and under rotten wood, under *Argusia* bark, in *Argusia* litter and under conditions in which the habitat of origin could not be determined. Most of this general search took place in December 1970.

Table 20. Degree of sharing of arthropods (excluding soil microarthropods) among various habitats on One-Tree Island, 1968–1972. Opportunistic collections are treated as a separate 'habitat'.

	All species		Established species	
	No. of species	% of total species on island	No. of species	% of total established species on island
Species restricted to **one** habitat	306	77.3	82	47.7
Species shared between **two** habitats	44	11.1	44	25.6
Species shared among **three** habitats	45	11.4	45	26.2
Species shared among **four** habitats	0	0	0	0
Species shared among **five** habitats	1	0.3	1	0.6
Total	396	100.1	172	100.1

species from the island were recorded from there. *Argusia* shrubs had the second largest numbers, *Pisonia* trees next and the beach had the fewest number of species (Table 21); species collected only opportunistically and which may have represented several habitats also had a large number of species.

The habitat with the highest proportion of its epigeal arthropod fauna restricted to it was the beach, with over 85% of its species having been found nowhere else. Following in descending order were the herb flat, *Argusia* shrubs, and the *Pisonia* forest (Table 21). The opportunistic collection probably contained a high proportion of transients and consequently had a high proportion of 'restricted' species.

Because of the large number of species in the herb flat and the moderately high degree of restriction of its species, it contained the greatest number of

Table 21. Degree of habitat restriction in arthropods from One-Tree Island, 1968–1972. Soil microarthropods are excluded.

	HF	AR	Pi	B	OP
Total species from habitat:					
number	265	120	70	7	72
% of total species for island	66.9	30.3	17.7	1.8	18.2
Number of species restricted to habitat	186	47	15	6	52
Restricted species as % of:					
total species for habitat	59.1	39.2	21.4	85.7	72.2
total species from island	47.0	11.9	3.8	1.5	13.1
total restricted species	60.8	15.4	4.9	2.0	17.0

HF = Herb flat; AR = *Argusia* shrubs; Pi = *Pisonia* forest; B = Beach; OP = opportunistic collections.

restricted species; over 60% of the total restricted species were found there. Although the beach had high proportional restriction it had few total species, and only two per cent of the restricted species occurred there (Table 21). The other habitats were intermediate. Thus, the beach was characterized by a few species with a high degree of restriction to that habitat, the *Argusia* and *Pisonia* areas by a moderate number of species that tended to occur in other habitats as well and the herb flat by a large number of species with a moderately high proportion not found elsewhere. The opportunistic collection (combining specimens from the *Pandanus* clump, the succulent mat and other habitats), had a moderately high number of species with a high degree of restriction.

It may be unrealistic to include transient species (for definition see Chapter 20) in data on habitat restriction as they were found on only one occasion and their occurrence in a particular habitat may have been fortuitous rather than indicative of the breadth of their habitat preference. This would be especially true of the opportunistic collection. If the transients were excluded from consideration, then slightly less than half of the species are restricted to only one habitat and slightly over half overlapped 2 or 3 habitats; there were still a negligible proportion occurring in more than 3 habitats (Table 21). Thus, even among established species there is a rather high degree of habitat specificity and the island is not populated primarily by extremely eurytopic species that occur in almost any habitat available. Rather, most are species that find appropriate habitats on One-Tree Island and occur in only one or a few of the available habitats which the island presents. Many of the transient species probably fail to become established because of lack of habitats on One-Tree Island which meet their requirements.

There was a tendency for certain habitats to share species with each other more than with other habitats. A means of expressing faunal similarity among habitats is by the Preston (1962) index of similarity (1–Z) which ranges from 0 (no similarity) to 1 (complete similarity). This index was applied to the 4 major habitats on One-Tree Island (Table 22). The two habitats with arborescent vegetation showed the greatest faunal similarity, and both had relatively great faunal similarity to the herb flat. Similarity with the beach fauna was low. The greatest number of species shared, however, was between the herb flat and *Argusia* shrubs, probably because of the great number of nectarivores shared. The greatest proportional sharing was between the *Argusia* shrubs and *Pisonia* forest with over 70% of the species from the former also occurring in the latter.

It can be concluded that although there is considerable habitat specificity among the epigeal arthropods of One-Tree Island and each habitat has its distinct fauna, there is broad overlap among the more highly vegetated habitats, especially between *Argusia* and *Pisonia*; the beach had few species and shared a low proportion of them with other habitats.

It was considered that the effective size of the island for particular species

Table 22. Faunal affinities of various habitats on One-Tree Island. X indicates inapplicable. Values on upper right are Preston's (1962) index of similarity (1–Z), those on the lower left are (1) the number of species shared (not in parentheses), (2) the percent of the *a* species shared with *b* (left set of parentheses) and (3) the percent of the *b* species shared with *a* (right set of parentheses).

a \ b	HF	AR	Pi	B
HF	X	0.36	0.33	0.07
AR	74 (61.7) (23.5)	X	0.47	0.05
Pi	46 (65.7) (14.6)	50 (71.4) (41.7)	X	0.03
B	1 (14.3) (0.3)	1 (14.3) (0.8)	1 (14.3) (1.4)	X

HF = Herb flat; AR = *Argusia* shrubs; Pi = *Pisonia* grove; B = beach.

might be the area of suitable habitat, the rest of the island being of little ecological consequence. If this were true, then the number of species in different habitats might be expected to be related to the total area of habitat available. The herb flat has the greatest number of species and the greatest area (Tables 1 and 21). However, the correlation did not hold up in detail, either when total species or only established species were used. The beach habitat had a low number of species for its area, probably because of the sparsity of vegetation on the beach. However, the vegetation cover of the beach was not measured and the areal value cannot be adjusted on that basis. The *Argusia* shrubs had a larger arthropod fauna than did the *Pisonia* forest even though the latter occupied a greater area of the island. This discrepancy is probably attributable to the fact that flowers were more abundant and seasonally continuous on *Argusia* shrubs than on *Pisonia* trees and hence the former was suitable for a larger number of nectarivores. This trophic group seemed to be especially abundant in the *Argusia* habitat. In general it appears that there is a loose relationship between habitat area and number of species but that this relationship is modified substantially by habitat quality.

Fauna of the small rubble cays

The small shingle cays on the northern edge of One-Tree Reef (Fig. 73) are unstable and shift with time; even the number varies from time to time (Hulsman 1977a). However, they are completely above water even during the highest tides for extended periods, and during the lowest tides are coextensive with wide expanses of the reef flat. Their substrate is much like the rubble beach on One-Tree Island. They contain no vascular plants although occasionally unicellular algae grow on isolated rubble chunks. Some terrestrial animals, probably often of One-Tree origin, occur there.

There were three cays in existence at the time we surveyed them (December 1970). The smallest was 7.5 m × 6.5 m × 40 cm high (above high tide line). The upper part had amorphous organic material, including guano, in the interstices and a green layer of unicellular algae coating some of the rubble. No seabirds were nesting. One coconut and 3 *Barringtonia* fruits were on the island, as well as various other flotsam and jetsam. Hand-collecting yielded in addition to intertidal isopods, 1 spider, several silverfish, 1 earwig (*Anisolabis maritima*), 1 mite, many collembolans and 1 fly, all either under blocks of rubble or in interstices between pieces of rubble.

The second cay was 31 m × 7 m × 1.3 m high. It was steep-sided but levelled off to a flat plateau 1 m wide on top. A film of algae occurred on some of the rubble, but no organic accumulation was present. There were 18 nests of *Sterna sumatrana*. Thirty-nine of these birds plus 1 Brown Booby and 3 Sooty Oystercatchers were observed. The same species of isopod, fly and collembolans as occurred on the smaller cay were present, as well as a second species of collembolan.

The third cay was 35 m × 5 m × 80 cm high and had little algal covering and no accumulated organic matter. A large butterfly landed on the island briefly and then left. Both previously mentioned species of collembolans were present as were the isopods and the fly. Fauna not observed on the other cays were a second species of fly, one small black beetle, and mites. The last were extremely abundant and were under nearly every block of rubble.

A total of 11 terrestrial arthropods (excluding the isopods) were recorded from these 3 small cays, a surprisingly rich fauna for such small and unstable islands, especially when it is noted that most of these animals are flightless.

Heatwole (1971) found that the numbers of species of terrestrial invertebrates on some Coral Sea unvegetated sand cays was related to island size. The maximum number on any such cay was 11 species. The rubble cays at One-Tree Island had faunal richnesses of the magnitude only found on the Coral Sea cays of 10,000–30,000 m^2. The richness of the fauna on the rubble cays on One-Tree reef is probably related to the close proximity of One-Tree Island. The cays in the Coral Sea had no nearby source for renewal of their biota.

15. Insect damage to leaves

Most trophic interactions are difficult to observe directly and consequently are difficult to assess. A notable exception is the consumption of foliage by chewing herbivorous insects. The leaf damage they cause is relatively conspicuous and usually distinguishable from other kinds of leaf damage. Seldom are entire leaves consumed.

Herb flat

In the herb flat different species of plants showed different degrees of damage by chewing insects. Four species (*Melanthera, Abutilon, Achyranthes* and *Plumbago*) had consistently low incidence of leaf damage. The mean loss to insects averaged 6% of total leaf surface area or less throughout the study (Figs. 62, 63). Values were remarkably constant for *Melanthera,* and *Abutilon* despite the temporal changes in leaf biomass and insect abundance over the time interval studied. *Achyranthes* and *Plumbago* showed somewhat more temporal fluctuation. *Achyranthes* did not occur in the sample plots initially and when it did make an appearance in mid-1969 there was no leaf damage evident. Thereafter, small and variable amounts of leaf damage were sustained. *Plumbago* had very low leaf damage for most of the study; values for early 1970 were somewhat higher than for other times.

Two species, *Canavalia* and *Ipomoea,* had much more extensive leaf damage and showed greater temporal fluctuations (Figs. 62, 63). Almost 30% of the standing crop of *Canavalia* leaves had been consumed by insects in the pre-drought period of 1968 (Fig. 63). Locusts were very abundant then (see Chapter 18) and probably accounted for much of the leaf damage to this species. Following the drought when the locust population density was low, leaf damage dropped to very low values which then progressively increased throughout the study, but never again achieved mean levels above 13.5%

Ipomoea had a leaf damage below 8% throughout the study except for one period, January–February 1970, when it slightly exceeded 20%. This was the

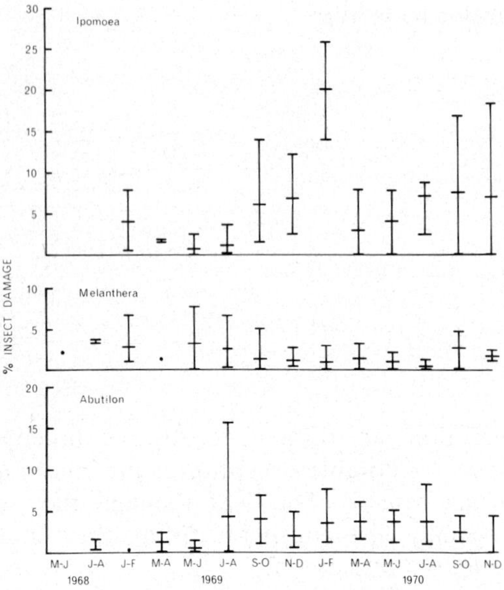

Fig. 62. Per cent of total leaf surface area lost to chewing insects by *Ipomoea, Melanthera* and *Abutilon.* Data obtained from cage plot samples from the herb flat. Horizontal line indicates the mean, the vertical line the range of values.

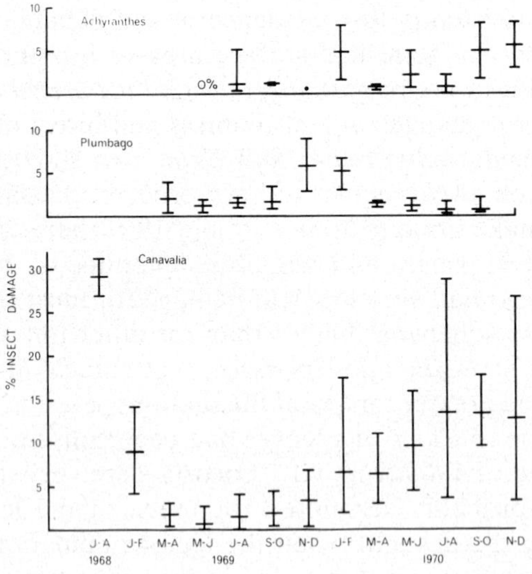

Fig. 63. Per cent of total leaf surface area lost to chewing insects by *Achyranthes, Plumbago* and *Canavalia.* Data obtained from cage plot samples from the herb flat. Symbols as in Fig. 62.

period just after the late 1969 peak in insects (Fig. 49) and the high damage to *Ipomoea* may have been a reflection of intensified grazing over the previous few months. An alternative, but not necessarily mutually exclusive, explanation for the high leaf damage in early 1970 might be that biomass was decreasing because of reduced formation of new leaves. Most of the leaf crop would consist of old leaves with concomitantly high incidence of damage. Unfortunately, leaf production estimates are unavailable for this period of time.

The relative damage suffered by *Canavalia* and *Ipomoea* was different from that reported for the same two species by Wilson (1977). He found that in Costa Rica *Ipomoea* suffered a great deal from insect damage, whereas *Canavalia* did not. He attributed these differences to arise from a greater toxicity of the latter species. On One-Tree Island both species had similar incidence of leaf damage (*Canavalia* often having greater damage than *Ipomoea*) which exceeded that of the other species present (Figs. 62, 63). If *Canavalia* is protected by toxins, then some of the other species may have even more effective protection. There are several possible explanations for the differences between our results and those of Wilson. One is that at One-Tree Island, the *Canavalia* leaves were older than those of other species and hence displayed greater accumulated damage. A second is that there may be regional differences between Australia and Costa Rica in the toxicity of these plants, and a third is that the two insect faunas involved are different in their responses to such protective mechanisms. Finally, the degree of error in using leaf-holes in estimating insect consumption may differ among the two species (see Chapter 7).

Solanum requires special mention. It was sufficiently rare during the early part of the study that it did not occur in the sample plots. Its first appearance in samples was in July–August 1969; leaf damage was 3.6%. In September –October it was 3.3% rising to 6.7% and 7.6% in November–December 1969 and January–February 1970, respectively. By March–April 1970 the incidence of leaf damage had risen to 12.7%; thereafter the species was no longer found in the sample plots. However, by September–October 1970 *Solanum* samples obtained elsewhere showed a damage of about 20%. This last estimate is believed to be an underestimate of the true damage. In contrast to samples of all other species, whole leaves of *Solanum* at that time had been consumed, leaving only the mid-rib in some cases. Some plants were nearly completely defoliated. Thus, the estimated leaf damage was the damage to the *leaves remaining*, not the total damage sustained by *Solanum*. Following this period the species was believed to become extinct on the island (see Chapter 18).

Other species were only occasionally assessed for leaf damage. *Tetragonia* sustained damage of 0.8–10.9% over the period from July 1969 to February 1970. In March–April 1970 these plants were rare and none that were observed (outside the sample plots) had any insect damage. *Sesuvium,*

Portulaca, *Tribulus*, *Commicarpus* and *Boerhavia* all sustained insect damage of less than 1%.

Some of the higher losses in leaf area (e.g. 20%–30%) sustained by the One-Tree species would seem to be significant in terms of decreased photosynthetic ability. However, Nielsen (1978) has pointed out that the time that loss is sustained is important. He suggested that a leaf damage of 20% may be expected to significantly affect primary production in an adverse way if it occurs in young, actively photosynthesizing leaves, but probably does not if it occurs in old leaves after their net assimilation has declined. Mattson and Addy (1975) suggest that insect grazing of 5–30% of the annual foliage crop (the range in which most One-Tree plants fell) does not usually impair annual primary production; light grazing may even stimulate growth or have a variety of indirect effects beneficial to plants (Chew 1974, Mattson and Addy 1975). Thus, depending upon the stage of leaf growth in which heavy grazing took place, One-Tree plants may or may not have suffered significantly from grazing herbivores. *Solanum*, which was severely defoliated, is a clear exception (see Chapter 18). Stored reserves may also influence the level of grazing that plants could temporarily sustain.

Because different species of plants in the herb flat had different leaf biomasses, and the relative contribution of the various species changed with time, estimates of per cent leaf damage does not directly indicate the relative contributions of different plant species to the total food of chewing herbivores. Furthermore, for different seasons and different leaf ages the leaf weight per unit area varied.

In order to assess the biomass of leafy material eaten by chewing insects in the herb flat, the weight–surface area calibration determined at each visit (see Chapter 7) was used to correct leaf surface area consumed to oven dry weights; the mean values for each species per m^2 of herb flat as well as the totals for all species were calculated. Because of a variety of uncontrollable sources of error the biomass of our values of insect consumption are probably overestimated (see Chapter 7).

In the pre-drought period *Melanthera* and *Canavalia* had both contributed a large amount of leaf biomass to chewing insects (Table 23). The former had a low incidence of insect damage which was more than compensated by its large standing crop; *Canavalia* had a small standing crop but extensive insect attack upon it. Throughout the rest of the study *Melanthera* contributed a variable but moderate amount of leaf biomass to the insects. After the drought, the biomass of *Canavalia* lost to chewing insects was low until the end of 1969 when recovery was nearly complete. Subsequently, large amounts of leafy material were eaten (Table 23). Neither *Abutilon* nor *Ipomoea* contributed a large amount of leaf biomass to insects in the pre-drought period as both standing crop and extent of insect damage were low then. During the recovery period the leaf biomass eaten increased and then dropped again after early 1970. *Abutilon* showed a similar initial pattern but with values

Table 23. Amount of leafy material consumed by chewing insects in the herb flat. Values are in g dry wt/m^2 of herb flat. Values in parentheses are kg dry wt for entire herb flat. Grasses are not included; their values were so small as to be negligible.

	1968	1969						1970						1978
	J–A	J–F	M–A	M–J	J–A	S–O	N–D	J–F	M–A	M–J	J–A	S–O	N–D	Sept.
Species														
Melanthera biflora	2.60 (79.3)	0.20 (6.1)	0.03 (0.9)	0.05 (1.5)	0.53 (16.2)	0.06 (1.8)	0.04 (1.2)	0.13 (4.0)	0.34 (10.4)	0.10 (3.1)	0.03 (0.9)	0.35 (10.7)	— —	0.49 (15.0)
Canavalia rosea	2.70 (82.4)	0.10 (3.1)	0.06 (1.8)	0.02 (0.6)	0.04 (1.2)	0.13 (4.0)	1.10 (33.6)	3.00 (91.6)	2.33 (71.1)	5.49 (167.5)	3.73 (113.8)	5.94 (181.3)	2.94 (84.7)	— —
Ipomoea macrantha	0	0.20 (6.1)	0.30 (9.2)	0.80 (24.4)	0.38 (11.6)	0.75 (22.9)	2.96 (90.3)	7.79 (237.7)	0.30 (9.2)	0.95 (29.0)	0.06 (1.8)	0.54 (16.5)	0.02 (0.6)	— —
Abutilon indicum	0.10 (3.1)	0.001 (0.03)	0.10 (3.1)	0.04 (1.2)	0.52 (15.9)	0.67 (20.5)	0.20 (6.1)	1.26 (38.5)	1.51 (46.1)	1.60 (48.8)	0.61 (18.6)	0.79 (24.1)	0.35 (10.7)	— —
All others	0	0	0.20 (6.1)	0.30 (9.2)	0.64 (19.5)	0.68 (20.8)	3.61 (110.2)	0.60 (18.3)	0.19 (5.8)	0.37 (11.3)	0.04 (1.2)	0.12 (3.7)	0.05 (1.5)	— —
Total	5.40 (164.8)	0.50 (15.3)	0.69 (21.1)	1.21 (36.9)	2.11 (64.4)	2.29 (70.0)	7.91 (241.4)	12.78 (390.1)	4.67 (142.6)	8.51 (259.7)	4.47 (136.4)	7.74 (236.3)	— —	— —

usually somewhat lower than for *Ipomoea*; values for *Abutilon* did not drop so markedly in 1970 and at that time they exceeded those of *Ipomoea*. The minor species of the herb flat collectively contributed only a small amount of leaf biomass to insects during the study except for November–December 1969 when their total value exceeded that of any of the major species (Table 23); *Solanum* was primarily responsible for that high value as it contributed 90% of the material eaten from leaves of the minor species at that date.

The amount of biomass consumed from all species collectively showed marked temporal trends (Table 23). Before the drought a large amount of leaf biomass had been eaten. After the drought, the amount eaten was low. During recovery the amount consumed increased and by the end of 1969 had exceeded the pre-drought values; thereafter the amount eaten varied, never dropped below 83% of the pre-drought level, and reached as high as 237%.

These values of biomass eaten are not rates of consumption, that is, they do not represent the amount eaten per unit time. Rather, they are the amount that has been consumed from the existing standing crop.

If leaf turnover is high, with new leaves being continually produced, maturing rapidly and falling early, the consumption of leaf biomass by chewing insects would appear to be lower than would be true for the same rate of grazing upon leaves with greater longevity and lower production. Seasonal changes within a species, and/or differences among species in leaf turnover and leaf longevity would influence apparent rates of leaf consumption. Furthermore, insects may prefer leaves of particular ages. Consequently, great care must be exercised in interpretation of leaf consumption data.

With this in mind, we proceed in a tentative way to relate leaf consumption to other events in the herb flat. Unfortunately, data on leaf production and leaf fall in the herb flat were obtained only toward the end of the study. Between July–August and September–October 1970 all four of the major herb flat species showed a net leaf production (rate of production of new leaves greater than that of leaf fall) (Table 13). During this period all four species showed an apparent increase rather than a decrease in amount of leaf biomass consumed by insects (Table 23). In *Abutilon* the leaf fall from September–October to November–December was the same as for the previous period but production of leaves was reduced by half and there was a net loss (Table 13); the apparent leaf consumption decreased rather than increased (Table 23). These results are all contrary to predictions of trends if changes in apparent leaf consumption were being influenced solely by differences in leaf production and turnover rates. It is probable that the actual grazing rates were also changing.

This supposition was initially strengthened by comparison of the patterns of leaf consumption with those of density of epigeal arthropod populations. As the insect populations built up in the post-drought recovery period, the total leaf consumption in the herb flat also increased (Fig. 64). The period following the peak density of arthropods exhibited the highest amount of leaf biomass previously consumed. Thereafter, the patterns of leaf consumption and arthropod density were similar (compare Figs. 49 and 64). However, the pattern of change in total number of chewing herbivores, a more relevant comparison, did not always follow trends in consumption or intensity of

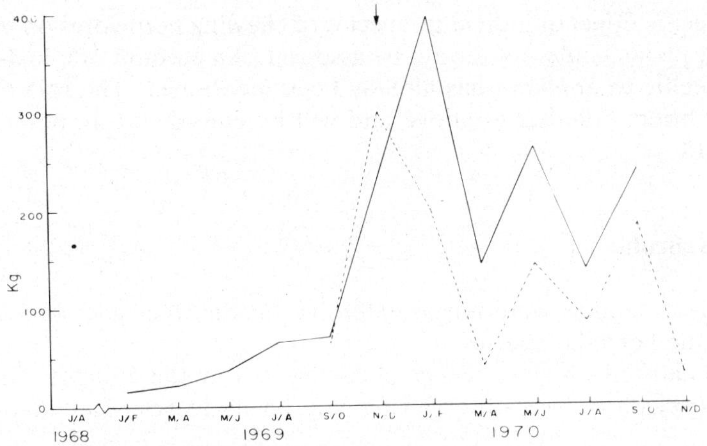

Fig. 64. Total consumption of leaf biomass (all species except grasses) by chewing insects in the entire herb flat (dot and solid line) and for *Pisonia* and *Argusia* combined (broken line). Arrow indicates period of peak density of epigeal arthropods.

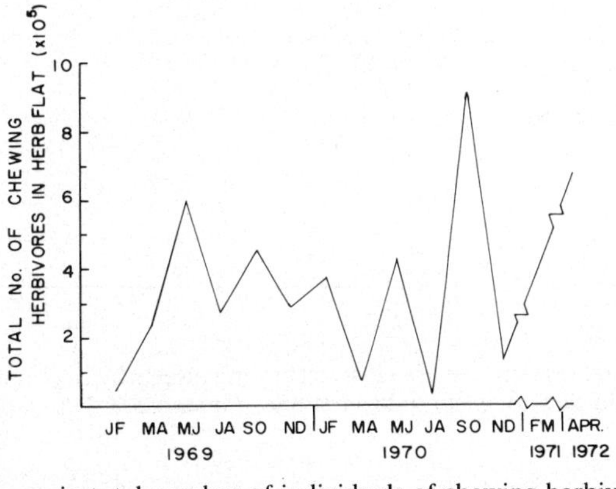

Fig. 65. Temporal change in total number of individuals of chewing herbivores in the herb flat.

damage in detail. There were 2 major peaks in number of chewing herbivores, one in March–April 1969 and the other in September–October 1970. Neither peak coincided with the periods of highest consumption of leaves (compare Figs. 64 and 65) or highest intensity of damage (compare Figs. 65 and 66); furthermore, the period of greatest consumption and intensity of damage (January–February 1970) was a period of only moderately high numbers of chewing herbivores.

During the entire year of 1970, however, the pattern of change in numbers of chewing herbivores and total leaf consumption and intensity of damage did follow similar trends in that all three increased and decreased together, albeit not at the same magnitudes.

The specific effect of individual species of chewing herbivores on particular species of plants could not always be assessed. An account of a leaf-chewing beetle specific to *Solanum* has already been mentioned. The role of locusts has been briefly alluded to above and will be considered in more detail in Chapter 18.

Trees and shrubs

Data on leaf damage were not as extensive for the trees and shrubs as they were for the herb flat species.

Pisonia showed a high incidence of damage before the drought. The locust, *Valanga irregularis,* was observed feeding on its leaves. The characteristic

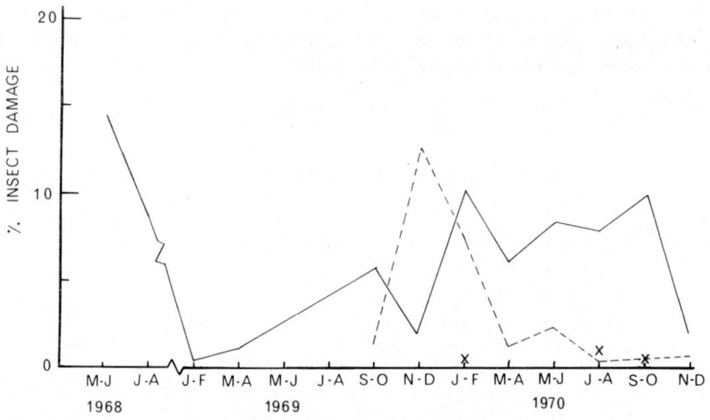

Fig. 66. Per cent of total leaf surface area lost to chewing insects by trees and shrubs on One-Tree Island. *Pisonia* = solid line. *Argusia* = broken line. *Pandanus* = X's.

shape of the eaten parts made locust damage identifiable and it accounted for much of the leaf damage to *Pisonia* in May–June 1968. After the drought, incidence of leaf damage was very low but later increased to variable but higher levels (Fig. 66). During the post-drought period it reached two peaks of about 10%, one in the period following the 1969 density peak of *Pisonia* leaf fauna and the other at the time of the 1970 leaf fauna peak (compare Figs. 59 and 66), the latter occurring despite a high net production of leaves at that time (Fig. 39). In March–April 1970 leaf damage was mostly by grasshoppers (see Chapter 18). In May–June of the same year there was heavy damage in a few localized areas but caterpillars were chiefly responsible. Thus, the prime agent of leaf damage may change over a short time.

Argusia showed quite a different pattern of damage. Sampling of this species commenced later. After a low value in September–October 1969, incidence of insect damage rose to a peak in November–December, after which it declined and remained low for the rest of the study. This pattern correlated with that of the abundance of its leaf fauna (compare Figs. 59 and 66). The peak of damage coincided with that of density of its foliage insects in late 1969. Foliage insect densities remained low throughout 1970, as did damage to leaves. Net production was high during the 1969 damage peak and again in late 1970 when damage was low (Fig. 40). It appears that incidence of damage reflects the changes in grazing by insects, and is not greatly altered by the changes in production.

Leaf damage to *Pandanus* was measured on only three occasions and each time it was extremely low (Fig. 64). Given the extreme longevity of leaves of this species (see Chapter 12), it must be remarkably immune to the ravages of chewing insects.

The leaf biomass consumed by insects was calculated as for the herb flat plants except that the standing crop values of leaves used in the calculations were the average of values obtained by the various methods. The pattern of change was similar to that of incidence of damage (Table 24).

Table 24. Biomass of leaves of shrubs and trees consumed by chewing insects.

	Pisonia		*Argusia*		Total	
Date	g dry wt/m^2 of canopy	For entire island (kg)	g dry wt/m^2 of canopy	For entire island (kg)	g dry wt/m^2 of canopy	For entire island (kg)
1968						
J–A	23.5	98.5	—	—	—	—
1969						
J–F	0.4	1.7	—	—	—	—
M–A	1.5	6.3	—	—	—	—
S–O	8.2	34.4	9.9	31.5	18.1	65.9
N–D	1.6	6.7	92.1	293.0	93.7	299.7
1970						
J–F	7.9	33.1	51.5	163.8	59.4	196.9
M–A	4.5	18.9	7.2	22.9	11.7	41.8
M–J	24.5	102.7	11.9	37.9	36.4	140.6
J–A	17.9	75.0	2.5	8.0	20.4	83.0
S–O	40.3	168.9	3.9	12.4	44.2	181.3
N–D	4.7	19.7	3.2	10.2	7.9	29.9

For the period for which comparative data were available, the combined biomass of *Pisonia* and *Argusia* was lower than the total for the herb flat except on one occasion; the general temporal pattern of change was similar for the two species (Fig. 64).

We believe the herb flat values are close to the total leaf consumption for the entire low vegetation on the island as insect damage to the succulents was very slight and biomass of beach vegetation was low. Similarly, *Pisonia* and *Argusia* combined, account for most of the arborescent contribution of leaf biomass to insects. There was very little *Scaevola* on the island and amount of damage to *Pandanus* was relatively small.

In conclusion, it appears that consumption of leaves is a complex phenomenon. In some cases it follows the pattern of change of numbers of chewing herbivores. However, changes in rate of leaf production and fall, and leaf longevity influence evaluation of grazing and introduces lag times which differ among species of plants. Consequently, extent of insect damage at any one time is not a precise indicator of the number of chewing herbivores present or *vice versa.* A further complication is that a particular species of insect may alter its grazing rate per individual under different environmental conditions, a variable not measured in the present study.

The incidence of leaf damage was sufficiently low throughout the study except for severe defoliation of *Solanum* on one occasion, that most of the primary production residing in the leaves is returned to the litter and goes into decomposers, rather than into herbivores.

Leaf damage is not entirely a one-way effect of action of herbivore upon the plant. Feedback may also occur, and plant responses may affect the grazer in subtle ways. For example, Haukioja and Niemala (1979) found that when lepidopteran and hymenopteran larvae were reared on birch leaves whose adjacent leaves had previously been damaged mechanically, the length of the larval period was protracted in comparison to that of larvae reared on leaves from previously undamaged foliage. This response occurred only during the period when leaves were young; after leaves had completed their growth they seemed to have lost the capacity to respond. Thus insect attack can lead to increased resistance to the attacking species, perhaps through increased production of leaf-phenolics. Under such a system, it is possible that an abundant, resistant herbivore might indirectly have an adverse effect upon a less common and less toxin-resistant herbivore by inducing plant resistance. Similarly, a plant which is favoured by a herbivore and heavily grazed, by responding to that grazing by producing toxin and becoming less palatable may shift the burden of herbivore pressure onto another plant species that had been previously less acceptable and more lightly grazed. The potential for subtle interaction between plant and herbivore, between herbivore and herbivore and between plant and plant becomes enormous. It remains to be discovered just how widespread grazing-induced resistance is among plants.

16. The soil microarthropods

One-Tree Island's litter and soil is inhabited by representatives of all major soil invertebrate groups. In a detailed study of the soil system, Done (1975) was able to characterize the soil microarthropod community of the island as a whole, and of three of its major vegetation areas – the herb flat, a *Pisonia* grove, and the succulent mat. The soil habitat has been described in Chapter 6.

The total number of species for the three areas collectively was over 130. The major taxa were represented by the following numbers of species:

Cryptostigmata	40
Mesostigmata	21
Prostigmata	34
Astigmata	7
	102 total mite species.
Symphypleona	2
Arthropleona	7
	9 total Collembola species.
Myriapoda	3
Insecta	13
Other	4[1]
	131 total species.

These numbers are underestimates, as some rare species were not identified past the ordinal level. A list of the species of Acari (mites) and Collembola, the two most abundant groups, is presented in Appendix III, and a list of the other taxa appears in Appendix IV.

[1] Includes the terrestrial snail *Gastrocopta* (*Sinalbinula*) *pediculus* (Shuttleworth) (family Chondrinidae), a leaf litter species.

Taxonomic study of the Mesostigmata (by D. C. Lee and E. E. Lindquist) and the Collembola (by Penelope J. Greenslade) revealed that the fauna consists of widespread species, frequently with pantropical or pantemperate distributions. Small collections from other islands in the Capricorn group showed great taxonomic overlap with that from One-Tree (Done, unpublished).

Several significant differences exist between the faunas of the three study areas on One-Tree. The species-individual curves in Fig. 67, each based on

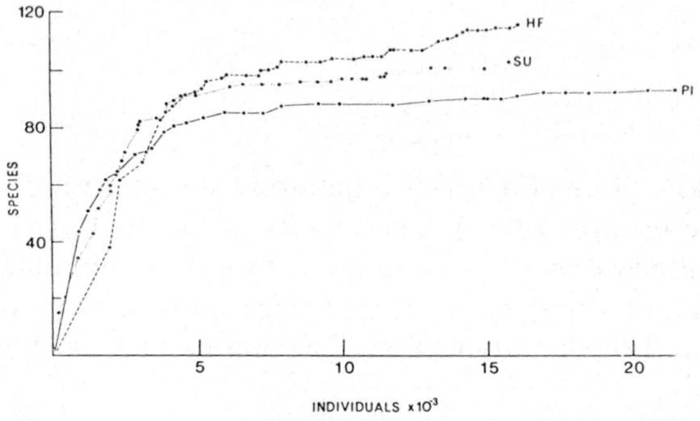

Fig. 67. Species-individual curves for the soil arthropods of One-Tree Island. Values for both species and individuals are cumulative totals. HF = herb flat. SU = succulents. Pi = *Pisonia* grove.

from 15,000–20,000 specimens, indicate the herb flat to have the greatest number of species (about 120), the *Pisonia* grove the least (about 95). The greater number of species in the herb flat reflects its relatively greater area and microhabitat variability and is in spite of its greater susceptibility to soil temperature fluctuations and desiccation (see Chapter 6). Conversely, greater microclimatic stability in the *Pisonia* grove was not accompanied by a greater species richness, as might have been expected.

The majority of species occur in all three of the study areas; however, some are exclusive or almost exclusive to only one (Table 25). Amongst this group is a disproportionate number of predatory species (8 out of 14 habitat-specific species, or 57%) in comparison to the overall proportion of predators to others (1 out of 5, or 20%).

The species which best characterize a fauna are those which are numerically dominant and widespread (constant). Those species comprising at least 5% of the total individuals and occurring in at least half the samples were defined as the 'dominant–constant' group. A comparison of the 'dominant–constant' groups of the herb flat and *Pisonia* grove appears in Table 26. Over the eight sampling periods, the size of this group ranged from 1 to 4 species in the herb flat (average 3.25) and from 2 to 8 in the *Pisonia* grove (average 4.50) reflecting a generally more equitable distribution of species abundance in the latter community.

Table 25. Species of mites restricted to particular habitats on One-Tree Island.

Site	Cryptostigmata	Mesostigmata	Prostigmata
Herb flat	*Ctenacarus* sp. *Haplocthonius* sp. *Zygoribatula* sp.		Pachygnathid II* Cryptognathid* Cheyletid* Stigmaeid VII*
Pisonia grove	*Ephilomannia* sp. Liodoid II	Parasitoid I* *Gamasiphis setosus**	
Succulents		*Gamasellus discutatus** *Digamasellus* sp.* Uropodid II	

* Predatory species.

Table 26. Species composition of the dominant-constant group of soil arthropods on One-Tree Island.

	1969			1970				
	Aug	Oct	Dec	Feb	Apr	June	Aug	Oct
Herb flat								
Nothrid I	+							
Sphaerochthonius sp.		+				+		
Coccetydeus sp.		+	+	+	+	+	+	+
Tarsonemid I		+			+			
Cosmochthonius sp.		+						
F. exiguus			+					
X. cf. *manusiensis*			+		+			
Synchthonius sp.				+		+		
Suctobelba sp. 2				+				
Oribatuloid II				+		+	+	
Ledermuelleria sp.					+			
Oribatuloid I							+	
A. major								+
Crypto, juv. 1								+
No. in group	1	4	3	4	4	4	3	3
***Pisonia* grove**								
Coccetydeus sp.	+	+	+	+	+		+	
Liochthonius sp.	+			+	+		+	+
F. exiguus	+		+					
Sphaerochthonius sp.	+		+		+	+	+	+
Eremelus sp.		+		+	+		+	+
Microzetes sp.		+						
Oribatuloid I		+						
B. stellaris				+				
Anoetid				+				
Carabodes sp. 1					+		+	+
Haplacarus sp.						+	+	
Tydeus sp. 1							+	+
Cultroribula sp.							+	
No. in group	4	4	3	5	5	2	8	5

The composition of the group is far less consistent in the herb flat than in the *Pisonia* grove. Only one herb flat species can be considered to be a more-or-less regular member of the group; *Coccetydeus* sp. occurred on 7 of the 8 occasions (in comparison to a maximum of 3 for the other herb flat

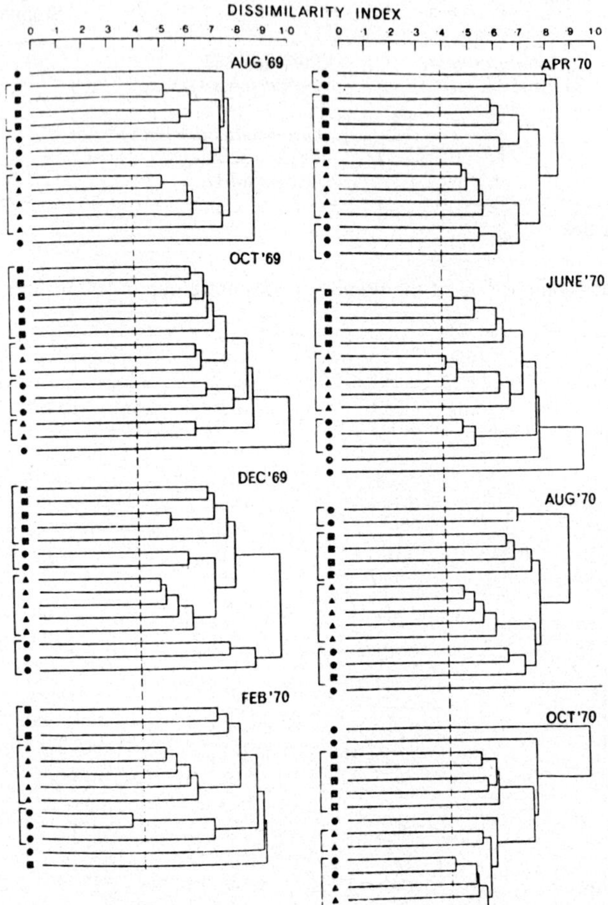

Fig. 68. Taxan 3 dendrograms produced from each site at each sampling date. Squares = herb flat. Triangles = *Pisonia* grove. Dots = succulent mat. Brackets indicate subjectively delimited clusters. For more detailed explanation see chapter 7, or Done (1975). The dissimilarity index is the Canberra Metric Dissimilarity Index.

species). By contrast, four *Pisonia* species are at least moderately regular – *Coccetydeus* sp. (6 occurrences), *Liochthonius* sp. (5), *Sphaerochthonius* sp. (6) and *Eremelus* sp. (5).

Subtle faunistic differences between the three sites are reflected in the classifications of the samples of each sampling period (see Chapter 7).

Despite the considerable overlap in species compositions of the three sites, samples from within sites group with each other preferentially on all sampling occasions (Fig. 68). However, samples from the succulent site consistently split into two groups. The dendrograms are interpreted to mean that the *Pisonia* and herb flat communities are faunistically homogeneous within

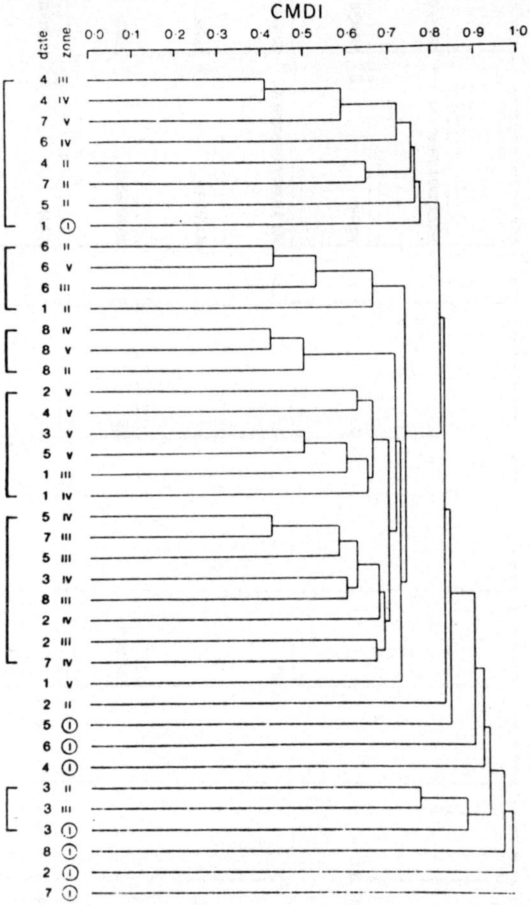

Fig. 69. Dendrograms of all sample units from the succulent mat. Numbers in 'date' column indicate 1st, 2nd, . . . 8th sampling occasion, Roman numerals indicate sampling zone (see Fig. 19). Brackets indicate subjectively defined clusters. For more detailed explanation see chapter 7, or Done (1975). CMDI indicates the Canberra Metric Dissimilarity Index.

themselves but different from each other; the succulent community is faunistically heterogeneous.

In a classification of all succulent samples combined (Done 1975), samples tended to group according to both their position along the slope by the pond and by date of sampling, reflecting changes in community composition in both space and time (Fig. 69). A further finding was that both species richness (S)

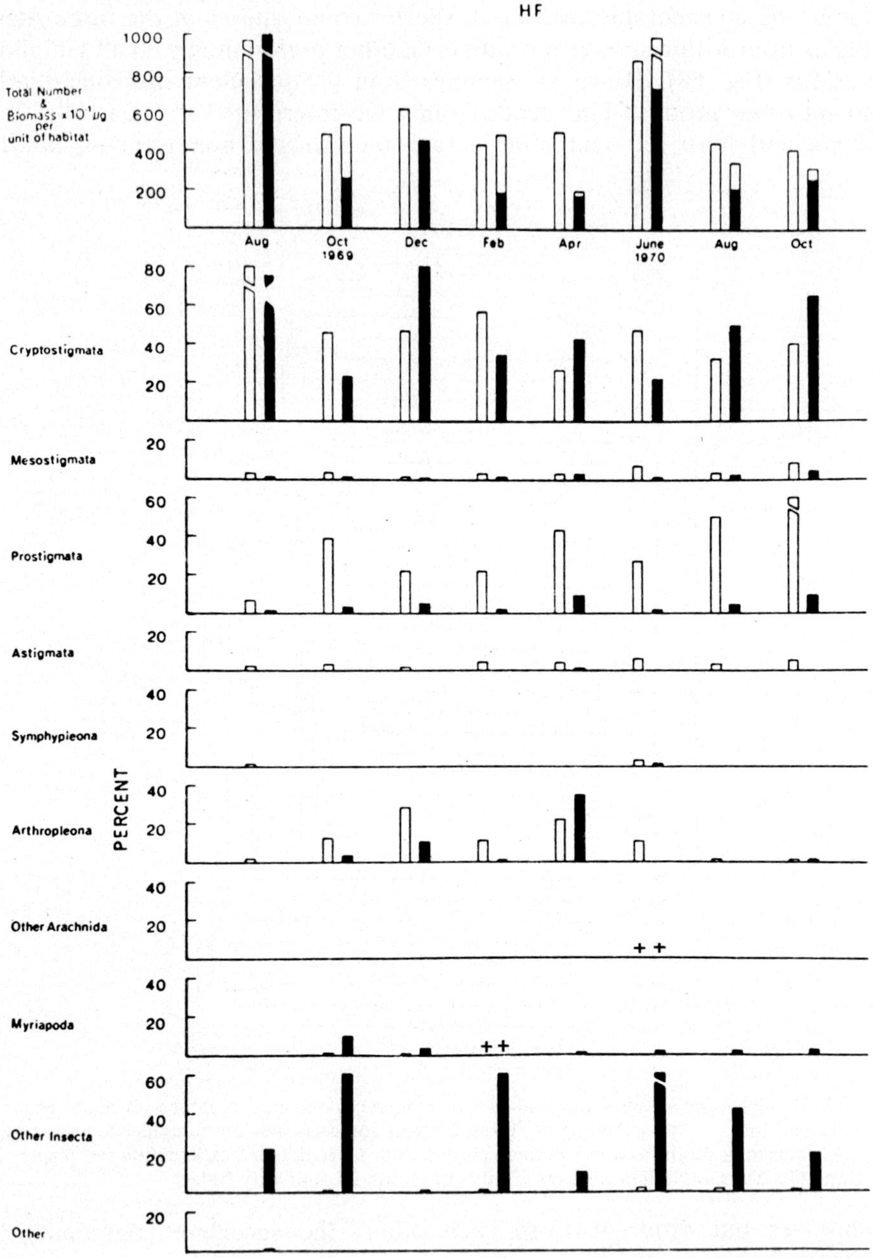

Fig. 70. Abundance (open histograms) and biomass (darkened histograms) of soil arthropods in the herb flat. Open top of darkened histograms refers to biomass of 'other insects'. + indicates less than 1% of total. Taken from Done (1975).

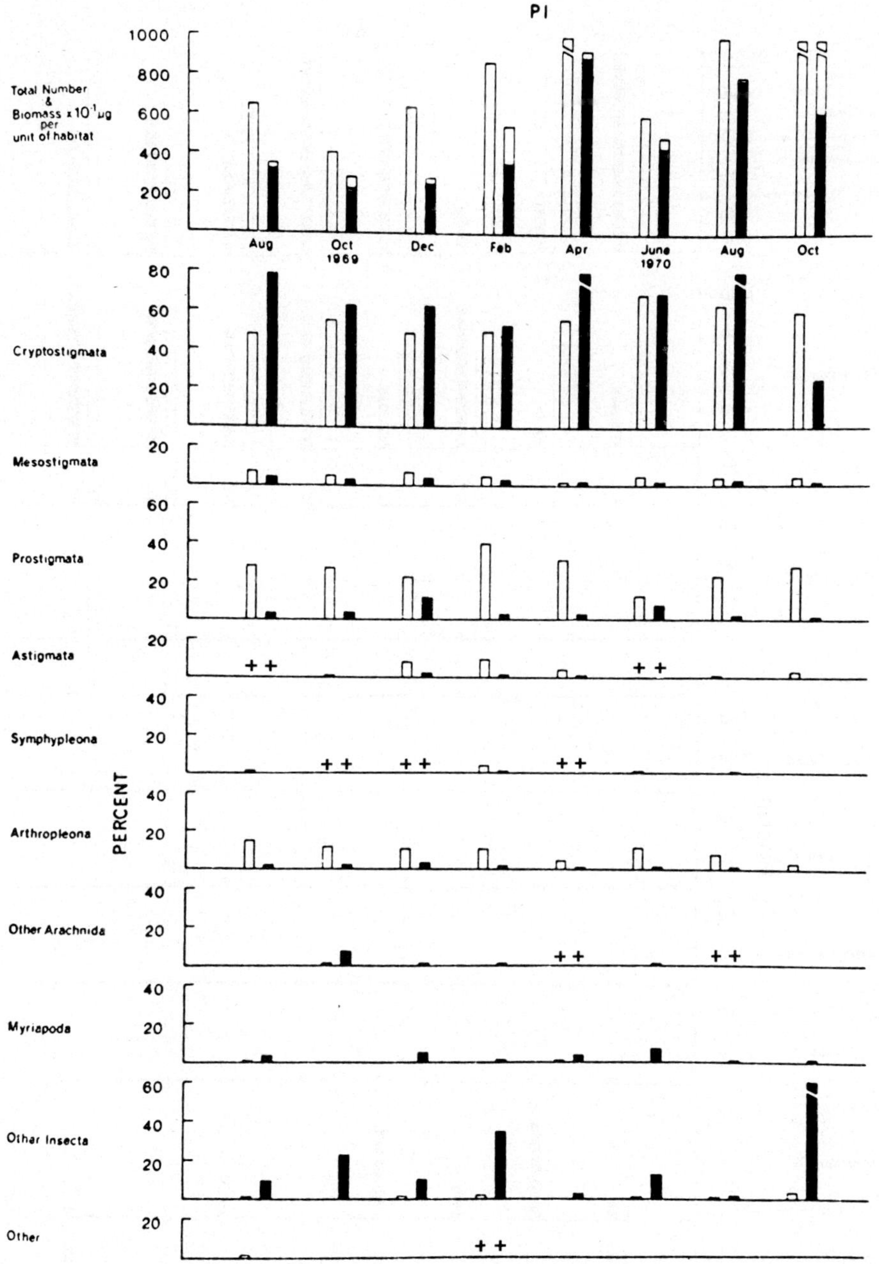

Fig. 71. Abundance and biomass of soil arthropods in the *Pisonia* grove. Symbols as in Fig. 70. Taken from Done (1975).

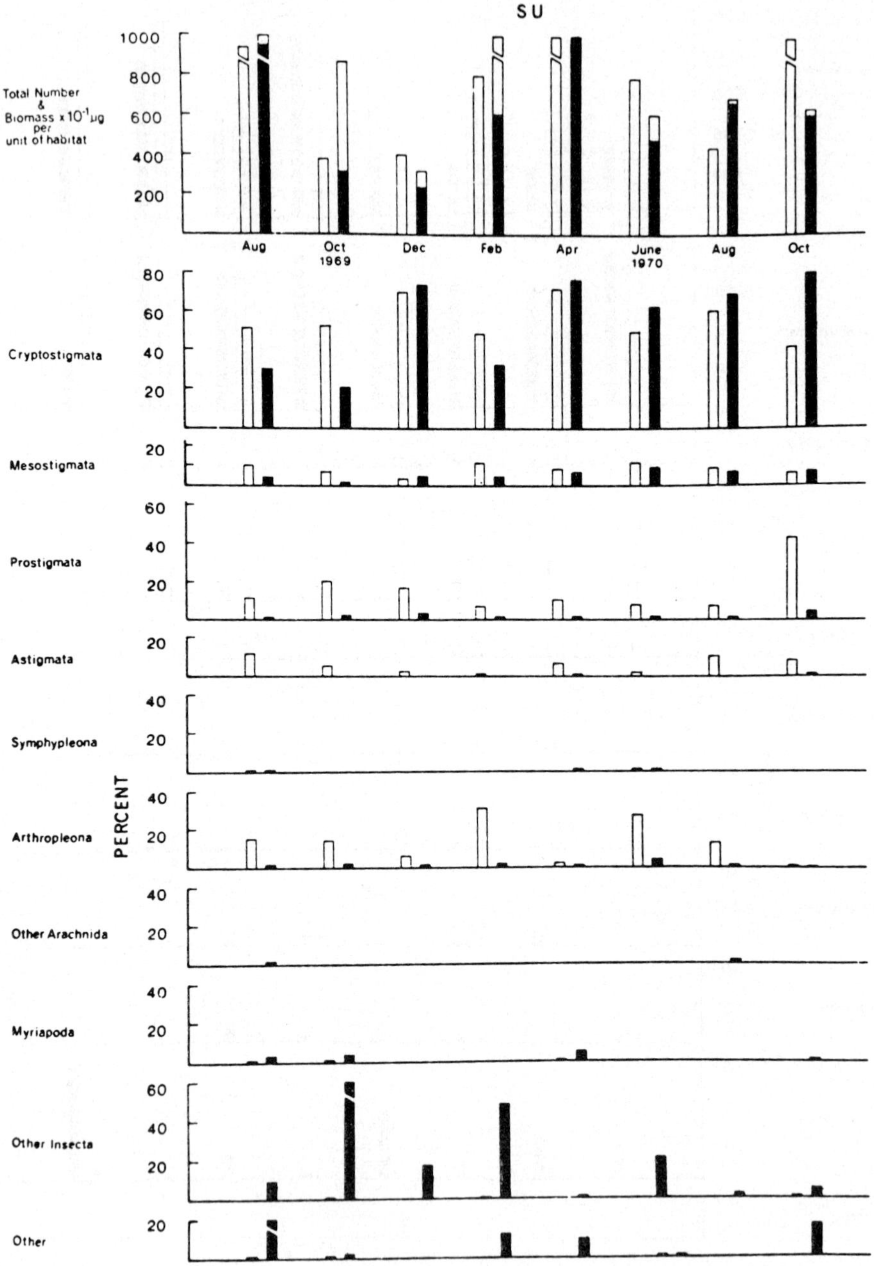

Fig. 72. Abundance and biomass of soil arthropods in the succulents. Symbols as in Fig. 70. Taken from Done (1975).

and Brillouin's Diversity (H) were positively correlated with distance from the pond; the susceptibility of the soil close to the pond to salt-water flooding makes it unsuitable for many species and causes the above trends.

Changes in the soil communities over time

Changes in the characteristics of the soil communities occurred during the study, the composition of the dominant–constant group (previously mentioned) being a case in point.

Estimates of total microarthropod abundance and biomass also varied considerably between sampling periods and suggested population fluctuations which did not have any obvious correlation with either the rainfall pattern or the seasonal changes in temperature (see Figs. 70–72). In all cases, the 1969 and 1970 patterns were quite different in month of occurrence of peaks and in the actual levels of population density reached.

In all three areas the Cryptostigmata were by far the most abundant group and accounted for the preponderance of the biomass at most sampling periods. Consequently, they influenced the total pattern much more than did other groups. The Prostigmata and the Arthropleona were the next two most important groups in terms of contribution to total numbers and biomass. All others were quite minor (except as noted for biomass of 'other insects').

Reproductive patterns

At the initiation of the study, it was not known to what extent reproduction in subtropical soil arthropods was limited to particular seasons. Of the seventeen individual species for which sufficient data were available, the vast majority had larvae and juveniles present year round; the data did not in any instance demonstrate a seasonal pattern to reproduction (Done 1975).

Convergence of community structure

Done's (1975) study provided evidence that qualitative changes were taking place in the soil microarthropod community of the herb flat which resulted in it tending to become more similar to the *Pisonia* soil community. The interpretation is linked closely to the changes in foliage cover of the former area, and its implied effect on the soil microclimate.

A net increase in similarity, in overall structure of herb flat and *Pisonia* microarthropod communities is indicated in Fig. 73 for the year 1970. This similarity is as measured by the Canberra Metric Index defined in Chapter 7. During this same period both Brillouin's Diversity (H) and within site homogeneity (intra-site Canberra Metric Index) increased, becoming comparable with levels in the *Pisonia* community.

An increase in similarity could equally as well result from the *Pisonia* fauna changing towards that of the herb flat as *vice versa*. That this was not so was demonstrated by comparing both sites with 'baseline' conditions, which were

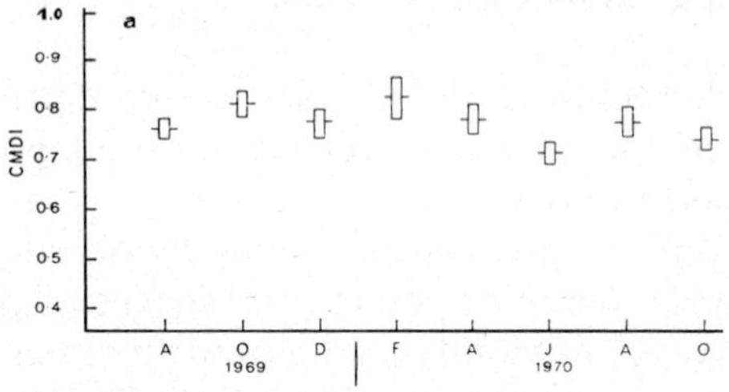

Fig. 73. Temporal changes in similarity between the soil arthropod communities of the herb flat and *Pisonia* grove. Horizontal lines represent the mean, and bars the 95% confidence limits of the Canberra Metric Dissimilarity Index (CMDI). Letters on the abscissa indicate months.

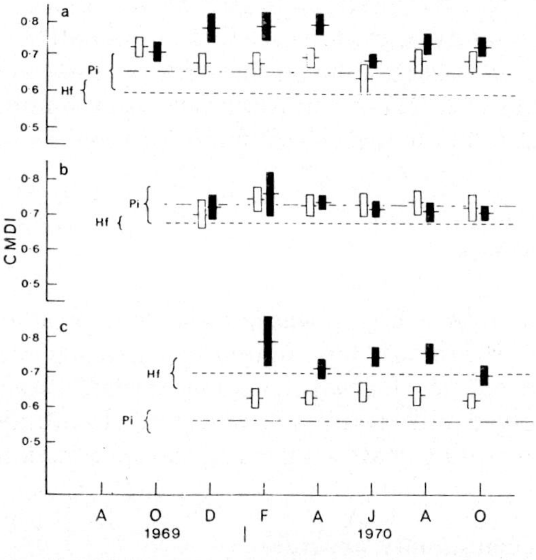

Fig. 74. Mean Canberra Metric Dissimilarity Index (CMDI) between baseline (broken line) and subsequent sampling dates. Brackets indicate 95% confidence limits of baseline; rectangles indicate 95% confidence limits of CMDI from baseline. Open rectangles = *Pisonia*; black rectangles = herb flat.

respectively defined as the community composition in August, October and December, 1969; all subsequent samples in a site were compared with the baseline condition using the Canberra Metric Index, and the results are presented in Fig. 74. In general, the *Pisonia* fauna did not diverge from the baseline condition whereas the herb flat fauna diverged initially, and returned to a similar condition some 10–12 months later. The herb flat fauna is thus shown to be the more labile of the two and it is change in its composition which causes the convergence between the two communities.

A 'pseudo-successional' interpretation may be made of the changes in the herb flat soil community. Done (1975) used the terminology that Wilson (1969) developed to describe refaunation of artificially defaunated islands. Defoliation caused by the drought of summer, 1968–69, is assumed to have increased insolation and caused great mortality among the herb flat soil arthropods.

The pre-drought and post-recovery conditions of heavy shading of the herb flat soil are assumed to create a more favourable (and more *Pisonia*-like) environment for the majority of soil arthropods. The convergence of the herb flat soil community towards the less labile *Pisonia* condition suggests the latter to be in a sense a *climax* type of community towards which the former progressed as the herb foliage increased.

From an immediate post-drought condition of low population levels, a sequence of non-interactive, interactive and assortative phases (after Wilson 1969) could lead to a *Pisonia*-like climax community in the herb flat, while undergoing the sorts of changes in homogeneity and diversity actually observed. These ideas are more fully treated by Done (1975).

The characteristics of the soil microarthropods described above may have wide application among insular communities. Spaull (1949) found that on Aldabra, the species composition of the soil arthropod fauna was similar at different sites and although there were exceptions, the distribution of species was largely independent of vegetation type. Thus, as on One-Tree Island there was little habitat specificity.

Spaull interpreted these results as arising from niche expansion of immigrant species, as influenced by the alternation of dry and wet years. During droughts, hardy species would be selected that could survive aridity, and others would become locally extinct, thereby reducing the number of competitors in the system and permitting niche expansion of the remaining species.

We would like to supplement this idea by pointing out that soil arthropods are probably not easily dispersed over water, and that those which do reach islands are probably able to do so only because they were already resistant to harsh conditions. Such species would tend to be eurytopic and occupy a broad array of habitats once they become established. The difference in these two views lie in the relative emphasis placed on immigration and establishment of soil arthropods which are not drought resistant, during years of high rainfall and their extinction during drought. The low species turnover among soil arthropods at One-Tree Island would tend to support the view that dispersal problems filtered out all but the hardy, eurytopic species, though of course the two theories are not mutually exclusive.

17. The vertebrate fauna

The only mammal (other than humans and one pig temporarily; see Appendix I) observed on the island was a bat (*Taphozous flaviventris*; family Emballonuridae) found lying on the beach on the morning of 31 March, 1970 after a night of strong wind squalls. It was examined and released. On the following day several bats of this species were observed on Heron Island.

No terrestrial reptiles or amphibians were ever observed on One-Tree Island.

Birds

A number of reports of the birds of the Capricorn–Bunker groups of islands have appeared over the years (Jukes 1847, Campbell and White 1910, Gilbert 1925–27, MacGillivray 1926, 1928, 1931, Edgell 1927, Cooper 1948, Moulton 1961, Shipway 1969, Booth 1970, Kikkawa 1970, 1973, 1976, Lavery and Grimes 1971, Fien 1971, Hawken 1972, Domm and Recher 1973, Chilvers 1975) and the avifauna is now well known. Some of these reports have included One-Tree Island or dealt with it solely. In addition, the biology of the terns and gulls (Hulsman 1974a, b, 1975, 1976, 1977a, b) and reef heron (Recher 1972, Recher and Recher 1972a, b) have been investigated at One-Tree Island.

Eighty-four species of birds have been identified on One-Tree Island by the present authors, Domm and Recher (1973), Hulsman (1973), MacGillivray (1928) and Chilvers (1975 and pers. comm.); 29 of these species were residents or regular visitors. Table 27 summarizes the taxonomic composition of the birds and a complete list appears in Appendix V. The most abundant birds on the island were terns, waders (in summer) and reef herons. The resident land birds were landrails, kingfishers, sylviid warblers, and silvereyes. Of the non-resident land birds, 16 were intra-Australasian north–south migrants, one was a trans-equatorial migrant and 11 were vagrants from the mainland. Only one genus was represented on One-Tree Island by more than one breeding and/or resident species, and this was *Sterna* – no fewer than 5 species occurred there. Two species, now both represented by museum specimens, were well north of their usual range (sooty albatross and dove prion) and were probably blown north by strong winds which lashed the coast just prior to their appearance on the island (Chilvers 1975).

Distribution of birds on the island

Resident and transient land birds, except for the sacred kingfisher (*Halcyon sancta*), were seen only in the vegetated parts of the island. Banded landrails (*Rallus philippensis*) were sighted most frequently on the ground within the *Pandanus* grove, round the pond, or close to one of the southern *Pisonia* groves. The silvereyes (*Zosterops lateralis*) frequented the herb flat in the vicinity of *Argusia* shrubs, and *Pandanus* and *Pisonia* groves. Golden-headed fantail warblers (*Cisticola exilis*) occupied the central herb flat. The kingfishers were usually seen in trees and shrubs on the perimeter of the herb flat, and occasionally round the central grove of trees, or on the shingle beach.

Many aquatic birds also used the vegetation of the island for roosting, nesting or feeding. Three species – little black cormorant (*Phalacrocorax sulcirostris*), little pied cormorant (*P. melanoleucos*) and white-capped noddy (*Anous minutus*) – roosted in *Pisonia* and *Argusia* but neither nested nor fed on the island. Reef herons (*Egretta sacra*) roosted and nested in the *Pisonia* groves and occasionally fed in and around the pond.

The bridled tern (*Sterna anaetheta*) nested entirely within the herb flat and its fringing shrub ring. Three other terns – the roseate (*Sterna dougalli*), crested (*S. bergii*) and lesser crested (*S. bengalensis*) – and the silver gull (*Larus novaehollandiae*), sometimes nested on the succulent mat around the pond, or on prostrate herbs on the perimeter of the herb flat. In other years they nested on strand vegetation or on bare coral rubble of the storm-ridged beach. Black-naped terns (*Sterna sumatrana*) restricted their use of the island to nesting colonies on the bare rubble beaches. The exact location of nesting colonies of the terns changed from year to year (Fig. 75). This may be a general characteristic of sea birds as crested terns and silver gulls are also prone to shift their nesting localities on islands off south-west Australia (Gillham 1961), and the distribution of nesting colonies of seabirds changed substantially with time on Aldabra (Diamond 1979).

The succulent mat and perimeter of the herb flat and strand vegetation also were used as resting areas by non-breeding terns and gulls. Some gulls, waders, and reef herons, fed in and around the pond, and the gulls foraged on the beach (Fig. 76) and visited the herb flat to feed on *Solanum* berries.

The terns fed almost entirely at sea, and the times and locations at which they rested on the island was influenced largely by tides except for the noddy tern which had a regular daily cycle (Hulsman 1977b). When they had young

Table 27. Summary of the One-Tree Island (O-TI) and Heron Island (HI) avifaunas. Data for One-Tree Island from the present study, Domm and Recher (1973), Hulsman (1973) and J. Chilvers (1975, pers. comm). Data for Heron Island from Kikkawa (1970).

	No. species		Residents or regular visitors		Total species in order	
	O-TI	HI	O-TI	HI	O-TI	HI
Procellariiformes					4	3
Diomedeidae	2					
Procellaridae	2	2	1	1		
Hydrobatidae		1				
Pelecaniformes					11	5
Pelecanoididae	1					
Sulidae	2	1	1			
Phalacrocoracidae	4	2	2			
Fregatidae	2	2				
Phaethontidae	2					
Ciconiiformes					5	3
Ardeidae	4	2	1	1		
Threskiornidae	1	1				
Anseriformes					1	1
Anserinae	1	1				
Falconiformes					3	1
Accipitridae	2	1	1	1		
Falconidae	1					
Gruiformes					1	2
Rallidae	1	2	1	1		
Charadriiformes					27	27
Haematopidae	2	2	1	2		
Charadriidae	4	4	2	3		
Scolopacidae	13	11	6	6		
Glareolidae		1				
Laridae	8	9	8	6		
Columbiformes					1	2
Columbidae	1	2		1		
Strigiformes					1	
Tytonidae	1					
Cuculiformes					4	2
Cuculidae	4	2				
Apodiformes					2	1
Apodidae	2	1				
Coraciiformes					3	2
Alcedinidae	2	1	1	1		
Coraciidae	1	1				
Passeriformes					21	22
Pittidae		1				
Hirundinidae	1	2				
Campephagidae	4	2				
Sylviidae	1	1	1			
Rhipiduridae	2	3				
Monarchidae	5	3				
Pachycephalidae	1	1				
Zosteropidae	1	1	1	1		
Meliphagidae	2	2				
Ploceidae	1					
Sturnidae	1	1				
Oriolidae		2				
Dicruridae	1	1				
Grallinidae	1	1				
Cracticidae		1		1 (in past)		
Total	84	71	27	25	84	71

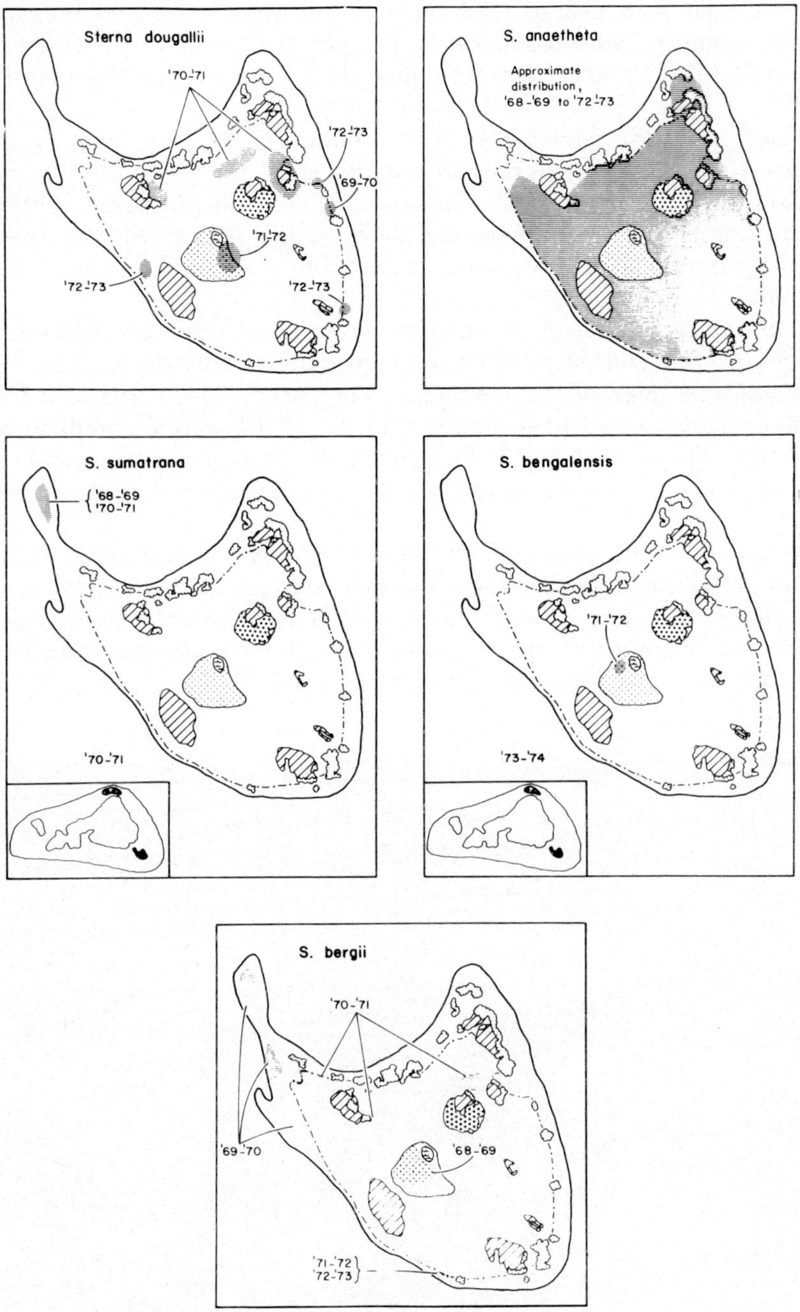

Fig. 75. Occasions of nesting colonies of seabirds near One-Tree Island. Inset map in middle two maps shows One-Tree Reef, One-Tree Island in black at the bottom, the three bare rubble islands (dots in black at the top).

they returned with fish to feed them, both intraspecific and interspecific robbing being of rather common occurrence (Hulsman 1976). Gulls sometimes robbed eggs and molested or killed chicks of the terns (Hulsman 1974a, 1976).

Waders used the island only to a limited extent; turnstones (*Arenaria interpres*) and oystercatchers (*Haemotopus* spp.) foraged along the beach, and sooty oystercatchers (*H. fuliginosus*) nested on the beach. Otherwise waders simply congregated on the shore of the island (without feeding), especially during high tides when their feeding grounds on the reef were covered.

White-breasted sea-eagles (*Haliaeetus leucogaster*) were seen intermittently during our visits and only visited the immediate vicinity of the eagle nest at the south-west edge of the *Pandanus*. They used a dead tree as a feeding perch; fresh carcasses and skeletons of birds and fish were found there nearly every trip. The nest (Fig. 77) located on the ground has been there since Juke's discovery of the island in 1843. It was not used for nesting during our study.

Five aquatic birds were seen on the surrounding reef and flying overhead, but not on the island itself; these were *Pelecanus conspicillatus, Phalacrocorax varius, Fregata ariel, Phaethon rubricauda* and *Cygnus atratus*. Carcasses of three other non-resident species were found above the high tide mark of the beach and thus ultimately contributed to the island's energetics; these

Fig. 76. A silver gull (*Larus novaehollandiae*) on the beach at One-Tree Island.

were the wandering albatross (*Diomedea exulans*), wedge-tailed shearwater (*Puffinus pacificus*) and brown booby (*Sula leucogaster*).

Nest sites

The species breeding on One-Tree Island are designated by a 'B' in Appendix V. The island carries major breeding populations of bridled and crested terns and minor colonies of roseate and black-naped terns (Lavery and Grimes 1971). The proximity of breeding colonies of four tern species on a small and structurally simple island, presented an opportunity to compare the nest site characteristics of each. The positions of nesting colonies of roseate, black-naped, bridled, crested and lesser crested terns, between 1969 and 1973 (data from Domm and Recher for the years 1971–73, are incorporated and shown in Fig. 75). Five nest characteristics of four terns are given in Table 28 and the variation in four other characteristics is illustrated in Fig. 78. The fifth species of tern, the lesser crested, has nested sporadically at One-Tree but breeding success has been low and no nest data were collected. Hulsman (1973) reported a large colony of this species nesting on one of the rubble banks of the island's reef crest in September 1973. One of these banks also was used by a portion of the nesting population of black-naped terns in the 1970–71 season (Fig. 75).

Bridled terns (*Sterna anaetheta*) selected the most sheltered nest sites, often shaded by vegetation, and in a natural crevice. Similar nest sites are described

Fig. 77. The sea eagle's nest located in the *Pandanus* grove in the centre of One-Tree Island.

in the literature, the recesses beneath *Pandanus* stilt roots, and tunnels in herbage being particularly favoured. However, on One-Tree Island in December 1927 the eggs of this species were 'often placed without any attempt at concealment on the rough shingle above high-tide mark' (MacGillivray 1928).

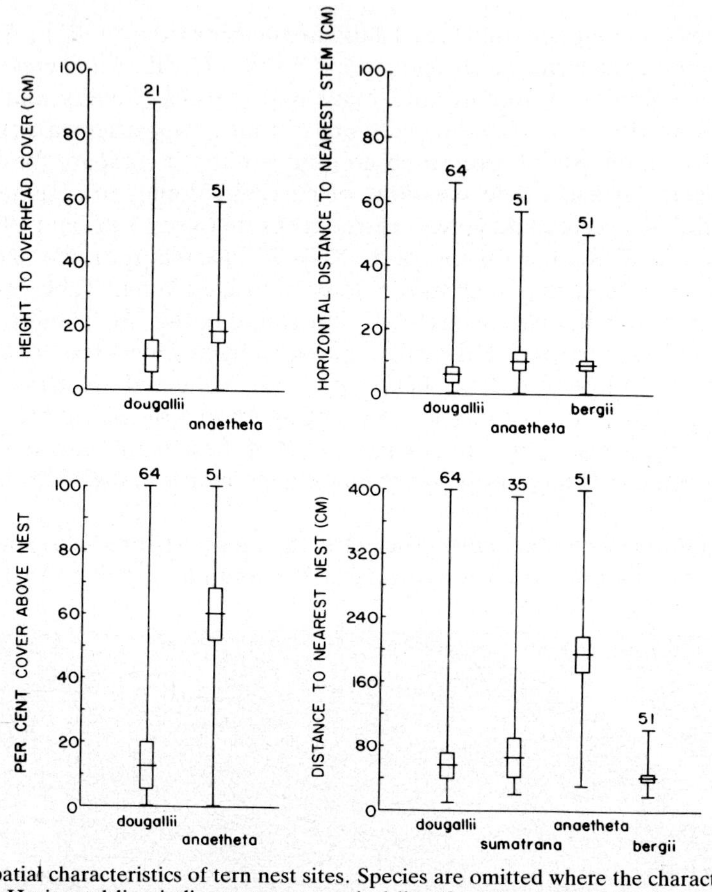

Fig. 78. Spatial characteristics of tern nest sites. Species are omitted where the characteristic did not apply. Horizontal line indicates mean, vertical line the range and rectangles two standard errors on either side of the mean. Numbers above symbols indicate sample sizes.

The various nesting sites of the crested tern (*S. bergii*) had in common an absence of overhead plant cover and a 'fine-grained' coral substrate overlaid with packed litter or succulent plants. Accounts in the literature refer to sand or grass substrates. Crested tern nest were the most closely grouped (Fig. 75); MacGillivray (1928) and Cooper (1948) estimated internest distances of some Capricorn colonies of crested terns to be 12″–18″ (30.5–45.7 cm).

Roseate nests (*S. dougallii*) were often situated amongst the herbs, but cover values were much lower than those of bridled tern nests (Fig. 78); the

Table 28. Characteristics of tern nest sites. Data collected November–December 1970.

Species	N	Location and vegetation type (% of nests)	% of nests with cover plant	Distance to sea (M) 0–10 (% of nests)	11–20	>20	% of nests on Fine substrate[a]	Medium	Coarse	Improvements
Sterna dougallii	64–66	Herb flat 83 Rubble storm-crest 17	72	17	17	66	1	38	61	Shallow depressions; approximately half the nests contained some plant material e.g. *Melanthera* and *Canavalia* sticks and leaves, apparently arranged.
S. sumatrana	35	Rubble storm-crest 100	0	100	0	0	0	50	50	Slight depressions. Nil.
S. anaetheta	43–80	Herb flat 79 Near *Argusia* shrubs 21	98	15	23	62	5	17	78	Slight depression, lined with finer rubble than surrounding substrate, often with accumulated plant debris in bottom.
S. bergii	50–51	Herb flat 37 Succulents 63	0	42	56	2	Packed soil and litter surrounded by rubble up to 25 cm diameter.			Stems and leaves of herbs, sometimes apparently arranged.

[a] Fine: 1–2 cm; Medium: 3.5–4 cm; Coarse: > 5 cm.

former species nested in discrete, dense colonies whilst nests of the latter species were scattered widely over the herb flat. Roseate tern nests described by MacGillivray (1926, 1928) and Gilbert (1925–27) were on shingle near vegetation, rocks or logs; 'all the sites that offered shelter from the . . . sun were availed of' (Gilbert 1925–27). Roseate terns trampled down the coarse grass on Bush Island (Torres Strait) to form depressions in which to nest (MacGillivray 1926).

Black-naped tern (*S. sumatrana*) nests were completely exposed and close to the water, being built on the storm crest above the beach. Gilbert (1925–27) and Hindwood *et al.* (1963) remarked that the shallow nesting depressions of Great Barrier Reef colonies of black-naped terns were often augmented with small pieces of coral or shell fragments, and substrates of sand and coral were used. The former author estimated the density of one colony to be 7 nests per square yard ($8.4/m^2$).

Differences in nest site characteristics of the terns are illustrated by the plant species growing closest to the nests (Fig. 79). Although bridled tern nests appeared to be scattered randomly over the herb flat, they were not associated randomly with plant species. The frequency of six forbs and two grasses on the herb flat (from releveé per cent cover estimates for November–December 1970) were compared with the frequency of their occurrence next to 39 bridled tern nests (the data illustrated in Fig. 79); the chi-square value was highly significant ($P < 0.01$). A higher proportion of nests were situated next to *Melanthera* than was expected on the basis of that plant's overall per cent cover value; likewise, more nests were close to *Canavalia* than was expected. It appears therefore, that bridled terns were selecting *Melanthera,* and to a lesser extent *Canavalia* plants as nest sites in preference to other species. Hawken (1972) has noted that bridled terns tended to nest 'under the vine-covered area' on Hoskyn Island as well as under bushes and *Pandanus* roots. *Abutilon* and *Plumbago* were the two species with low values of association with nests; these species are erect woody herbs or shrubs, whilst *Melanthera* and *Canavalia* have a weak-stemmed, semi-decumbent, vine-like habit. *Melanthera* and *Canavalia* were important herbs, in cover and biomass terms, for most of the two-year study of the herb flat.

Many roseate nests were close to *Lepturus* and *Ipomoea*, species which are early colonizers of disturbed and bare areas and behind the storm ridge. The nest characteristics reviewed above indicate that although terms which breed in north-eastern Australian waters exhibit considerable latitude in their nest site requirements, some marked species differences do exist. Similar variation is described for some other Australian sea birds (see especially Hindwood *et al.*, 1963); MacGillivray (1928) remarked that sites of nests of silver gulls in the Capricorn Group varied from island to island depending on the material and shelter available.

Reef herons built stick platforms in the *Pisonia* groves (Fig. 80) and occasionally in the shrub species (Table 29), generally in a vertical fork or on a horizontal branch. Materials were those readily available at the nest site, a nest in *Pisonia* containing a number of *Pisonia* sticks but few *Argusia* sticks, and vice versa. Herb species provided the bulk of nest material, however (Fig. 81). *Melanthera,* which was one of the dominant herbs on the island, occurred in all of 14 nests in which the plant material was identified.

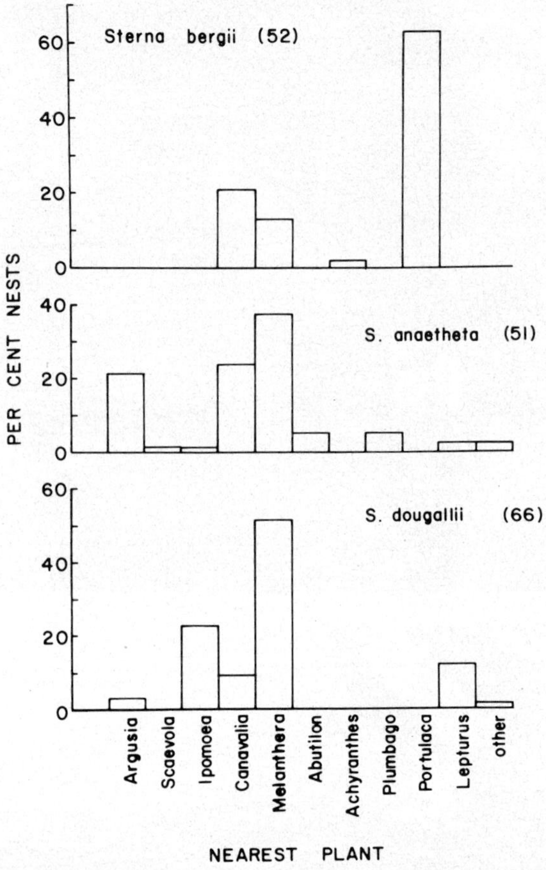

Fig. 79. Plant species associated with tern nest sites. The values are the percentage of total nests which were closest to the plant species indicated. Numbers in parentheses are the total number of nests. Data from November–December, 1970. Other = rock, log, or dead unidentified plant material.

The sea-eagle nest built on the ground on One-Tree was first described in 1843 by J. B. Jukes, naturalist on the 'Fly'. Figure 77 illustrates the appearance of the nest in 1970 and its dimensions are given in Table 29. The nest was not used during the period of the study, nor in the time that Domm and Recher recorded bird data on One-Tree Island (i.e. from 1968 to 1973). Nine other nests of the white-breasted sea-eagle have been described in the Capricorn Island group; data are summarized in Table 29.

Fig. 80. Nests of the reef heron, *Egretta sacra* on One-Tree Island. Top: adults (white phase) attending a nest in a *Pisonia* tree. Bottom: two chicks in nest.

Two empty silvereye nests were found and were made entirely of local materials (Table 29). Nests of the banded landrail, fantail warbler and sacred kingfisher were not found, although the former two species are known to breed on the island (Domm and Recher 1973, and personal observations of juvenile landrails). The kingfisher may not breed on One-Tree Island. MacGillivray (1928) described nest sites of this species on three other Capricorn islands as hollows bored in soft *Pisonia* tree trunks 2–10 metres above the ground, just inland from the beach. Such sites are available on One-Tree Island.

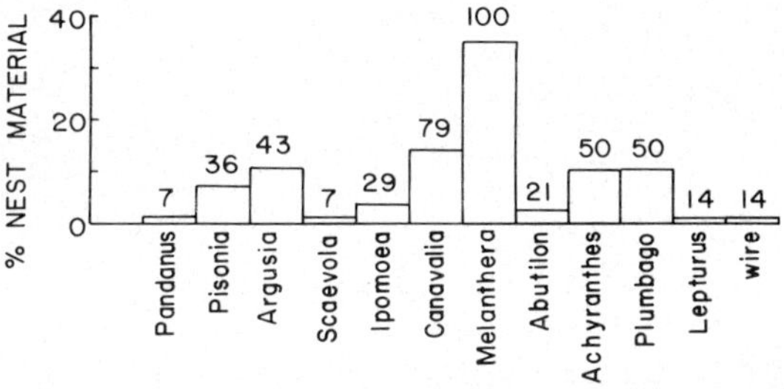

Fig. 81. Relative importance of plant species as materials in reef heron nests. Numbers above histograms indicate the per cent of nests which contained the indicated type of plant material.

Population changes and species composition

Domm and Recher (1973) published a detailed account of the One-Tree Island avifauna for the years 1971 and 1972 and October 1968, and Hulsman (1973) supplied a brief addendum of 1973 data. Table 30 lists the populations of birds which they recorded and our own observations for 1969 and 1970, May and July of 1968 and March of 1971. Records are thus available for six consecutive years. The minimum number of reef herons we counted (55 in May 1968) is almost certainly an underestimate of the population; unusual tidal conditions at the time of that census meant that the birds were less conspicuous than usual. Domm and Recher's estimate of the *Cisticola* population probably applies also to the 1968–70 period; continuous sampling activities in the herb flat made it difficult for us to observe these birds when they were not disturbed, and our estimate is probably low.

Apart from these two discrepancies between our data and those of Domm and Recher, numbers of resident birds in the two columns of Table 30 generally agree. There seemed to be a rather stable resident avifauna, although the numbers of breeding pairs of terns appeared to drop in 1971–72.

Hulsman (1976, 1977a) gives the numbers of terns and gulls on One-Tree Island during the breeding season (time of maximum numbers) for two seasons beyond the above data (1973–74, 1974–75) with a few notes on the 1975–76 season. *S. bergii* was at about the previous level (358 birds) in

Table 29. Characteristics of the nests of three resident bird species on One-Tree Island. Data are included from other islands of the Bunker Capricorn Group. Values in parentheses represent sample sizes.

Species	Plant in which nest was built	Mean % cover	Mean clearance (cm)	Distance to nearest other nest (cm)	Entire Nest: Mean height (cm)	Entire Nest: Mean diameter (cm)	Bowl of nest: Mean depth (cm)	Bowl of nest: Mean diameter (cm)	Height above ground (cm)	Construction materials
Egretta sacra One-Tree Island	*Pisonia* 72%, *Argusia* 19%, *Pandanus* 8%, *Scaevola* 2% (51)	71% (45)	54.4 (43)	50% > 5m 50% < 5m of which x̄ = 2.6m	14 (45)	31 (45)	—	—	x̄ = 1.8 (49)	See Fig. 81.
Other islands[a]									0.5–4.6	
Haliaethus leucogaster One-Tree Island	On ground among *Pandanus* (1)			No others on island	235	180	14	78	0	Mostly sticks (*Argusia, Pisonia, Pandanus,* bamboo) 0.3–1.4 m long and 1–3 cm in diameter; *Pandanus* leaves; 1 piece of sea fan. Lining: shredded *Pandanus* leaves; small sticks up to 10 cm long and 1.3 cm in diameter.
Other islands[b]	*Pisonia* 63%, *Casuarina* 37% (8)				69, 100, 120 210		10		9–13 (4) On ground	—
Zosterops lateralis	*Pisonia* (2)	100%	3.0			8.7		7.0	2.1	Mostly fibres of *Melanthera* and *Pandanus*; 23 feathers. Lining: rootlets, grass, *Pandanus* fibres.

[a] MacGillivray (1926), Domm and Recher (1973).
[b] MacGillivray (1926, 1928), Edgell (1927), Cooper (1928), Kikkawa (1970).

1973–74 but despite poor nesting success, was unusually numerous (864 birds) the following year. The numbers of *S. bengalensis* were much higher than the values in Table 30 for both years (240 and 180[1]) but included those on the small rubble cays on the outer edge of the reef which we were not able to count adequately in our study. *S. anaethetus* was slightly more numerous but of the same order of magnitude in 1973–74 and 1974–75 (500–1,000 and 500–800 respectively) as previously. Gull numbers did not change appreciably. *S. dougalli* showed a considerable decline to 80 birds in 1973–74 but was at full numbers again the following year; *S. sumatrana* showed a similar pattern. It would appear that in general there were fewer terns during the 1973–74 nesting season than was usual for previous years or were present in 1974–75.

The causes of lack of nesting success were mainly predation by silver gulls (up to 83%) and flooding of nest areas (up to 51%). In the 1975–76 season, a severe storm ('Cyclone David') prevented adults from foraging for 3 days and heavy mortality of chicks through starvation occurred. Only 20 survived. Hulsman (1977a) concluded that the breeding success of 2 species, *S. sumatrana* and *S. dougalli* were too low to maintain their present size at One-Tree Island. At the present time it is not known how much the size of the breeding colony for any one year reflects nesting success on the island and how much it reflects year to year differences in movement of adults from one breeding area to another.

Seasonal fluctuation in bird numbers was great, varying by a factor of 10. The mid-winter population of residents averaged about 300, whilst summer residents numbered close to 3,000. Intra-Australasian migrants were most commonly seen in September and October (when mainland records indicate they are moving south to breed) (see Table 31). Of the remaining six land birds which do not have a documented regular migration, the magpie lark (*Grallina cyanoleuca*) and the common starling (*Sturnus vulgaris*) were the most frequent visitors to the island. The former was recorded on five different occasions in two years, and the latter appeared three times in as many years. Neither species remained on the island for more than a week or so, and neither is a resident of other Capricorn islands. No successful establishment of new bird species was witnessed during the study.

Of the 84 species of birds observed at One-Tree Island, only three species (a maximum of about 46 individuals), were entirely dependent upon the island for all resource requirements. These were *Rallus, Cisticola* and *Zosterops,* and even then, exchange of individuals occurs between the One-Tree and Heron Island populations of *Zosterops* (Domm and Recher 1973, quoting Kikkawa). We observed silvereyes on One-Tree which had been banded on Heron. Kikkawa mist-netted 14 silvereyes on One-Tree Island in October 1972, 3 of which had been banded on Heron Island.

[1] Misprinted as 80 in Hulsman's (1976) paper.

Table 30. Avian populations on One-Tree Island.

Location and taxon	Domm & Recher (1973) Min.		Max.	Present study Min.	Max.
I. Residents					
A. Occupying vegetated part of island					
Phalacrocorax melanoleucos	10		25	0	34
Egretta sacra	200		275	55*	284
Haliaetus leucogaster		≥1		1	2
Rallus philippensis		~4		1	4
Larus novaehollandiae	25		60	14	101
Sterna bergii	10		300	12	460
Anous minutus		150		1	104
Halcyon sancta	1		2	1	3
Cisticola exilis	10		12	1*	8
Zosterops lateralis	6		14	7	30
	417		843	93	1,030
B. Periphery of island and reef crest					
Sula leucogaster		~6		1	22
Haematopus fuliginosus		≥2		1	3
Sterna sumatrana	10		250	6	310
S. bengalensis		20		23	26
Waders (Charadriidae and Scolopacidae)	20		80	21	102
	58		358	52	463
II. Regular summer visitors					
A. 8 Waders (Charadriidae and Scolopacidae)	60		600	68	615
B. Terns					
Sterna dougalli	200		300	100	680
S. anaetheta	300		500	375	650
	560		1,400	543	1,945
III. Irregular visitors					
A. Aquatic Feeders and Raptors					
Diomedea exulans					1
Puffinus pacificus			>1		2
Pelecanus conspicillatus					5
Sula dactylatra					1
Phalacrocorax sulcirostris			12		27
P. varius			6		
Fregata ariel			≥1		7
Phaethon rubricauda			1		
Ardea novaehollandiae			10		
Cygnus atratus			1		
Accipiter novaehollandiae			1		
Falco peregrinus			1		
Haematopus ostralegus			2		
9 Waders (Charadriidae and Scolopacidae)			4–18		?
Sterna albifrons			105		2
B. Land Birds (Mostly Passerines)					
Chrysococcyx plagosus			1		1
Eudynamys scolopacea			1		
Hirundapus caudacutus			1		
Eurystomus orientalis			2		1
Hirundo neoxena					2
Coracina novaehollandiae			1		2
Lalage sueurii					2
Rhipidura fuliginosa					2

Table 30. (continued).

Location and taxon	Domm & Recher (1973) Min.	Max.	Present study Min.	Max.
Myiagra rubecula		1		
M. cyanoleuca		1		2
Monarcha frater		1		
M. leucotis		1		
Pachycephala rufiventris		1		1
Meliphaga fasciogularis		1		1?
Passer domesticus		1		
Sturnus vulgaris		1		1
Dicrurus bracteatus		1		1
Grallina cyanoleuca		1		2

* Values probably unrealistically low, see text.

J. Chilvers (pers. comm.) has supplied recent data on the landrails. In 1973 when she first took up residence on the island she believed that there were only a few pairs on the island. She fed them regularly and by 1978 the population had increased to an estimated 70 birds. The caretaker who took up residence in 1979 did not artificially support these birds and the population appears to be declining again (Sale, pers. comm.).

Species numbers

Table 27 compares the avifauna of One-Tree Island with that of its nearest neighbour, Heron Island. No doubt some transient species which visit these islands have been missed because of the discontinuity of ornithological records. Heron Island, which is three times the size of One-Tree, recorded 10 fewer species than the latter, but observations on both islands have been spasmodic. The species composition differed more than the table suggests. Of 11 species which have been recorded breeding at Heron, only six were common to the 12 breeding species of One-Tree.

Differences in substrate and vegetation of the two islands probably account for much of the difference in resident avifaunas. The rubble substrate of One-Tree Island was unsuitable for shearwater nesting burrows and we suggest that the low stunted habit and closed canopy growth of *Pisonia* there, do not provide suitable nest sites for the noddies, which nest elsewhere in the Capricorns, in tall open *Pisonia* forests. On the other hand, the populations of nesting *Sterna* species were much larger than on Heron. As Table 28 shows, some of these terns preferred open nesting situations and much of the undeveloped portion of Heron Island is forested. Human disturbance would have a major effect on ground nesting birds on Heron Island, and in particular on beach-nesting species.

Table 31. Frequency of sightings of intra-Australasian migrants (1927*, 1968–1973).

Species	Number of years in which seen						
	J–F	M–A	M–J	J–A	S–O	N–D	Total
Chrysococcyx spp.			1		2		3
Eudynamys scolopacea					1		1
Hirundapus caudacutus						1	1
Eurystomus orientalis		1			3		4
Hirundo neoxena		2					2
Coracina novaehollandiae			1	1	2		4
Lalage sueurii			1		2		3
Myiagra rubecula					1		1
M. cynaoleuca	1		1		2		4
Monarcha frater					1		1
Pachycephala rufiventris		1			1		2
Dicrurus bracteatus			2		1		3
	1	4	6	1	16	1	29

* MacGillivray (1928). Observations made on 7–8 December 1927.

Breeding land birds not shared by the two islands were the fantail-warbler (only on One-Tree) which is a species of 'long grass and other tangled vegetation' (Slater 1974) and the bar-shouldered dove, *Geopelia humeralis* (only Heron Island) which is a species of 'dense vegetation . . . along seashores' and open forest, among other habitats (Slater 1970). The distribution of these birds on the two islands is probably determined largely by the differences in vegetative cover.

The species composition and variety of birds on One-Tree Island may be influenced more by the topography and vegetation of the island than by its size or distance from the mainland. Up until 1967 human influence on the island was slight, but since then, activities by scientific parties have probably had a deleterious effect on nesting success, or even whether nesting occurred at all (e.g. sea eagle).

Trophic relations

The important role of the sea birds and waders of coral cays in transferring marine-derived energy and nutrients to the island system has been noted by Davis *et al.* (1938), Airy-Shaw (1952), Heatwole (1971) and Domm and Recher (1973) and is considered in Chapter 20.

In addition to this role the gulls probably had a direct effect on the terrestrial flora and fauna. Gulls eat a variety of seeds and fruits (Lea and Gray 1935a) and are effective carriers of seeds in their crop, especially annuals (Gillham 1961). It is likely that they have assisted the land birds in bringing some of the plants to One-Tree Island. The almost certain role of the silver gull in the immigration of *Solanum* to One-Tree Island and its subsequent spread throughout the island is detailed in Chapter 18.

Of less widely realized significance is the insectivorous tendencies of silver gulls. Their stomachs often contain insects (Lea and Gray 1935a), and they have been observed hawking insects on various occasions (Wheeler and Watson 1963, Green 1966). One such observation occurred on One-Tree Island. In February 1974, Hulsman (1974b) observed mantids being assisted in their flight by a 12 knot westerly wind, and moving across the One-Tree lagoon from the direction of the mainland toward the island. He made his observations from 0743 hrs to 1900 hrs during which time mantids arrived singly, flying at a height of 6–9 metres. Gulls ate these insects, either catching them directly in the air, or in the attempt, knocking them into the water and catching them there. A number of insects were involved as Hulsman indicated that 20 of 22 attempts to catch them were successful.

These observations are important, because they indicate that silver gulls may become directly involved in terrestrial chains by hawking single, large conspicuous insects and that they can constitute a significant risk for a potential immigrant while it is still over the water and near its destination; gulls may thereby influence immigration rates. Arrival may well be the most hazardous part of the trip for large, slowly flying insects.

These observations also indicate that under appropriate weather conditions, a number of individuals of the same species may arrive at an island nearly simultaneously, thereby improving chances of establishment. It is not known how many mantids escaped the attention of the gulls. During our study we only observed this group on One-Tree Island once and then only one individual was encountered.

On One-Tree Island, most of the visiting land birds were insectivorous and were observed feeding during their stay. A bronze cuckoo (*Chrysococcyx* sp.), grey fantail (*Rhipidura fuliginosa*) and rufous whistler (*Pachycephala rufiventris*) ate *Utetheisa* caterpillars and/or moths, and a black-faced cuckoo-shrike (*Coracina novaehollandiae*) was seen to hawk a locust once. None of these birds were seen for more than a few days at a time, and their visits were concentrated in a few months of the year. Only a few individuals at a time were involved. They are therefore not important predators in the system.

Utetheisa moths were eaten by the fantail-warblers which presumably consumed a wide variety of herb flat arthropods; literature indicates they are primarily insectivorus (Lea and Gray 1935c, Cayley 1946).

Silvereyes eat a variety of insects, spiders, seeds and fruits (Lea and Gray 1935d). They caught insects on the foliage and flowers of *Pisonia* and *Argusia* and on herbs. They also ate *Solanum* berries, *Pisonia* flowers, and *Stenotaphrum* seeds, and drank nectar from *Abutilon* flowers. Wyatt (pers. comm.) listed a wide variety of insects (aphids to katydids and ants), fruits of *Scaevola* and *Argusia*, and perhaps nectar of *Pisonia* and *Scaevola* flowers, as some of their food on Heron Island. The silver gull was another important consumer of *Solanum* fruits on One-Tree Island. The role of these two species in the dispersal of *Solanum* over the island has been discussed elsewhere (see Chapter 18). Silvereyes probably disperse a variety of plants (Gannon 1934).

The landrail often fed around the pond; according to Lea and Gray (1935a) and Cayley (1946) this species normally feeds on insects, freshwater molluscs, seeds and other vegetable matter. J. Chilvers (pers. comm.) observed it eating *Solanum* fruits. The sacred kingfishers fed mainly on the shore and out on the exposed reef at low tide, so it is assumed they resembled the sea-birds and herons in their trophic position, although Lea and Gray (1935b) indicated this species also takes insects (Orthoptera, Coleoptera).

The remaining birds which fed in the interior of the island, concentrated their activity on the succulent mat around the pond, and in the pond itself. Reef herons, several wader species, silver gulls and magpie-larks (*Grallina cyanoleuca*) were observed there at some time. We collected small snails from the pond, and mosquito larvae occurred sporadically (Chapter 19); these were probably some of the food items sought by the birds.

The sea-eagles' perch was checked each visit for remains of their prey. These food items are summarized in Table 32. The biomass of food introduced by the eagles' predation alone, was considerable. Fish and sea-birds composed the bulk of the material. MacGillivray (1928) lists sea snakes and fish below a nest on Lady Musgrave Island and sea snakes, fish and birds below the one on One-Tree Island during a visit in 1927. Gilbert (1925–27) reported sea eagles preying on diseased and weakened wedge-tailed shearwaters, and mentioned remains of birds and fish around a nest on North-West Island. Thus, the observation by Campbell and White (1910) that sea eagles 'did not seem to prey on the birds of the locality [Capricorn Group], but on sea-snakes and fish' is only correct in part.

Sea turtles

The green turtle (*Chelonia mydas*) and the loggerhead (*Caretta caretta*) both nest throughout the Capricorn–Bunker Islands and inhabit the waters of the surrounding reefs; Wreck Island is the main breeding site in the area for the loggerhead (Great Barrier Reef Committee 1978). Other species of sea turtles inhabit the Bunker–Capricorn waters but do not nest in the area. Turtles are quite abundant and indeed, in the first three decades of this century, a canning factory for turtle soup was in operation 3–4 months per year on Northwest Island (Musgrave 1925–27a, Nebe 1928, Limpus 1978); one was also reported to be on Heron Island (Moorehouse 1933, Cooper 1948), perhaps erroneously.

Most of the islands in the area have sandy beaches and the literature indicates that the loggerhead habitually nests in sand (McAllister *et al.* 1965, Hughes 1974a, b). Caldwell (1959) indicated that physical features of the

Table 32. Food items collected from below the Sea-Eagles' perch

	Pre M–J 1969	M–J 1969–F–M 1971
Mollusca		
Cephalopoda (cuttle fish)	1	
Crustacea		
Decapoda (crab)	1	
Pisces		
Acanthuridae (surgeon fish)		1
Balistidae (*Balistapus undulatus,* trigger fish)		1
Carangidae (*Trachinotus*, dart)		1
Diodontidae	1	
Fistuleriidae (*Fistularia petimba*)	1	
Hemiramphidae (half-beaks)		2
Labridae (*Choerodon* spp. + others, tusk fish, wrasse, etc.)	4	9
Muraenidae (reef eels)		1
Scaridae (parrot fish)	1	3
Scombridae (mackerel)		4
Serranidae (sea basses and rock cod)		1
Siganidae (rabbit fish, *Siganus lineatus*)		1
Sparidae (sea bream)	1	
Tetraodontidae (toad fish)		1
Tylosuridae (*Tylosurus strongylurus*)	1	11
Indeterminate species	2	3
Reptilia		
Chelonidae	1	
Hydrophiidae	1	1
	15	*40*
Aves		
Puffinus pacificus	1	2
Phalacrocorax sp.		4
Egretta sacra	2	7
Larus novaehollandiae		1
Sterna anaetheta		1
S. sumatrana		4
Sterna spp.	1	15
Anous minutus		1
Eurystomus orientalis (immature)		1
Indeterminate species		1
	4	*37*
	19	**77**
Total birds	**41**	

beach are the most important factors determining nesting activity in loggerheads. He gives the preferred nesting habitat as high sandy beaches backed by rounded dunes or vegetation. A few exceptions are known, however. Carr (pers. comm.) indicates that loggerheads 'often dig in shell fragments, . . . and wherever pebble, cobble, or shingle beach adjoins sand you may find a certain amount of straying on to stony places . . . probably most nests dug in rocks are abandoned without receiving eggs'. One-Tree Island with its beaches of coarse rubble would not seem to be a likely place for this species to nest, yet it does so at least in small numbers (Fig. 82).

Nesting success of loggerheads on One-Tree Island was low in November 1970, the only time during our study that we directly observed turtles nesting there. One female (recognized by paint marks placed on her at the first emergence) was observed emerging on land 18 times, during which she made 13 attempts to nest; no eggs were deposited (it is common for turtles that are unsuccessful in nesting to return repeatedly to the beach; Caldwell *et al.* 1959b). On one occasion she made 4 unsuccessful attempts to dig a nest cavity, destroying each before finishing. On the six occasions when the total time the turtle was out of the water was known, she was on land from 27 to 115 minutes (mean = 69 minutes). A second female emerged once and successfully laid 140 eggs, breaking 3 while covering them up with rubble. The site she selected had been previously excavated by the other turtle. She spent 100 minutes ashore.

Both turtles had difficulty in nesting in the rubble. The rubble tended to cave in on the egg chamber; often only a token depression was excavated as a body pit. Limpus (pers. comm.) indicates that 'the depth of a body pit is a function of the type of substrate . . . it varies according to the moisture content/type of sand/grass/type of sand/root content etc.'

It seemed to be difficult for the turtles to loosen the rubble in order to remove it. Two turtles on One-Tree digging in rubble respectively required 5–8 scrapes (mean = 6.5; N =11) and 3.5 scrapes (mean = 3.5; N = 10) of the flipper to loosen enough rubble to get a flipper-load of material to throw out of the cavity. The turtle with the lower number of scrapes was digging at a site previously excavated and loosened by another turtle. Comparable values for two loggerheads digging in sand on Heron Island were 1–2 scrapes per flipper-load (means =1.5, 2.0; N's = 10). On two occasions a turtle on One-Tree had a hind flipper injured and bleeding as a result of attempts to dig in rubble. Other than that the flippers of the One-Tree turtles appeared normal in all respects.

Caldwell (1959) indicated conditions affecting digging as responsible for the abandonment of primary excavations without ovipositing. They were: sand too dry or soft (resulted in cave-ins), sand packed too hard, layers of oyster shells, and an abundance of tough plant roots. He reports a case of a female digging egg holes without laying eggs. On One-Tree Island the substrate not only was difficult to dig but also resulted in cave-ins.

Another feature of the One-Tree loggerheads was that a number of emergences were in daylight. Carr (1967) states that with a few exceptions, virtually all marine turtles nest at night and Caldwell *et al.* (1959a) indicate that most loggerhead activity takes place 4–5 hours after dusk and report a few instances to the contrary as exceptional. Hughes (1974b) indicates that daylight emergences are rare in the loggerhead. However, Bustard and Greenham (1968), Bustard *et al.* (1975) and Bustard (1979) indicate that

Barrier Reef *Caretta caretta* usually avoid the heat of the day, but may nest in the late afternoon if the tide is high. Limpus (pers. comm.) indicates that a small proportion of loggerheads will attempt to nest at midday on Heron Island if tides are suitable.

On One-Tree Island 18 of 19 emergences (95%) took place at least partly in daylight. Four of these attempts may have started before or near dawn but many were near midday. There were high tides during the day during our observation periods. Limpus (pers. comm.) indicates that 'daytime beaching is often associated with previously unsuccessful [nesting] attempts'. Parmenter (pers. comm.) has made similar observations.

The attempts to nest on One-Tree Island were not unique to the year 1970. We observed 89 excavations that appeared to be made by sea turtles. Of those, 58 (65%) had the coral in the bottom of the pit discoloured like surface coral elsewhere, and were at least a year old. The remaining 31 (35%) had coral in the bottom of the pit which had not yet become darkened by exposure and were probably of the 1970 season. Only two (3.5%) of the old nests and 3 (9.7%) of the new ones were judged to have been possibly successful (error in making such assessments is probably considerable).

Observations by Ted and June Chilvers (pers. comm.) indicate nesting of loggerheads on One-Tree Island in 1974. One turtle laid eggs at 0900 hours on 21 January 1974 in a pit excavated by the turtle and assisted by the Chilvers. What was perhaps the same turtle (same pattern of carapace barnacles) returned on 9 February 1974 and nested again about 3 metres from the previous site. The first nest yielded 80 hatchlings on 8 April 1974 and the second one 48 hatchlings on 6–18 May 1974. This result indicates that conditions in rubble can be suitable for development of sea turtle eggs. However, it is not known whether hatchlings could dig out of the nest chamber unassisted, because the Chilvers helped dig them out. The 'incubation' times were longer (77–98 days) than those reported for this species by Caldwell (1959) (mean = 55 days, range 49–62 days) and the young may have been imprisoned in the nest chamber.

Loggerheads usually (Caldwell *et al.* 1959b) but not always (Bustard and Limpus 1971, Le Buff 1974) return to the same beaches repeatedly to nest. It is possible that the nestings on One-Tree are only of a few individuals returning repeatedly and at different years to nest; such animals may have hatched on One-Tree themselves. Alternatively they may represent 'strays' from populations normally nesting on other Capricorn or Bunker Islands.

Nesting primarily occurred on the point of land on the northwestern corner of the island and the shores of the crescent-shaped bay along the northern edge. A few nests were on the north-western point. Only one other nest was found elsewhere; it was on the western edge of the island (Fig. 83). The almost exclusive use of the northern part of the island for nesting is probably related to the fact that it has easy access from the lagoon. All other areas require traversing extensive areas of reef flat in order to reach the beach.

Fig. 82. Loggerheads (*Caretta caretta*) ashore on One-Tree Island. Note coarse texture of the coral rubble. Upper left: emerging on to beach. Upper right: beginning the body pit. Lower left:

digging the egg chamber. Lower right: nesting under an *Argusia* shrub; the egg chamber is being dug; its subsequently filled with eggs. Two of the photos courtesy of David Vleck.

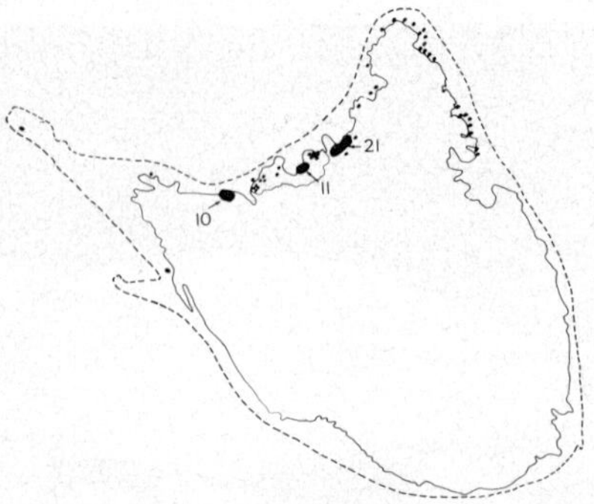

Fig. 83. Map of the locations of nest excavations of sea turtles (successful and unsuccessful) on One-Tree Island. Data from 1970 (old and recent nests) and 1974 (nests with eggs). Single dots indicate location of individual nests. Larger black areas indicate sites of a number of nests, the number of which appears on the map with an arrow pointing to the location. The dotted line indicates the outline of One-Tree Island. The solid line indicates the extent of the vegetated area. The smaller area along the bay on the northern edge enclosed by a solid line indicates the patch of *Argusia* shrubs under which a number of nests occurred.

All of the observed nesting on One-Tree was by loggerheads (only one green turtle was seen ashore and there was no evidence of it have attempted to dig) and consequently it is probable that most, if not all, of the nest cavities mapped on One-Tree are of loggerheads.

Most of the nests on the northwestern point were well above the high tide line but outside the heavily vegetated zone. By contrast most of those near the bay were in the vegetation and even under *Argusia* shrubs (Fig. 83). Table 33 summarizes the characteristics of the nests observed in 1970. Low incidence of nesting is emphasized and frequent attempts at nesting in or near vegetation is evident.

Table 33. Characteristics of nest excavations (successful and unsuccessful) observed on One-Tree Island in November 1970. Values are means with ranges in parentheses.

Characteristic	Old nest sites (dug prior to 1970 season) N = 58	Recent nest sites (probably 1970 season) N = 7	New nest sites (freshly dug, certainly 1970 season) N = 24
Distance to nearest vegetation (m)	0.6 (0–5)	2.9 (0–10)	22 sites in or next to vegetation. Others 1–2 m distant.
Distance to high tide line (m)	8.3 (3–26)	16.0 (3–22)	8.1 (2.5–25)
Altitude above high tide line (m)	2.1 (1.5–2.5)	1.6 (1–2)	1.6 (1–2.5)
Diameter of body pit (m)	—	1.2 (1–1.6)	1.1 (1–1.6)
Depth of body pit (cm)	—	15 (10–25)	14 (5–25)
Depth of egg chamber (cm)	—	5	14
Number of sites with egg chamber	—	2	14
Probable successful sites	?	1?	1, perhaps 2

The latter characteristic undoubtedly has an ecological effect on the island. Digging a nest cavity can be destructive of vegetation through ripping out or breaking low plants and exposing the roots of trees or shrubs. The nesting attempts we observed sometimes resulted in uprooting of herbaceous vegetation and in one instance branches of *Argusia* were broken off for a distance of 3 metres.

Plants may have an effect on nesting loggerheads. Caldwell (1959) reports that 5.3% of the eggs laid among *Uniola paniculata* were destroyed by its roots. The reasons were: 'The hair roots formed thick mats around the individual eggs, eroded the shells, and desiccated them; often the sharp-pointed stolons pierced the eggs. A small number of the turtles that pipped their eggs were unable to completely escape the shell and died. Still another small percentage escaped from the shell successfully but were caught in the nest and died. This was caused by their inability to climb through the tightly packed deposit of hatched eggs or matted vegetation roots, or their tendency to burrow horizontally into hard sand instead of perpendicularly to the surface.' Similar observations have been made on Great Barrier Reef islands (J. Miller pers. comm.).

Another effect the turtles have on the island is that the hatchlings may be briefly available to local predatory birds (such as gulls) on their way from the nest to the sea. Sand crabs are known to be predators of turtle hatchlings (Caldwell 1959); on One-Tree with its rubble beaches, however, these are absent.

18. Selected case histories

The most important species, especially the plants, birds, insects and soil arthropods, have been treated in the various preceding chapters. Many of the epigeal arthropod species were transient or rare and had relatively minimal impact on the system and have appeared in the text only as contributing to the total numbers of species or individuals. Others were important at least temporarily, and yet it is beyond the scope of this book to treat each in detail. However, certain species or other taxonomic groups illustrate particular points especially well, or were of particular interest and this chapter presents a more detailed treatment of those than could be afforded to every species.

The Gall-Fly (Diptera, Tephritidae, Tephritinae; *gen. et sp. nov.*)

During the pre-drought sampling of 1968 it was observed that the *Melanthera* was infested with a stem or petiole gall containing a dipteran larva or pupa. Sample plots in the herb flat yielded mean estimates of 16.6 galls/m^2 and 29.5 galls/m^2 in May 1968 and July 1968 respectively[1]. In one plot 23 of the 24 galls were empty (adult emerged) and one had a dead pupa. In the other plot 30 galls were empty, 1 had a dead pupa and 4 had live pupae.

An independent sample of 61 green galls consisted of 23% with live larvae, 61% with pupae (33 live, 1 dead, and 3 parasitized by a hymenopteran), 13% with the adult fly already emerged, and 3% with adults in the process of emerging. Larvae ranged from 1.1 mg to 13.0 mg dry weight, and pupae averaged 11.3 ± 0.36 mg dry weight (range = 8.4 mg to 15.3 mg), with 3 empty pupal cases having a mean weight of 0.8 mg. Two adult flies averaged 8.1 mg dry weight.

The frass produced by a growing larva accumulated in the gall and on the average its dry weight was 119% of the larval dry weight. Any gall with a pupa contained the entire fecal production of the previous stages. These were

[1] Similar galls were noted on *Melanthera* on Heron Island during the same periods, as well as later.

collected from 37 galls and weighed. The total frass production by the larval stages of the gall fly averaged 7.2 mg dry weight or 64% of the dry weight of the pupa, or 89% of that of the emergent adult.

Adults were abundant in May and July 1968 and it was anticipated that a bioenergetic study would be carried out on this species. However, by the next sampling period at One-Tree (January–February 1969) the drought had intervened and the host plant had been reduced in biomass. Neither adults nor new galls of the gall-fly were found, nor were any subsequently observed at any time throughout 1969 or in 1970 until November–December. At that sampling, adults were present in low numbers ($0.01/m^2$); no galls were obtained in the sample plots and only two were observed anywhere on the island. Both were brown and on the same *Melanthera* plant; one had a dead pupa and the other had become occupied by ants. No galls were observed in the single sampling period of 1971. By 1972 green galls were found in 4 of the 10 sample plots for a mean density of $2.3/m^2$ (range = $0–10.4/m^2$) although adult density was still estimated at $0.01/m^2$. In 1976 galls were extremely abundant; all sample plots had large numbers and the density ranged from 46.8 to 181.8 per m^2 (mean = $108.8/m^2$). Clearly the species had become well established again. By 1978 it had again declined somewhat (there had been another drought between 1976 and 1978) and only half of the sample plots contained galls. The mean density was only $4.7/m^2$ (range = $0–13.0/m^2$).

We interpret these results to indicate extinction of a dense population of gall flies by the drought, an absence of the species from the island for almost 2 years, followed by immigration and re-establishment.

The Locust, *Valanga irregularis*, and Grasshoppers

It has been mentioned briefly already that the locust, *Valanga irregularis*, was extremely abundant during the pre-drought 1968 period and was responsible for considerable defoliation, especially of *Pisonia* (Fig. 84). Key (1970) has noted that this species feeds on the leaves of trees and shrubs in the moister tropics and subtropics and as a result may be of local horticultural importance. Estimated mean population density of flying adults (by the tape-dragging method) was 80 per hectare for the open herb flat and 50 per hectare for the sparser ecotonal parts of the herb flat away from arborescent vegetation. This represents an underestimate as the locusts roosted in shrubs and trees at night and were most abundant in and near arborescent vegetation even during the day; *Pisonia* leaves seem to be a favourite food. Consequently, most of the population was not amenable to the methods used.

Following the drought (February 1969) the value had dropped to 10 per hectare and by May 1969 no locusts were found either by dragging the tape or during the releveé sampling along the transect lines in the herb flat. An intensive search of the island for locusts revealed only 7 adults, 3 near the

Pandanus grove and 4 in the herb flat. There were also 2 juvenile locusts and 2 adult grasshoppers observed. Thus, the population had reached extremely low levels. Thereafter, for the next 2 times (July–August 1969, September–October 1969) tape-dragging did not yield any locusts and intensive searches near arborescent vegetation revealed only 9 and 5 locusts and grasshoppers collectively for the 2 respective dates. The populations seemed to have picked up subsequently as searches near trees and shrubs for November–December

Fig. 84. Pisonia foliage showing damage by the locust, *Valanga irregularis*.

1969 yielded 24 locusts and grasshoppers and January–February 1970 yielded 30. Tape-dragging in the herb flat on both occasions gave an estimate of 20 per hectare.

In March–April, 1970, 41 locusts and grasshoppers were observed in the vicinity of *Argusia* shrubs, 36 near *Pisonia* or *Pandanus* clugs, and 25 while running the releveés (total 102), and the population seemed to be increasing. However, by the next visit (May–June 1970) only 12 locusts and grasshoppers were seen near *Argusia*, 3 near *Pisonia* and 5 near the herb flat releveés (total 20). The population seemed to have declined again, a suggestion supported by the finding of 4 dead adult grasshoppers, each at a different site (2 in herb flat, 1 in *Pandanus* and 1 in *Pisonia*). This was the only time that grasshopper corpses were ever found on the island.

In July–August 1970 none were observed during the tape-drag and only 15 at all sites (3 near *Argusia*, 11 near *Pandanus* or *Pisonia*, and 1 in the open).

The last two sampling periods of 1970 yielded no orthopterans by dragging the tape and only a total of 6 and 5 observed at all sites for September–October and November–December, respectively.

In early 1971 only 10 were observed, but most of these were juveniles suggesting that maybe the population was on the upturn.

The above data indicate that the populations were very high in 1968 before the drought but declined to very low numbers throughout the rest of the intensive study, with a slight temporary increase in late 1969–early 1970. The original high density was almost entirely due to *Valanga irregularis* as other species were not seen while dragging the tape. Once the locust population crashed, however, it did not recover during the intensive part of the study. Adult locusts observed on each occasion numbered only 3–4 or fewer, the remainder of the count being made up of other species; the temporary increase beginning in late 1969 was entirely due to species other than the locusts. It is possible that *Valanga* may even have become locally extinct for a brief period.

After the intensive part of the study, locusts did reach a high population density again at least once. The Chilvers (pers. comm.) indicated a 'plague' in 1973 when large numbers came into the research station buildings.

In 1978, the population was estimated to be 31 per hectare in the herb flat, the highest we had directly observed since the 1968 decline and may have been building up toward another 'plague'.

Musgrave (1925–27a) mentions that in 1925 grasshoppers were attacking the leaves of *Argusia* on North-West Islet and had nearly defoliated some of the trees.

In conclusion, grasshopper populations were generally low but that of the locust *Valanga irregularis,* fluctuated markedly; peak densities occurred at about 5-year intervals (1968, 1973, 1978). At the highest density (1968) the estimated number on One-Tree Island would have been 238 adult *Valanga* in the herb flat plus an additional large number around the shrubs and trees. Feeding experiments with 20 individual *Valanga* on Heron Island indicated that adults consumed an average of 0.92 g dry weight of *Argusia* leaves per day and presumably values would be similar for other plant species, or perhaps even more for more preferred foods. Thus, at peak densities *Valanga* would consume about 213 g dry weight of foliage in the herb flat per day and a large but unknown amount of tree and shrub foliage.

The Moth, *Utetheisa pulchelloides*

The moth, *Utetheisa pulchelloides* feeds on *Argusia argentea* and on One-Tree seems to do so exclusively, as it was never seen on any other plant. The young larvae seem to prefer buds whereas the larger caterpillars consume mostly foliage. The adult feeds on flower nectar. Woodell (1979) has suggested a pollinating role of the moth, although *Argusia* is pollinated by a variety of other insects as well.

The larvae were sampled as part of the foliage fauna (see Chapter 14) but the adults were not amenable to any of the sampling methods previously described. Unlike most other species they did not appear in the quadrats of the foliage in proportion to their true abundance. They would fly up from the shrubs at the approach of a person and not settle again until the investigator had gone some distance away. Consequently a special census method for sampling had to be devised (see Chapter 7)[2].

The size structure of the larval component of the population was ascertained by measuring the length of the first 50 individuals encountered during a search of *Argusia* foliage, using those shrubs on the northeastern corner of the island. On some occasions when larvae were not abundant, fewer than 50

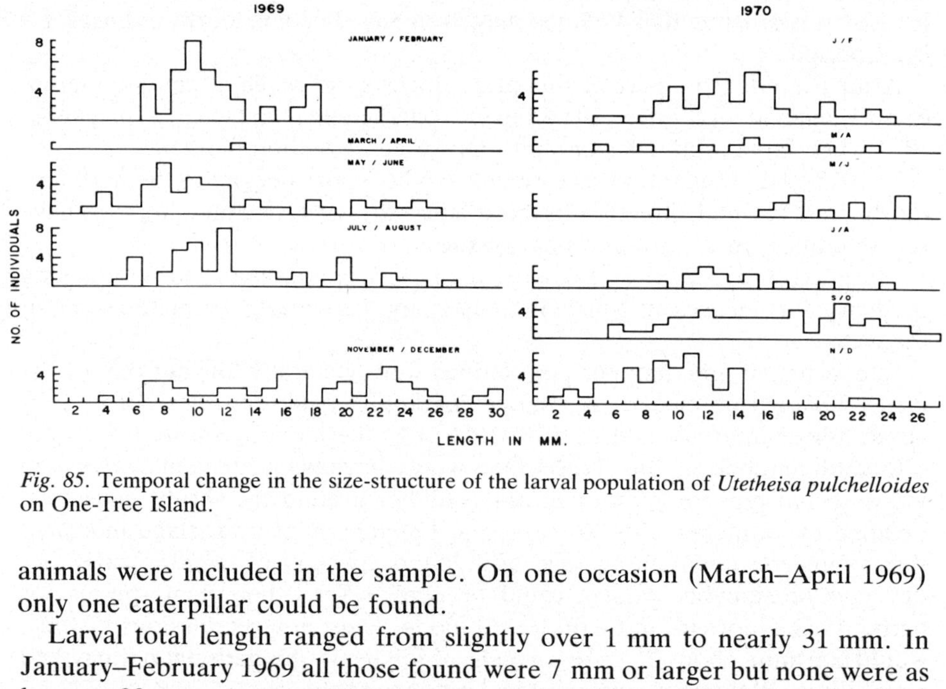

Fig. 85. Temporal change in the size-structure of the larval population of *Utetheisa pulchelloides* on One-Tree Island.

animals were included in the sample. On one occasion (March–April 1969) only one caterpillar could be found.

Larval total length ranged from slightly over 1 mm to nearly 31 mm. In January–February 1969 all those found were 7 mm or larger but none were as large as 23 mm; most were between 7 mm and 13 mm (Fig. 85). In June the distribution was similar but there were some larger ones, and some very small ones had also appeared. The situation hadn't changed markedly by August. No measurements were made in September–October. By early December 1969 (the time of the peak in adult population density; Table 34), the size frequency was relatively uniform throughout. There were some individuals of maximum size. The year 1970 was very different. The January–February population had larger individuals than for the same month of the previous

[2] An indication of the extent of underestimation of adult *Utetheisa* by the foliage quadrat method can be gleaned from comparing 2 methods at the peak of adult population density (November–December 1969). Foliage quadrats estimated 1,591 individuals; the direct census gave 5,000.

year; most individuals were between 10 and 17 mm. By August, most were larger than 17 mm; a relatively uniform spread of sizes occurred in October, but by December (again the peak in adult numbers) most were 14 mm or less. Comparisons on a month by month basis indicates that never was the size structure similar for equivalent seasons in the two years. Thus, size structure seems to neither remain stable nor to show regular seasonal trends.

It must be stressed that the observed patterns may be restricted to the northeastern corner of the island where the sampling of size-structure occurred. Toward the end of the study it became apparent that the entire *Utetheisa* population was not homogeneous. In fact each *Argusia* shrub differed in the characteristics of its *Utetheisa* population. As noted in Chapter 11 there was some flowering of *Argusia* at all times of the year but there were peak flowering periods (Fig. 35). Each shrub seemed to have its own phenology; at a given time one would be heavily in flower, another would have scarcely any. This seemed to influence the *Utetheisa* population. Adult moths seemed to be attracted to shrubs that were heavily flowering and consequently tended to oviposit there with the result that subsequent to the flowering peak the shrub would be relatively heavily infested with caterpillars. When these metamorphosed it left the shrub relatively bare of caterpillars until after its next flowering peak. Consequently, adjacent shrubs might have quite different stages of the lepidopteran population cycle, and localized sampling would be biased accordingly.

In essence the moth population of each shrub was a somewhat separate entity and would require individual sampling. By the time this situation was appreciated qualitatively, it was too late to incorporate appropriate techniques into the sampling of size-structure to ascertain the differences between shrubs at different phenological stages. In any event, the time required for analysis on that level would have been prohibitive in view of the requirements of the rest of the sampling programme. The population dynamics of *Utetheisa* in relation to the flowering of its host plant, on a shrub by shrub basis could well be the topic of a detailed ecological study and should be extremely rewarding. This we leave to future investigators.

The sampling of numbers of *Utetheisa* on the island was not subject to the same bias as that of the size-structure; the larvae were sampled over a wider area and the adults were censused for the whole island.

In the early part of 1969 numbers of adults ranged from 236 to 714 (Table 34). When caterpillars were first sampled in September–October they were estimated to be over 19,000 for the entire island. The following visit many of these had metamorphosed as adult numbers had increased to 5,000 (the peak for the entire study). Caterpillar numbers declined steadily from the September–October 1969 value until March–April 1970 when density was so low they didn't appear in the foliage sample quadrats. Thereafter they increased slightly and had reached high densities again by the end of 1970.

After the November–December 1969 peak, the number of adults dropped

Table 34. Temporal changes in total number of caterpillars and adults of *Utetheisa pulchelloides* on One-Tree Island. – indicates present but not quantitatively sampled. + indicates present but not in foliage quadrats.

Date	No. of caterpillars	No. of adults	Total
1969			
March–April	—	714	
May–June	—	236	
July–August	—	490	
September–October	19,086	210[a]	19,296
November–December	13,360	5,000[b]	18,360
1970			
January–February	7,316	37	7,353
March–April	+	19	
May–June	318	10	328
July–August	318	18	336
September–October	1,272	62	1,334
November–December	7,953	928	8,881

[a] Of which 6 inadvertently killed.
[b] Of which 25 inadvertently killed.

precipitously from 5,000 to 37 (Table 34) and remained well below 100 individuals until November–December 1970 when the value rose to over 900.

There must be heavy mortality among adults in the population (they are probably short-lived) and/or extensive migration away from the island. The peak number of adults was only slightly more than one-fourth that of the immediately preceding peak number of caterpillars (Table 34). Even more significantly, at the sampling period following an estimate of 13,000 caterpillars there were only 37 adults. Similarly, following a population of over 7,000 caterpillars there were only 19 adults. Except possibly on one occasion (March–April 1970) adult population size was greatly below both the previous and the simultaneous caterpillar population size. In 1969 caterpillars were abundant during much of the period of heavy flowering of *Argusia* and the peak of adults coincided with the end of the flowering peak (compare Table 34 and Fig. 35); in 1970 the peak in numbers of adults also occurred in December, but was only about one-fifth as high as in the previous year. It occurred when flowering was still relatively heavy, and in the sampling period just following the peak of flowering of *Argusia*. Thus, on an island-wide basis the population seems to undergo a rather regular pattern of seasonal change with numbers of adults low most of the year but abruptly increasing to a peak in December, a time generally coinciding with the end of the flowering peak of *Argusia*. It is not known if this is a fixed seasonal rhythm which has developed in response to selection for peak numbers of adults at a time when their food supply is abundant. This hypothesis seems reasonable in view of the close association of moth and plant (this species of *Utetheisa* appears to be restricted to *Argusia*). However, there is also the possibility that adults from other islands may migrate to One-Tree when *Argusia* flowers are abundant there. The differences among islands (if any) in the flowering cycle of *Argusia* has not been assessed. The caterpillars are chewing herbivores feeding exclusively on *Argusia* leaves; food was continuously available as the standing crop of *Argusia* foliage did not change greatly in 1969–70; however, the high densities of caterpillars mostly occurred in periods when net production of new leaves was high and hence new succulent foliage most available (compare Table 34 and Fig. 40).

Thus, it appears that the life histories of the moth and the shrub coincide in ways maximizing food supply, either in terms of quantity (nectar) or quality (leaves) for both the larval and adult stages.

At the peak abundance in November–December 1970, and at the end of the detailed sampling of this species, a series of 38 small caterpillars (mostly 2–4 mm in length) and 75 medium sized ones (mostly 12–14 mm in length) were collected, oven-dried and weighed. The smaller ones had a mean dry weight of 1.23 mg per individual, the larger ones a mean of 5.90 mg per individual. Nine pupae averaged 10.33 mg per individual and the largest caterpillars would probably have a similar weight. Forty-nine adult moths had a mean individual weight of 9.25 mg. These data in conjunction with the data on numerical abundance provide a crude assessment of biomass. At lowest population densities, the biomass for the adults of the entire island would have been approximately 0.09 g and at highest population densities about 46.3 g. Judging from the size structure of the population of caterpillars in July–August and November–December, 1969, the mean size of caterpillar in September–October of that year would have been about the value of the intermediate sized individuals used for dry weight determinations. Using that value, the estimated biomass of the caterpillars at that time (peak population) for the whole island would be about 113 g. At lowest population density, it would have been below 2 g.

Ants

As pointed out previously, the appearance of ants in various kinds of samples was determined by their activity patterns as well as by their actual abundances. They live in nests, often in sites that are not amenable to sampling by these methods and only that variable portion of the colony active above ground is subject to sampling at any given time; furthermore, the fact that they live in colonies means that their distribution tends to be highly clumped. Occurrence of a nest in a sample (e.g. in the base of a plant in a cage plot) would give a highly inflated estimate of population density for that time. For these reasons little reliability was placed on density estimates of ants in the previous analyses and they hardly indicate more than an 'order of magnitude'. Nevertheless, ants of at least one species were abundant (see Chapter 14), were an important part of the fauna associated with *Argusia* flowers, and occurred often in significant numbers on the plants of the herb flat, and on the foliage of trees and shrubs. Two species were known to scavenge on dead vertebrates (see Chapter 20). It seems reasonable to present at least a brief evaluation of the role of ants in the One-Tree Island community.

There were eight species recorded from One-Tree Island (Table 16). Of these at least four are tramp species that have reached a large number of places via the agency of humans. *Pheidole megacephala* is a native of Africa but has been spread by commerce to almost all the humid parts of the tropics, including many of the Pacific islands. It is present on all of the Hawaiian group, where it is one of the commonest species and is generally attributed as being one of the agents most destructive to the native Hawaiian insect fauna (Taylor 1967, Wilson and Taylor 1967). Wherever it has become established it has aggressively displaced native species of ants. However, it is not always dominant as in turn it has been displaced by a similarly aggressive and expansive tramp species (*Iridomyrmex humilis*) from South America, when the latter arrived in Madeira; *I. humilis* is now in the process of replacing *P. megacephala* on Bermuda (Haskins and Haskins 1965, Crowell 1968).

Cardiocondyla nuda is also probably African in origin and has been spread by commerce throughout the warmer part of the world (Taylor 1967, Wilson and Taylor 1967). *Tapinoma minutum* occurs in Australia, New Guinea, the Solomons, Samoa, the Cook Islands, Niue and Micronesia (Taylor 1967, Wilson and Taylor 1967). Clearly it is a species subject to inter-island transport and its presence on One-Tree is not surprising. We also found it on Heron Island in 1978.

Iridomyrmex glaber is a tramp species of Australian origin which has been transported inadvertently by man, and has reached New Zealand; in Australia it usually lives in savannah woodland and in gardens and other domestic situations (Brown 1958).

Iridomyrmex glaber was by far the most abundant species. It was clearly an omnivore. It was recorded visiting the flowers of *Melanthera, Canavalia, Scaevola, Plumbago, Ipomoea, Pisonia,* and *Commicarpus* and appeared to be feeding on nectar in all cases. It was also attracted to green *Pandanus* fruits and was observed transporting dead insects, including thysanurans, 3 species of flies, a bug and an unidentified insect. It also scavenged on reef heron faeces and vertebrate carrion, as detailed below, and was attracted to sardine baits. This species was clearly dominant. It accounted for nearly all the individual ants captured by sweeps and cage plots, was obtained by sweeping the succulent mat, was a prominent member of the fauna associated with *Argusia* flowers, occurred on *Argusia* and *Pisonia* foliage and was collected under loose *Pisonia* bark, in *Pandanus* litter; winged reproductives were frequently observed or captured.

Camponotus sp. was also an omnivore as it was observed to feed on *Argusia* nectar, gull and reef heron faeces and vertebrate carrion. It occurred in the plot and sweep samples, and on *Argusia* and *Pisonia* foliage.

Tapinoma minutum was collected only in the cage plots in the herb flat; workers and winged reproductives were both taken. *Cardiocondyla nuda* occurred primarily in sweep samples of the herb flat. It also appeared twice in plots, and was captured under loose *Argusia* bark, in *Abutilon* fruits, and on *Pisonia* foliage.

Pheidole megacephala was occasionally taken in sweep or plot samples from the herb flat.

An unidentified species occurred only in the succulent area; it was attracted to sardine baits (see below).

There was a second, unidentified species of *Iridomyrmex* and a third species that was not identified at all. The former was not abundant and the

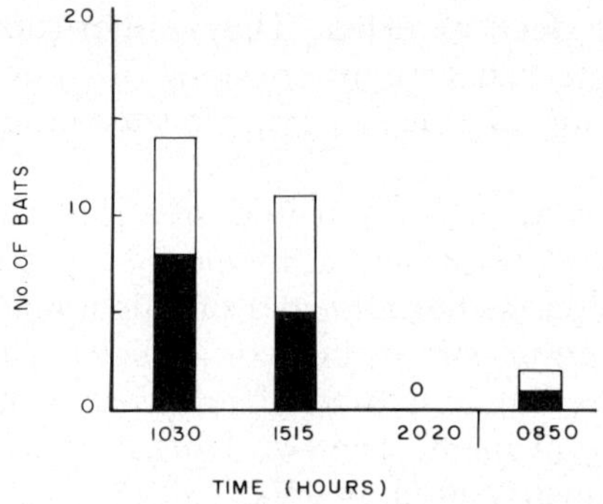

Fig. 86. Activity of the ant (*Iridomyrmex glaber*) at various sardine baits in the herb flat, 16–17 September 1978. Dark part of histogram indicates ants that were feeding, open part of histograms indicate ants foraging in the vicinity of the bait.

latter was known from only one winged reproductive (perhaps a newly arrived individual?).

In the herb flat, only one species of ant (*Iridomyrmex glaber*) came to sardine baits. At the season when baits were laid (September) they were most active during the daylight hours. The highest numbers occurred in the morning, with a slight decline by mid-afternoon; no ants were at the baits at the night reading, but by the following morning they had begun to return (Fig. 86).

Not all of the baits in the herb flat were visited. Four (20%) were never observed to have an ant visit them or forage nearby (within 0.5 m), six (30%) had ants at only one of the observation periods, nine (45%) had them at two periods, and one (5%) had ants at three; none of the baits had ants at all four observation periods.

In most cases when ants were feeding they were also carrying food back to their nests. It was difficult to follow their trails in the herb flat and no nests were found by trailing them. However, during other sampling several nests were found. They were usually at the woody base of a plant.

The only bait in the *Pisonia* that was visited by ants was one of those on a tree trunk. *I. glaber* were feeding on it at both morning observation periods, or foraging nearby in the afternoon but were absent at night. The other baits in the tree and all those on the ground always lacked ants.

In the *Argusia* area both baits up in shrubs were fed on by *I. glaber* at all observation periods, even at night. By contrast, none of the ground baits were visited by ants at all; both did have collembola at them in abundance at night, and a cricket was present on one occasion. The ground baits readily disappeared and were apparently eaten by something other than ants.

At one of the *Argusia* stations in a shrub the ants were trailed from the bait to their nests. The ants carried pieces of sardine along two separate trails from the bait. These in turn diverged to go along different branches. There were three nest sites, all in dead branches. Thus, either separate colonies of this species fed at the same bait without engaging in combat over food, or else there were three subunits of a single colony located in separate dead branches on the shrub.

The succulent mat was the only baited area that yielded more than one species of ant. *I. glaber* was present at the periphery, coming only to the bait near the ectone between the herb flat and succulent mat. It was feeding there at all daylight observation periods but not at night. The second species was unidentified. It was feeding in numbers by day on the bait at the edge of the pond; it was not seen at night. The two baits in the middle portion of the succulent mat were never visited by ants.

In summary, *Iridomyrmex glaber* forages on the ground and low herbs in the herb flat and on shrubs and trees. It nests at the woody bases of plants. In the areas with shrubs and trees it forages above ground, but doesn't seem to be active on the ground very much. It penetrates into the edge of the succulent mat. The latter habitat, at least close to the pond is occupied by an unidentified, also diurnally active species.

When feeding on nectar *I. glaber* was active at all times of day or night (see Fig. 56). At least during the 2 days of sardine baiting, this species was predominantly diurnal. The only sardine baits visited by *I. glaber* at night were in *Argusia* shrubs. The ants were probably recruited from those visiting flowers on the same plant.

One of the unusual features of the One-Tree Island ant fauna is the coexistence of several species of tramps that often aggressively displace other species. *Pheidole megacephala* is notorious for its destruction of ant faunas, yet on One-Tree Island it coexisted with the more dominant *Iridomyrmex glaber*. It is likely that these tramps have severely altered the composition of whatever ant fauna may have been present before. However, several species are present on the island which are not known as tramps and which may represent part of the original fauna.

In summary, most of the species of ants on One-Tree Island are tramps, two of which are well known as destructive of native faunas when they reach new locations; yet some presumably native species still persist on this tiny island. How the two tramps coexist with each other and with the non-tramp species has not been made clear from the present study. A detailed analysis of the interactions of ants on One-Tree Island should be a most fruitful and rewarding study.

The role of the ants, especially the two more aggressive tramps, in influencing the composition of the rest of the insect fauna also remains unevaluated. Majer (1976) has shown that one or a few dominant species of ants can markedly influence the composition of entire insect faunas.

Trachymene procumbens vel. aff.

This umbelliferous plant was found on the open rubble beaches where it was the dominant species of the sparse beach vegetation (Figs. 6, 23). Only once was it found elsewhere (November–December 1969) and then on rather exposed rubble in the more sparsely vegetated part of the herb flat. Because it occupied an exposed habitat, grew in discrete clumps, and was sparse in numbers of individuals it could be censused by direct counting and lent itself well to a study of vegetation dynamics. Such censuses were made regularly during 1969 and most of 1970.

The first census occurred in January–February 1969. There were less than 300 plants on the island and almost one-third of those were dead (Fig. 87), probably as a consequence of the drought which had occurred previously. The dead plants were of a wide range of sizes with a slight predominance in the smaller size groups (Fig. 88). By contrast the live ones were mostly small, suggesting that the previous population had been mostly killed by the drought and that the current population of live plants was made up primarily of those which had sprouted and grown after the breaking of the drought. However, many of these young plants had reached maturity and were flowering and fruiting (Fig. 87), with an overall average of 2.8 umbels per plant. At the next observation period (March–April 1969) other activities did not allow time for the *Trachymene* count to be carried out and it was merely noted that they were abundant and were flowering and fruiting. By May–June 1969 the population had increased to over 1,600 individuals with only a few dead ones to be seen. The previous dead ones had apparently been blown away by the wind. Almost all of the plants were seedlings (80.5% were less than 2 cm in diameter, and only 0.9% were greater than 15 cm in diameter). Thus even some of the larger live plants of the previous periods had died and the population consisted mostly of seedlings; there were only four umbels in the entire population (mean of 0.0025 per plant). The population peak was reached in July–August 1969 when there were 3,456 plants on the island, about 15 times the number just after the drought. Almost all of these plants were young and umbels were not abundant, there being on the average only

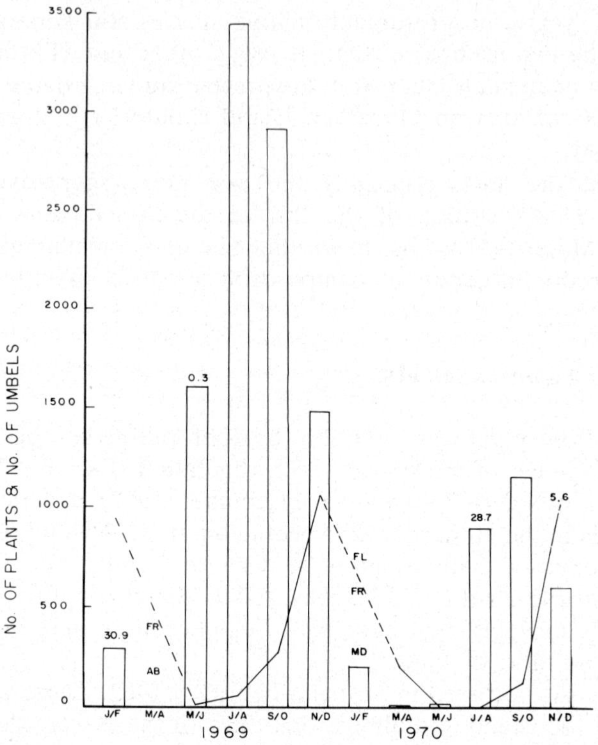

Fig. 87. Temporal changes in total number of individuals (histograms) and the numbers of umbels (flowering plus fruiting) (line graph) of *Trachymene* on One-Tree Island. Numbers above histograms indicate the percentage of the plants that were dead. AB indicates plants that were abundant but uncensused. FL and FR and dashed line indicate that flowering and fruiting were occurring but that the umbels were not counted. MD indicate there were many dead plants but that live and dead plants were not tallied separately.

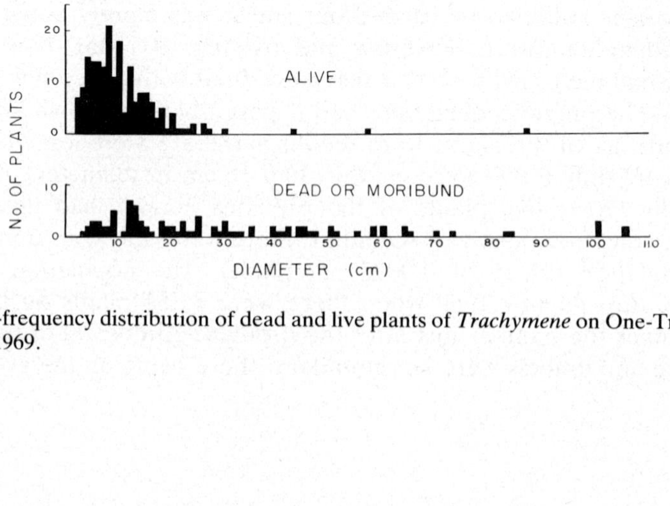

Fig. 88. Size-frequency distribution of dead and live plants of *Trachymene* on One-Tree Island on 6 February 1969.

0.016 per plant. Over the next two sampling periods the numbers decreased markedly. Many of the seedlings had died and had been blown away. However, others matured and the number of umbels of flowers and fruits increased (Fig. 87) reaching respective mean values of 0.09 and 0.71 per plant. By January 1970 many more plants had died and many dead ones were observed still on the beach. The remaining live ones were mostly greater than 5 cm in diameter and many plants were mature. Most of the umbels observed had ripe fruit; there were few flowers. This trend continued into March–April 1970 when the nadir of the population decline was reached. There were only four *Trachymene* plants on the island. However, these were in reproductive condition, collectively bearing 3 umbels of flowers and 200 of fruits. The following period (May–June 1970) there was only one mature plant and it had no flowers or fruits (this period and the subsequent one were the only two times in the study that no fruiting or flowering was evident). There were a few new plants growing; 16 individuals less than 5 cm in diameter were present. By the next sampling period their number had markedly increased, and seedlings were abundant, although more than one-quarter were dead. Of the live plants, 95.6% were less than 2 cm in diameter.

In September–October 1970 the population had increased and was still made up mainly of young plants although 1.2% were now greater than 10 cm in diameter, and both flowers and fruits were evident (Fig. 87) (mean = 0.10 umbels/plant).

By December 1970, the last time quantitative data were obtained, the number of plants had decreased and a number of dead ones were present. There must have been considerable mortality since the previous time as 86% of the plants were very young (less than 10 cm diameter). The larger plants were flowering and fruiting heavily with more than 1,000 umbels present on the island (mean = 1.46 umbels/plant, or 10.13 umbels per plant greater than 10 cm in diameter).

Although no more quantitative data were collected, it was noted that the species persisted until the last visit to the island in 1978.

The population fluctuations followed very different patterns in the 2 years and consequently cannot be a regular seasonal cycle. For example, in 1969 the peak population density (3,456 individuals) occurred at about mid-year. In 1970 very low densities occurred in mid-year. These changes are more likely attributable to immediate weather conditions. Rainfall was probably one contributing factor, though there was no close correlation of precipitation and plant abundance. The low population density in early 1969, as indicated previously, was probably a consequence of the previous drought. The massive production of seedlings that subsequently occurred was probably a response to the high rainfall of March and May and the large numbers of seeds produced in January–February (Fig. 87). Rainfall was high, at least for most alternate months for the rest of the year, but gradually declined throughout most of 1970, a time when *Trachymene* populations were at low levels. In July plants began to grow again although rainfall remained at a very low level. The total precipitation was 990 mm in 1969 and 752 mm in 1970; the former, wetter year had a much higher population of *Trachymene* than did the latter. It can be concluded that although there are some correlations between rainfall and *Trachymene* population levels, the two are not closely related and

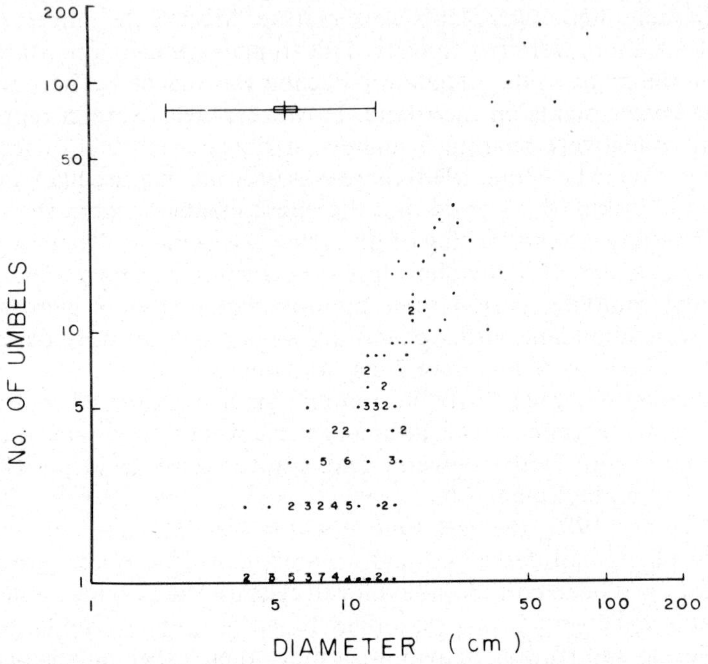

Fig. 89. Relation of number of umbels to diameter of flowering and fruiting *Trachymene* plants on One-Tree Island, February–March and April–May, 1969. Dots indicate a single value; numbers indicate multiple ones. Symbol in upper left corner of graph indicates the mean diameter (central vertical line), range (horizontal line) and two standard errors either side of the mean (rectangle) of the plants which lacked umbels.

factors other than precipitation must be exerting an important effect on the plant.

Trachymene occupies the harshest environment on One-Tree Island. The open beach is exposed to the highest intensity of wind and the greatest amount of salt spray. It is also the hottest and driest habitat on the island. *Trachymene* was often abundant on the eastern beach where it would be exposed to the full brunt of wind and spray. Clearly, this species has high physiological tolerances to a variety of factors, but does seem to succumb to major drought.

The low point of the population during the 1968 drought was not observed. The species went precariously close to extinction in 1970 when only four plants were left. This is probably more apparent than real, however, as many seeds probably persisted in the beach and were available to restore the population under favourable conditions. Indeed, the 1969 recovery from the drought was probably implemented in part by sprouting of seeds produced before the drought.

It is likely that the persistence of *Trachymene* in such a harsh environment is in part attributable to its pattern of reproduction. Important aspects are: (1) It is capable of flowering at all times of the year if conditions are favourable. Although the flowering and fruiting cycle appeared to be more similar between the two years than were population fluctuations, and could represent a regular seasonal phenomenon, the same reproductive pattern could equally well be explained as being generated by the changes in population size and age structure. In early 1969, although population density was low, many of the plants were mature and seed production was high (Fig. 87). Later when there were many plants, most were immature seedlings and seed production was low. A lag in reproduction was imposed by the time it took these seedlings to mature and hence the apparent seasonality of flowering and fruiting may be artifactual. More data are required to clarify this point. However, reproduction did occur to some extent in every month, at least once over the study period. (2) An abundance of seeds are produced. For example, large plants can produce up to 165 umbels (Fig. 89). Umbels contain a mean of 161 mericarps each (N = 4). Consequently, production can be very high. (3) Plants set fruit when they are very small. Although individual plants grow up to slightly over 1 m in diameter, some begin flowering when they are only 4 cm in diameter. This permits them to be opportunistic and take advantage of temporarily favourable conditions to produce large numbers of fruits. No plant greater than 13 cm in diameter was ever seen without umbels on it (Fig. 89). The greater the size of the plant, the greater the number of umbels (and hence fruits) it produces.

The Nightshade, *Solanum nodiflorum* subsp. *nutans* and an Halticine Beetle

The small herb, *Solanum nodiflorum,* illustrates just about every aspect of the dynamics of insular ecology. Almost everything that could possibly happen to an insular plant in fact did!

Only one individual plant was found on the first trip (July 1968) and unfortunately it was collected. It occurred on the western edge of the northwestern point on bare rubble and isolated from all other vegetation. In February, April and June 1969 none were found in the herb flat but 3–10 small plants were observed growing in the coral rubble just above the high tide line at a site where silver gulls commonly roosted; some were flowering or fruiting on each occasion, but others were moribund. In June 1969, a few scattered individuals had spread to the sparse vegetation nearby but had not extended beyond the base of the northwestern point. By October 1969, the species had spread over much of the narrow peninsula on the northwest corner of the island. It subsequently spread rapidly throughout the herb flat until it accounted for over 16% of the total herb biomass there (in October 1969; see Table 6). At that time silver gulls were observed actively feeding on the ripe berries throughout the herb flat. At no other time were these birds ever observed feeding in that area. Silvereyes also were feeding on *Solanum* fruits in October, 1969, and the blue-stained faeces of both species contained undigested *Solanum* seeds. The biomass of *Solanum* then declined rapidly and the plants were observed to be infested by a small blue unidentified species of halticine beetle which was severely damaging the plants, in some cases completely defoliating them. It had not been seen on the island before and it was never observed feeding on any plant but *Solanum*. *Solanum* was very rare in February 1971 and was still being heavily damaged by the beetle. By April 1972 the plant was not noted on the island and only one beetle was found.

By 1976, *Solanum* was again present on the island, but was restricted to the edges of paths between buildings which were erected between 1972 and 1976. In 1978 the condition was much the same. On neither of the last two occasions were any of the beetles found and the *Solanum* appeared undamaged.

From these facts probable events can be reconstructed. In the beginning *Solanum* was not abundant on One-Tree and probably became extinct there during the drought. If there were any plants left, they were very scarce and were certainly not on the exposed beaches where they would have been most clearly visible to birds. The fact that gulls feed heavily on *Solanum* fruits when they are available (even going to places they do not otherwise visit) and that the first plants after 1968 were found at, and subsequently spread from, the spot on the beach where the gulls habitually roosted, suggests that these *Solanum* were transported to One-Tree as seeds in the digestive tracts of the gulls flying from nearby Heron Island where *Solanum* was abundant throughout the study. The first clump of plants to grow was located in a conspicuous place and when they fruited the gulls probably ate the fruits and distributed the seeds in the general area of the roost. As the plants became more abundant the silvereyes, which normally feed in the trees and in shrubs in the herb flat (but were seen to eat *Solanum* fruits on the beach) were attracted to them and rapidly spread the seeds, via their droppings, throughout the island and into places not normally inhabited by the gulls. When *Solanum* became abundant inland, the gulls were attracted to it and aided the silvereyes in contributing to its further spread. There were few species that fed on the *Solanum* leaves at that time; the beetle was probably not on the island. If it chanced to arrive prior to the presence of *Solanum* there, it would probably have failed to become established because of lack of food. For the same reason is probably followed its host plant into local extinction. The subsequent appearance of *Solanum* around buildings probably resulted from introduction by humans. The beetle had not yet reached the island again by September 1978.

We believe that we witnessed the extinction, immigration, establishment, and again an extinction and re-immigration of *Solanum* on One-Tree Island.

Only longer studies will tell whether these events were unique or whether this process recurs with repeated immigrations of *Solanum* followed by the beetle and the eventual extinction of both. However, we suspect that this has occurred a number of times. MacGillivray and Rodway (1931) visited One-Tree in 1927 and reported *Solanum* to be present then.

19. The pond

The pond on One-Tree Island (Figs. 5, 6, 9) deserves a detailed study in itself. It is a small body of water, subject to periodic drying, that is well isolated from other such habitats. Unfortunately, because of time limitations we were unable to carry out a detailed study of the aquatic community inhabiting it and only a few cursory observations were made. The limited information we do have is presented in the hope it will stimulate, and serve as a basis for, future investigation. The only other published information on small ponds or pools on coral islands of which we are aware are the studies of Donaldson and Whitton (1977a, b) and Whitton and Potts (1977).

The pond, when full, is about 20 m in diameter and 75 cm deep. Hulsman (1976) referred to it as a 'tidal pond', but its water level is not related to exterior tidal levels in any simple way. At times pond level fluctuates widely over short periods without any intervening rainfall; at other times the level remains relatively stable for considerable periods. Examples of pond level fluctuations can be gleaned from some of our field notes:

'23 May 1969, pond nearly dried up, about 1 m in diameter. After rains it was completely filled. Water extremely clear.'

'30 November 1969, pond 12 feet in diameter. By 1 December it had emptied but mud was wet. Began to fill on 5 December, probably influenced by onset of spring tides and possibly by rainfall during the week. By 8 December water had reached edge of succulents and by 9 December it covered succulents 2 metres away from the usual pond border.'

'On 28 January 1970, pond full, fell to only 1/4 full on about 31 Jan. 1970; full again on 3 February 1970. 5–7 February 1970, pond overflowed to succulents surrounding it.'

'23 March 1970 pond about 3/4 full; the level fell gradually throughout the week. On 31 March it was only about 1/4 full.'

'21 May 1970, pond full; overflowed during several nights and was always full or overflowing until 28 May 1970.'

'On 29 September, pond 7 m in diameter. Disappeared on October 4. Only wet mud left. This despite rain. No obvious relation to tides.'

Almost every visitor to One-Tree Island has speculated about the factors controlling the pond level, and numerous theories have been proposed. The favourite one seems to be that the island has a saucer of hard material beneath the rubble that is raised sufficiently that only the spring tides are high enough to flow over it and fill the pond[1]. We leave this intriguing problem to future investigators.

Water clarity varied greatly and ranged from very clear to murky green; the former was the more usual condition.

Surface temperatures of the water in daytime were: 28.0–28.3° C (March 1970), 23.3° C (May 1970) and 26.5° C (July 1970); in the One-Tree lagoon, January temperatures are bout 28°–29° C and June ones are about 22°–24° C (Ludington, pers. comm.).

Hulsman (1976) indicated that the pond varied from brackish to salt. Certainly the alternate effects of rain and extensive evaporation, combined with the contribution of salt spray and the direct influence of the sea, must have caused rather large fluctuations in the salinity. In May 1969 we measured the osmolarity of a sample of pond water. It was 977 milliosmoles or about 98% that of average sea water.

Oxygen concentrations of the pond water measured on several occasions were: 8.5–9.0 ppm (March 1970), 6–6.5 ppm (May 1970), and 8.4 ppm (December 1970), or near saturation for water of that temperature and salinity (Green 1968).

The bottom of the pond was mostly soft deep mud with occasional pieces of coral rubble or other objects sticking up above it. When dry the mud would crack and leave deep fissures. Several cores revealed dark areas several centimetres below the surface of the mud. These had a sulphurous smell suggesting subsurface anaerobic conditions.

The aquatic plant life consisted only of algae. The greenish caste of the water on occasions was probably due to blooms of unicellular algae, and there may have been a variety of other planktonic organisms as well. On one occasion (May–June 1970) heavy growths of filamentous green algae occurred in the shallow water at the edge of the pond but by the next visit (July–August 1970) the water level in the pond had dropped and the filamentous algae had died and bleached and was left as a decaying mat on the exposed mud. There was no conspicuous growth of filamentous algae at other times.

Animals seen in the pond included mosquito larvae and three species of marine snails. A variety of terrestrial invertebrates (ants, isopods, beetles, bugs and dipteran larvae) were found under the dead algal mat at the edge of the pond in July–August 1970.

The snails were *Bembicium kielmanseggi* (Littorinidae), *Clyreomorus moniliferum concisus* (Cerithiidae) and *Amphinerita polita* (Neritidae). They

[1] Eschmeyer and Eschmeyer (1971) state: 'During neap tides no addition of water occurs and the salinity gets higher, killing all animal life which may have been introduced with the high tides.'

were seen tracking across the surface of the mud on the pond bottom, on pieces of rubble sticking out of the mud, on emergent mud at the edge of the water, and during periods of low water or complete drying under an algal mat or in fissures of the mud.

A number of factors made sampling of the snails difficult and only very approximate data were obtained. Individuals were not always distributed evenly over the pond bottom. On one occasion most were in a 4 m^2 area where there was considerable rubble with the mud, and on another (dry) period they seemed attracted to the dead algal mat. Probably there was also some vertical migration into fissures during dry periods.

In May 1968, it was merely noted that the snails were present in rather large numbers. Later a more quantitative approach was attempted and in May 1969, 10 underwater quadrats (each of 20 cm on a side) were examined for snails. The number on the surface and in the upper 3 cm of mud were collected and sorted into live and dead categories. On occasional later visits similar quadrats, varying in number and size depending on immediate conditions, were used.

Bembicium was the most common snail. In May 1969 it was estimated that there were 115 live animals and 47.5 dead shells per m^2 of pond bottom. Thus about 71% of the animals found were alive; some were observed copulating. This was the largest estimated population at any time during this study. At the following sampling period there were only 33.7 live snails per m^2 and 14.5 dead shells per m^2 (70% live). By January 1970, the estimated number of live animals had dropped to 22 per m^2 and only 32% of the total shells found contained live animals. On the 4 subsequent samplings (March, July, September and December 1970) no live *Bembicium* were obtained in quadrats indicating the population to be very low. They were not extinct as an occasional live individual was seen when not sampling. Density of dead shells was often quite high and there were 144 per m^2 in December 1970.

Occasional live individuals of *Clyreomorus* were observed but occurred live in the samples only once (July 1969) when they reached a live density of 3.3 per m^2; 73% of all shells seen contained a live animal. Dead shells were found in the samples in low numbers (1.2–24.0 per m^2) at each sampling period except July and September 1970 when neither live nor dead animals of this species were found.

Amphinerita was relatively rare and was never encountered live in the samples. Dead shells never reached densities higher than 4.0 per m^2, and in over half the sampling periods neither live animals nor dead shells of this species were encountered.

The above data suggest that there was a decrease in the numbers of the initially abundant *Bembicium*, with the other two species remaining relatively uncommon throughout.

The frequent drying of the pond meant that any permanent aquatic residents would need to exist for long periods in a resistant dormant stage.

Other species that became established would probably be those easily dispersed in or on the bodies of birds. Some of the seabirds nested around the pond (Chapter 17) and sometimes swam in it. Other seabirds and especially shore birds visited the pond and were frequently seen wading in shallow water.

It is known that aquatic organisms including algae and crustaceans can be transported viable in the digestive tract of birds (de Vlaming 1967, Proctor *et al*. 1967) and it has been demonstrated that snails can be transported on the feet of shore birds (Malone 1965). Dragonflies occasionally visited the pond; Revill *et al*. (1967) has indicated that viable algae and protozoa of many species are dispersed by insects such as midges and craneflies; dragonflies may also be dispersal agents.

20. Interaction of marine and terrestrial communities

The sea-land interface is sharp, changing only slightly with tidal fluctuation, and the boundary between insular and reef communities is more sharply demarcated than is often the case between adjacent terrestrial ones. Many species are restricted entirely to one medium or the other.

Despite this physical discreteness, there may be a significant ecological interdependence between islands and reefs. In a very real sense a coral cay is an extension of the reef ecosystem. The functioning of the above-water community is greatly influenced by the subtidal communities, exchanges energy with it, and shares some of the same materials (Davis *et al.* 1938, Heatwole 1971, Alexander 1979).

On most cays, whether of rubble or sand, the land originated from the living coral, foraminiferans and shells of the reef. Upon this substrate the terrestrial vegetation roots and grows and many animals shelter or burrow there.

The island receives not only its substrate from the marine environment but some nitrogenous compounds and other plant nutrients as well. Organic matter arising from decomposition of marine organisms occurs as a thin film on the surface of the sea and/or is adsorbed on small bubbles. It reaches the atmosphere as a thin surface film around small droplets ejected into the air, especially when bubbles break and where the water is agitated in surf or whitecaps. These droplets with their contained inorganic ions and their organic surface films are carried by wind to land. The amount has been estimated to be about 0.3–0.7 micrograms of organic material per microgram of sea salt (Blanchard 1968), and it has been suggested that it might provide a significant nutrient supply for the large amount of life near the intertidal zone (Wilson 1959).

Another source or organic matter is by direct wash-up on the beach of marine organisms such as seaweed or dead fish. Seldom can one walk around the One-Tree beach without encountering marine carrion of some type. This carrion is available to terrestrial scavengers and may contribute a significant proportion of the total food base of the insular community; on very small islands it may be a major source (Heatwole 1971).

The sea also influences input of materials to an island by transporting terrestrial flotsam. Beaches are often littered with dead logs or branches of trees of species not occurring on the island alive. Such material may become available as shelter for terrestrial animals or food for wood-borers.

At present, jetsam contributes a special form of input for islands near shipping lanes. A vast amount of organic material and food is cast or discharged from ships. Some of this material floats to islands and becomes incorporated into the insular food web. Even food in sealed jars or tins eventually becomes available when lids rust through.

In many cases the indirect transport of materials from the sea to an island is more important than is direct wash-up. The former is expedited by 'transfer organisms'. Some of these may be intertidal species such as crabs, but most of the transfer is performed by sea birds. Sea birds prey upon marine animals and hence are top-level predators in the marine ecosystem. However, their requirements for shelter, resting, and reproduction are terrestrial (insular). They obtain food from the sea and transfer part of the materials therein to the insular community in the form of guano or their own dead bodies or those of their offspring.

An attempt was made to quantify some of the transfer of materials from the sea to the shores of One-Tree Island. Flotsam, jetsam and carrion were recorded once each visit during systematic inspection of the entire perimeter of the island.

Carrion

On the average there were 7.6 dead bodies of sea birds or marine animals on the beach at any one time (Table 35). Since on each visit only new carrion was counted, and large animals tallied on the previous visit but still not completely decomposed were ignored, the actual standing crop would be somewhat higher. Not all bodies were available intact to terrestrial scavengers, as some had been partly decomposed in the sea and only a fragment had reached the shore. This was especially true of the sea turtles; of the eight individuals found during the study, only two (one adult and one hatchling) were whole.

Sea birds accounted for over half (54%) of the dead bodies on the beach. This was not an estimate of the total avian contribution of carrion to the island. During the study and apart from the bird carcasses below the eagle's perch, two land rails, seven terns (*Sterna sumatrana*) and two silvereyes were found dead inland; doubtless others remained undiscovered.

Eggs and regurgitated or dropped prey of the nesting terns, gulls and herons further raise the avian contribution of organic input to the island. For example, in July 1968, a *Sterna sumatrana* dropped a fish (*Pranesus capricorniensis*, family Atherinidae), weighing 1.32 g (0.66 g dry wt.) on the One-Tree beach. On other occasions we saw small fish on the beach, presumably

Table 35. Marine carrion and dead sea birds on One-Tree beach during 1-day census periods. For detailed list see Table 40.

Date	Sea birds and shore birds	Land birds	Fish	Sea snakes	Sea turtles	Sponges	Squids[a]	Sea cucumbers	Jelly fish	Crustaceans	Unidentified	Total birds	Total marine	Total
1968														
May	1	1	0	0	0	0	0	0	0	0	0	2	0	2
July	3	0	0	0	1	0	0	0	0	0	1	3	2	5
1969														
February	15	0	4	0	0	0	1	2	0	0	2	15	9	24
March	0	0	1	0	0	0	0	0	0	3	0	0	4	4
May	9	0	2	0	0	0	0	0	0	0	0	9	2	11
July	9	0	2	0	1	0	0	0	0	1	0	9	4	13
September	11	0	4	1	1	0	2	0	0	0	0	11	8	19
November	2	0	1	0	1	0	0	1	0	0	0	2	3	5
1970														
January	1	0	0	0	0	0	0	0	0	0	0	1	0	1
March	0	0	0	0	0	0	0	0	0	0	0	0	0	0
May	0	0	0	0	0	1	0	0	0	0	0	0	1	0
August	0	0	0	0	0	0	0	0	Many[b]	0	0	0	0	0[b]
September	3	0	3	0	1	0	0	0	0	0	0	3	4	7
December	4	0	1	1	3	2	0	0	1	2	2	4	12	13
Total	58	1	18	2	8	3	3	3	Many[b]	6	5	59	48[b]	107[b]
Mean No. per visit	4.1	0.1	1.3	0.1	0.6	0.2	0.2	0.2	—	0.4	0.4	4.2	3.4[b]	7.6[b]
% of total	54.2	0.9	16.8	1.9	7.5	2.8	2.8	2.8	—	5.6	4.7	55.1	44.9	100.0[b]

[a] Excluding washed up cuttlefish skeletons, of which there were usually many. These were probably free of flesh before reaching the island.
[b] Jellyfish not included in totals, means or percentages as they were too numerous to count in August 1970; the beach was littered with them.

dropped by birds. Recher and Recher (1972a) note that noddy terns often regurgitate or drop fish, especially when disturbed.

Less than half the species and individuals of sea birds found dead on the beach nested on the island. Of the 13 identified species only 5 or 39% (*Egretta sacra, Sterna sumatrana, S. anaetheta, S. bergii,* and *Larus novaehollandiae*) nested on One-Tree. These plus the 16 terns that were too decomposed to identify to species (but which were probably species that nested locally) accounted for only 25 (43%) of the individual marine birds in the carrion.

Table 36. Birds and marine organisms found dead on the beach at One-Tree Island during 1-day censuses, at about 2-monthly intervals between May 1968 and December 1970. Large animals persisting for more than one census period were only tallied on the first census.

Vertebrates	
Sea birds and shore birds	
1	*Diomedea exulans* – Wandering albatross
16	*Puffinus pacificus* – Wedge-tailed shearwater
6	*Sula leucogaster* – Brown booby
1	*Phalacrocorax melanoleucos* – Little pied cormorant
3	*Egretta sacra* – Reef heron
1	*Charadrius* sp. – Dotterel
1	*Numenius phaeopus* – Whimbrel
1	*Larus novaehollandiae* – Silver gull
3	*Sterna sumatrana* – Black-naped tern
1	*Sterna anaetheta* – Bridled tern
1	*Sterna bergii* – Crested tern
16	*Sterna* spp. – Unidentified terns
3	*Anous minutus* – White-capped noddy
4	Unidentified sea and shore birds
Land birds	
1	*Zosterops lateralis* – Grey-breasted silvereye
Reptiles	
6	*Chelonia mydas* – Green sea turtle
2	*Eretmochelys imbricata* – Hawksbill turtle
2	Hydrophiidae – Unidentified sea snakes
Fish	
1	Myliobatiformes – Ray
1	Mugilidae – Mullet
1	Sygnathidae – Pipe Fish
1	Balistidae – Trigger fish
1	Batrachoididae? – Frogmouth?
1	*Diodon* sp. – Porcupine fish
12	Unidentified fish
Invertebrates	
3	Sponges
3	Squids
3	Sea cucumbers
2	Crabs
1	Hermit crab
3	Crayfish
Unidentified	5

Thus, there is a significant input beyond that contributed by resident species.

Most of the non-avian carrion consisted of fish, with sea turtles and crustaceans also well represented; sea snakes, squids, sponges, sea cucumbers and jellyfish provided relatively minor inputs (Table 36).

There is a bias towards large species in the recording of carrion. Small animals could reach the beach and decay or be completely consumed between visits and would not be tallied, whereas larger species would persist and *would* be counted. Another difficulty in evaluating input from carrion is the quantification of biomass, as on a particular visit different cadavers are in different stages of decomposition and weighing them would not be very instructive; the degree of decomposition suffered in the sea before being washed ashore cannot be determined.

Despite these difficulties and uncertainties, however, it is obvious that both sea bird carrion and marine carrion are important sources of organic input to the island. Further, they are rather continuous sources of organic matter. In only 2 of the 14 censuses of the beach were no dead bodies found at all (Table 35).

Carrion enters the terrestrial food web through the scavengers which eat it. In addition to the decomposer microorganisms (which were not studied), there were several macroscopic species that utilized carrion. Table 37 lists the arthropods observed to be associated with carrion on One-Tree. Some of

Table 37. Invertebrate fauna associated with carcasses of 23 birds and a parrot fish (Scaridae), and a sea turtle

Taxon				Numbering of times occurring
Pseudoscorpionida				1
Araneida	Lycosidae			1
	indet. species			2
Acarina	Ticks			2
	Mites			2
Insecta	Collembola			4
	Dermaptera			6
	Orthoptera: *Teleogryllus oceanicus*			2
	Hemiptera			1
	Coleoptera: Staphylinidae			1
		Dermestidae: adults		2
		larvae		7
	Diptera: *Atherigona* sp.			1
		Lamprolonchaea metatarsata		2
		Phoridae		1
		indet. species: adults		1
		pupae		2
		larvae		4
	Hymenoptera: Formicidae: *Iridomyrmex glaber*			1
			Camponotus sp.	1
			indet. species	1

them were not scavengers but were predators (pseudoscorpions, spiders), ectoparasites attached to the bird before its death (ticks) or were probably attracted to the cover or microclimate provided by the carcass. However, the earwigs, beetles, flies and ants listed in Table 37 are known scavengers (also see Chapters 14 and 18).

Hermit crabs are abundant scavengers on many islands, and in some cases there are other terrestrial crabs that play an important role in energy exchange between islands and the sea (Alexander 1979). On One-Tree they were rare. *Coenobita perlata* were recorded on three occasions. A number were observed in November–December 1969, and one each on two other occasions. In 1979 Elvish (pers. comm.) found two unidentified species. Two of one species were found, one on the beach and the other in the *Pandanus* grove. Many of the second species were also seen, all in the *Pandanus* (5 collected); they were under the roost sites of reef herons and were observed to feed on fish regurgitated by the birds.

The earwig (*Anisolabris maritima*) lived in very dense populations in the interstices of the coral rubble of the beach just above the high tide line. An attempt was made to sample this population. Empty tins were buried in the beach so that the opening was slightly below the level of the coral. A flat piece of coral was placed over the tin to shade it. Four tins were placed at 1 metre intervals in a line slightly above and parallel to the high tide line. Each tin was baited with a sardine enclosed in a wire mesh cage so that the earwigs could not actually feed on it. The tins were placed in late evening of 4 October 1969 and observed periodically until 6 October 1969. At each observation, each animal was marked with a dot of non-toxic plastic paint and released (except for a few killed and partly eaten by other earwigs). Few earwigs were trapped during the day but many were trapped at night.

The morning after the baited pitfall traps were set a large number of earwigs were present as well as a few silverfish and flies and 1 collembolan

Table 38. Results from baited pitfall trapping in the One-Tree beach. Values are total numbers of individuals trapped since last observation period; values in parentheses indicate the range in number of individuals trapped per tin.

Taxon	Morning 5 Oct. 1969	Evening 5 Oct. 1969	Morning 6 Oct. 1969
Dermaptera			
Anisolabris maritima	60 (5–26)	7 (0–3)	56 (6–25)
Thysanoptera			
(1 species)	6 (1–2)	0	0
Collembola			
(1 species)	1 (0–1)	0	0
Diptera			
(1 species)	3 (0–3)	0	0
Total	70	7	56

(Table 38). During that day there were only a few earwigs trapped but numbers were high again on the morning of 6 October. Although earwigs are found in carrion during the day they seem to move around in search of new food sources principally at night.

Earwigs were seldom observed elsewhere than in the beach rubble, or eating anything other than carrion. However, one individual was seen eating a fungus in a *Pisonia* grove.

The attempt to estimate the population density by mark-recapture methods was unsuccessful as no marked individuals were ever recovered.

Also living in the interstices of beach rubble was the carnivorous centipede

Scolopendra sp., probably the chief predator on the earwigs. This general predator is found elsewhere on the island but seems to reach its greatest abundance in the upper beach. However, no quantitative assessment of population densities proved feasible.

Flotsam and jetsam

An attempt was made to quantify the flotsam and jetsam cast upon the shores of One-Tree Island; at each visit we walked around the island and noted the wash-up.

A variety of flotsam and jetsam was found on the beach; on the average there were 132 such items per visit (Table 39). The largest component (mean about 73%) of the organic material was usually dead wood, with various fruits and seeds being second (about ¼ of the items). Live plants or their vegetative parts accounted for only a small amount of beach drift. The most common plant portions to wash up were roots (unidentified species). Bamboo and sugarcane were the only other items more numerous then three (Table 40).

Table 39. Flotsam and jetsam on the One-Tree beach during 1-day census periods, March 1969 to December 1970.

Month of census	Terrestrial plants (or parts)	Large seeds or fruits	Boards, sticks or logs	Total organic flotsam-jetsam	Manufactured jetsam[a]
1969					
March	3	10	6	19	36
May	0	15	32	47	67
July	4	8	69	81	58
September	4	25	31	60	67
November	5	55	144	204	97
1970					
January	0	25	47	72	55
March	2	45	70	117	81
May	0	36	53	89	58
August	2	32	46	80	60
September	0	85	515	600	122
December[b]	11	28	39	78	43
Total	31	364	1,052	1,447	744
Mean No. per visit	3	33	96	132	68
% of total organic material	2.1	25.2	72.7	100	—

[a] Plastic, rubber, shoes, light bulbs, tins, jars, etc. Four of the jars and tins had food in them.
[b] All flotsam-jetsam collected in late September 1970; thus the December values include only the net accumulation since then.

Twenty different species of large fruits or seeds were found on the beach. Numbers on particular visits ranged from 8 to 85 (mean 33). Only two seeds, both *Pandanus* sp., were of a species established on the island. Clearly, establishment for many species is a problem. A similar phenomenon has been noted on Aldabra Atoll (Wickens 1979b), Cocos-Keeling Atoll (Guppy 1891), New Zealand (Mason 1961), South Africa (Muir 1937) Krakatau (Docters van Leeuwen 1936) and Western Australia (Kenneally 1972) and as a general case (Safford 1921). Alexander (1979) suggests that consumption of seeds by crabs may prevent some species from becoming established; in the absence of such crabs on One-Tree Island other factors must be involved.

The manufactured jetsam has little impact on the biotic community as most

Table 40. Terrestrial plants, vegetative parts, seeds, and fruits on One-Tree Beach during censuses, March 1969 to December 1970.

Whole plants or vegetative parts	Fruits and seeds
1 *Cakile* sp.	51 *Barringtonia asiatica*
1 Umbellifer stem	39 Mangrove planulae
1 aroid	78 *Cocos nucifera*
2 Palm patioles	2 *Pandanus* sp.
6 Bamboos	7 *Cerbera odollam*
5 Sugar cane (*Saccbarum*)	187 Unidentified seeds belonging to 15 species[a]
1 Onion	364 Total
2 Grapefruit peels	
3 Lemon pieces	
1 Dried peach	
1 Piece of bark	
7 plant roots	
31 Total	

[a] Tentative identifications based in part on Kenneally (1972) and Gunn and Dennis (1976) include *Entada* sp., *Caesalpinia bonduc, Heritiera* sp., *Aleurites* sp., and *Calophyllum inophyllum*.

could not serve as a food source. However, several tins or glass jars containing some food (e.g. jam) were found on the beach. If such items were not washed away again before rusting of the lid or tin occurred, the contained food would become available to the terrestrial community. Some of the organic drift was almost certainly jetsam (onions, dried peaches, grapefruit peels, lemon pieces).

Although the amount of flotsam-jetsam on the beach was quite large, much was not usable by the One-Tree fauna. A great proportion of the drift was either wood or large, hard woody fruits and seeds. There were few, if any, animals on the island equipped to penetrate the tough outer coverings of these fruits and seeds and no wood-borers were ever collected except those inhabiting live *Pandanus* seeds. Consequently, the woody flotsam was not cycled in the community. It merely lay on the beach for a time, usually to be washed away again. The edible component (some of the plants and plant parts, vegetables, fruits and possibly food in containers) was a very small proportion of the total flotsam and jetsam and was negligible in comparison with plant material produced by the island's vegetation.

Woody debris can serve as shelter for animals, however. A large hollow seed on the beach in December 1970 contained 4 terrestrial species (6 silverfish, 4 collembolans, 17 beetle larvae and 1 thrip).

Transfer

On some very small cays vascular plants are lacking altogether. Despite this lack of a terrestrial autotrophic food base, a number of terrestrial animals are present including some predators such as centipedes and spiders. The entire food base sustaining this community is marine-derived; much of it is sea bird carrion and guano (Heatwole 1971). Small cays with green plants have an endogenous energy supply and can support herbivores in addition to scavengers and predators; a more complex community is possible. The larger, and more heavily vegetated an island, the greater is its independence from the marine community. There is thus a gradual progression from complete marine-dependence on the part of unvegetated sand cays to nearly complete autonomy of energy and food on large, forested islands. Most cays exist at some intermediate location along this spectrum, and no insular community is completely emancipated from marine effects. Diagrams illustrating food webs on progressively more complex unvegetated sand cays and the flow of energy and materials on a vegetated one are presented by Heatwole (1971, 1976).

Plants also may benefit from marine-derived carrion and guano. Some species, such as the *Pisonia* tree, seem to depend on high levels of nitrogen and phosphorus in the soil and can only become established where there have been numerous sea birds (see Chapter 8). Thus, the type and condition of vegetation on coral cays may be directly influenced by a marine-derived mineral.

The defecation of up to more than 3,000 birds on the island in summer contributes large quantities of phosphorus, nitrogen and organic matter to the island substrate. Alexander (1979) considered sea birds to eat about 10% of their body weight in food per day and produce 1/16 of that value per day in guano. Table 41 summarizes data on defecation rates of 6 species of sea birds, the reef heron and the silvereye. The high value for the silvereye and the low one for the heron probably reflect different body sizes and concomitant metabolic rates. The phase of the bird's daily activity cycle would influence its defecation rate, so interspecific comparisons should only be made with reservations. It is questionable whether the rates should be lumped for each species for this reason. However, Recher and Recher (1972a) indicated that typically, reef herons feeding close to shore leave the water to defecate on land, so most of their faeces would have been dropped on One-Tree Island. Other herons also exhibit this behaviour (Brackbill 1976) and although they often feed in shallow water, their faeces and regurgitated food are most often deposited on land.

Sea birds often defecate while flying over water and it does not seem reasonable to try to quantify the faecal deposition by them for the island as a whole until more is known of the habits of the individual species involved. However, the yearly transfer of marine materials to the island through guano deposition by birds is clearly great. Even some species not actually landing on One-Tree Island may deposit guano on it. On one occasion we saw a frigate bird defecate on One-Tree Island while flying overhead.

The greatest faecal deposition is probably by species which nest on the

Table 41. Defecation rates of some bird species which occur on One-Tree Island.

Species	Total minutes of observation (bird mins)	Total defecations	No. of defecations/HR
Egretta sacra	232.1	2	0.516
Larus novaehollandiae	2,097.0	60	1.716
Sterna dougalli	107.0	0	—
S. sumatrana	7.0	0	—
S. anaetheta	190.5	0	—
S. bergii	248.0	0	—
Anous minutus	3,373.5	28	0.498
Zosterops lateralis	32.8	3	5.490

island. We placed a large plastic sheet under a heron nest for 3 days. The mean dry weight of faeces reaching the ground per day at the nest site, including contributions by the adults and nestlings was 11.21 g. Comparable figures for a noddy tern nest (on Heron Island) was 0.95 g. The oven dry weights of individual droppings were 4.3 mg for the silvereye and 2.04 g for the reef heron.

Diamond (1979) estimated that on Aldabra sea birds brought about 105 tons of guano into the atoll ecosystem each year.

PART FOUR

Synthetic characteristics

21. Immigration, extinction and the equilibrial number

Many of the epigeal species on One-Tree Island were sampled on only one occasion, and only a few species were recorded continuously throughout the study. This statement applies to all the methods of sampling, including sweeps and cage plots (Fig. 46), foliage quadrats and counts at flowers (Fig. 90). In

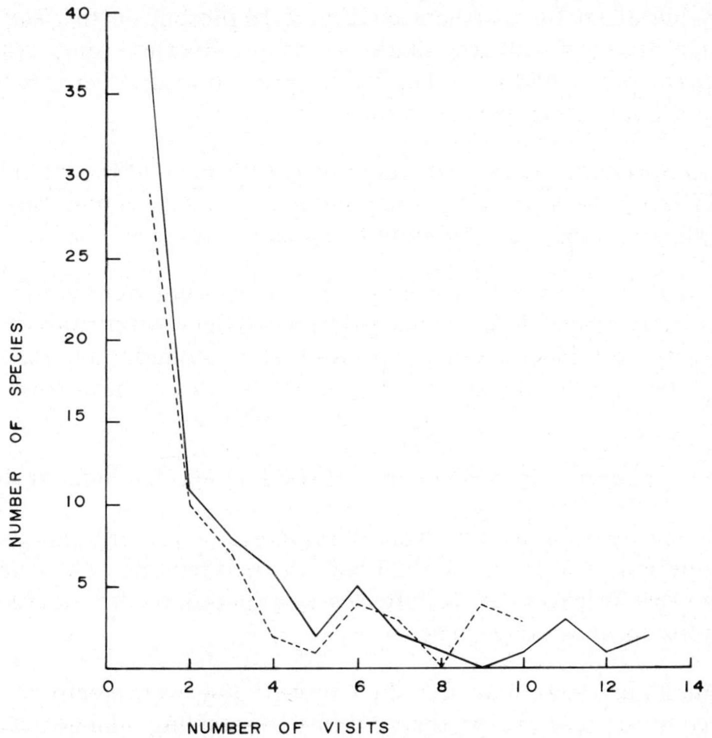

Fig. 90. Frequency of species captured at different numbers of sampling visits. Solid line = arthropods associated with *Argusia* flowers. Broken line = arthropods on *Argusia* and *Pisonia* foliage.

all cases the most numerous category by far contained the species sampled only once during the study; there was a rapid decrease in number of species captured more frequently.

The absence of a species from a particular type of sample may be due to two causes in addition to extinction from the island. Species could have moved from one habitat (e.g. the shrub foliage) to another one (e.g. the herb flat) seasonally or as conditions changed, and a species registered extinct by one sampling method may have become 'extinct' only in a particular habitat, not for the island as a whole. Therefore, a table of presence or absence of species in the combined sampling methods was analyzed to provide island-wide estimates of extinction and immigration. Absence from all the samples collectively may also in some cases be due to sampling error, rather than extinction. The probability of a species being missed depends on its abundance and ease of sampling. Rare species are more likely to be missed than abundant ones, and consequently will have a greater number of spurious gaps in their sampling record. We recognized that a gap of a single sampling period was unlikely to represent a true absence if the species were present both before and after that period. Two consecutive absences in samples would be a less probable event for a species continuously present on the island, three consecutive absences still less likely, *et cetera*. Because some criterion of relative permanency had to be applied in order to analyze species turnover, we set up the following working guidelines:

Permanent Species—species that were present throughout the intensive study (1968–72) except for gaps in the sampling record not exceeding two consecutive sampling periods, i.e., 4–6 months apparent absence.

Resident Species—species that were not permanent but were present for two or more sampling periods not separated by more than two periods of absence. This category would include those species that emigrated to the island during the study, became locally extinct during the study, or both (one or more times).

Established Species—the permanent and resident species collectively.

Transient Species—species that were observed to be present only during one sampling period, or if for more than one, the two periods were separated by more than two absences. In the latter case, a species was considered to have been a transient more than once.

These guidelines were not slavishly applied but were tempered by our knowledge of the seasonality, susceptibility to sampling, and general biology of particular species, e.g., a species that occurred only in the beach rubble and was observed frequently there throughout the study but with gaps of more than two consecutive periods would be considered a permanent species as the beach fauna was not subjected to regular sampling. We believe that our assessment is as accurate as the data permit in the absence of specific information on the reproductive success of all species.

Using the above guidelines, the total time span for each species on the island was indicated by a line on a linear calendar chart. The number of immigrations and extinctions were tallied by each taxon for the island as a whole. The transients were probably greatly underestimated as many probably visited the island briefly between sampling periods; some very rare species may have been missed even when they were present during research visits. The established species, generally having larger populations would be less subject to underestimation. Simberloff (1976) has pointed out that such species turnover is probably 'pseudoturnover' as it involves transients, and Lynch and Johnson (1974) have pointed out that transient species, i.e., those not breeding on an island, should not be considered in assessing immigration and extinction rates. We concur, and such a treatment is consistent with the terminology set out in Chapter 4. Diamond and May (1977) have pointed out that long intervals between censuses result in an underestimate of species turnover. Our intervals were short (two months) and this effect would be minimal.

Of the 396 species sampled during the study, 224 (or 56.6%) were transient (Table 43). The average proportion of transients at particular sampling periods was about 21% (Table 42). Of the 172 species which were established at least once, only 26 were permanent (15.1% of the established species or 6.6% of the total species recorded).

The proportion of the total species that were transient or that were established did not vary greatly with time (Table 42), greatest differences being only slightly more than two-fold throughout the intensive study (12.8% to 26.6% for transients). During periods when there were few established species there tended to be few transients and *vice versa*. This suggests that the same conditions which favoured the persistence of established species also permitted transient individuals to live long enough to be included in samples, even though they did not become established breeding populations.

Different taxa differed markedly in the extent to which they occurred as transients. The group with the highest proportion of transient species were the beetles (Table 43); less than 20% of the species of beetles recorded on One-Tree ever became established, whereas for all other groups more than 30% of the species became established at some time or other, and for some taxa the values reached as high as 70%. It would appear that the disproportionately low representation of beetles on One-Tree is partly because a large number of the species that reached the island failed to become established. Also, fewer beetles may have reached the island in the first place (see p. 160).

With the exception of spiders, wingless groups constituted a small proportion of the transient category. Many wingless species were permanently

Table 42. Dynamic Characteristics of the Epigeal Arthropod Fauna of One-Tree Island. I = immigration; E = extinction; $\hat{S}$ = equilibrial species number during enduring equilibrial phase (March 1970 to March 1971) = 79.3 species.

	1969						1970						1971	
	J–F	M–A	M–J	J–A	S–O	N–D	J–F	M–A	M–J	J–A	S–O	N–D	F–M	$\bar{X}$
Transients														
% of total species for date	20.0	26.6	20.8	20.0	26.0	24.3	12.8	14.6	24.7	23.2	19.4	20.6	18.4	20.9
% of established species for date	37.5	50.0	32.0	29.8	50.8	47.8	20.0	23.0	46.2	39.3	32.3	34.9	30.5	36.5
% of $\hat{S}$	15.1	26.5	20.2	21.4	40.4	41.6	17.7	17.7	30.3	27.7	25.2	27.7	22.7	25.7
Established species														
% of total species for date	80.0	73.4	79.2	80.0	74.0	75.7	87.2	85.4	75.3	76.8	80.6	79.4	81.6	79.1
% of $\hat{S}$	60.5	73.1	76.9	85.8	114.8	129.9	119.8	103.4	92.1	92.1	104.7	107.2	100.9	97.0
I/E	7.00	3.00	1.75	11.0	3.67	1.27	0.39	0.31	0.62	3.25	1.63	0.83	0.50	2.02*

* Omitting the value for J–A 1969.

Table 43. Transient and established species in different taxa of arthropods (excluding soil microarthropods) from One-Tree Island, 1968–1972.

Taxon	Transient species		% established species		Total
	No.	%	No.	%	
Araneida	18	56.3	14	43.8	32
Orthoptera	3	30.0	7	70.0	10
Hemiptera	19	48.7	20	51.3	39
Coleoptera	27	81.8	9	18.2	36
Diptera	56	60.2	34	39.8	90
Lepidoptera	39	68.4	18	31.6	57
Hymenoptera	52	53.6	45	46.4	97
All other winged taxa	7	43.8	9	56.3	16
All other Apterous taxa	3	15.8	16	84.2	19
Total	224	—	172	—	396

present on the island, the low number of transients probably being due to their difficulty in dispersing overwater. Spiders contributed significantly to the transient group, probably because of the many species which disperse aerially using gossamer nets.

Arrival of epigeal arthropods to the island was relatively frequent, and never dropped below 20 species added since the previous visit (slightly less than a two-month interval) and on one occasion exceeded 50. A number of these were transient species which did not contribute to the established fauna. However, the arrival rate of species that subsequently became established never dropped below 10 species per visit to the island and once exceeded 20 (Fig. 91). At any one time, the new arrivals usually included more transients than immigrants that would become established, sometimes the transients outnumbering the others by more than a factor of two. This is almost certainly an underestimate as many transient species were probably overlooked. The sole exception was in January–February 1969 just after the drought. On that occasion, the subsequently established immigrants slightly exceeded the transients.

Arrival of new species showed a broadly seasonal trend (Fig. 91) and the pattern was similar for both transients and subsequently establishing species. The lowest frequency of arrival at One-Tree was May–June in 1969 and March–April in 1970. The peak periods of arrival of epigeal arthropods at One-Tree was more extended in 1970 (May to December) than in 1969 (September to December). However, the 1969 peak was considerably greater than in 1970 (Fig. 91). It appears that 1969 was characterized by an intense but short-term movement of arthropods to the island in spring and early summer, and that 1970 had a less intensive but seasonally more prolonged period of high incidence of arrivals (late autumn through winter and spring to early summer). Such yearly differences could reflect differences in the availability (due to differences in activity or abundance) of arthropods on the mainland, or to differences in the conditions influencing dispersal from mainland to island. Information needed to assess the former was not at hand; however, wind directions in 1969 and 1970 were available for analysis.

Although there was not a complete correspondence between numbers of days of winds of favourable direction and rates of immigration, the peaks were closely associated with favourable winds. The two highest peaks of total arrivals (I + T; Fig. 91) were in September–October 1969 and July–August 1970 and the two highest frequencies of favourable winds occurred in September 1969 and July–August 1970 (Fig. 92). Furthermore, the sampling

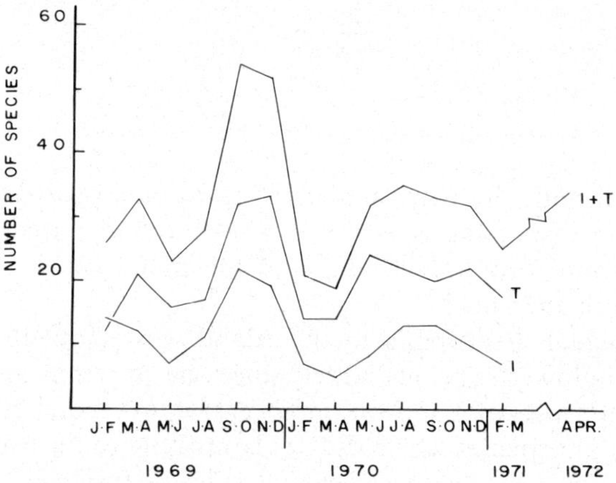

Fig. 91. Arrival of epigeal arthropods to One-Tree Island (all habitats). Values represent number of speices that had arrived since the previous visit (slightly less than a two-month interval). I indicates arriving species that subsequently became established. T indicates arrivals that did not become established and had disappeared before the next visit. I + T indicates total arrivals.

period following two of the three periods of unfavourable winds (March 1969, February 1970; Fig. 92) showed few new immigrants (Fig. 91). The poorest correlation between frequency of arrivals and favourable wind directions occurred in October 1969 (few days of favourable winds yet many new arrivals present at the next sampling) and July 1969 (high incidence of favourable winds but no immediate rise in immigration rate). In general it seems that frequency of new arrivals and frequency of favourable winds are related but that there are exceptions. Other modifying variables may be wind strength and duration of bouts of favourable winds (4 days of continuously favourable wind directions may have a different effect than 4 bouts of one day each, separated by a considerable time interval). More precise analysis of factors affecting immigration rates are not warranted by our data. However, a study designed to test the above variables might be extremely rewarding.

The pattern of extinction was very different from that of immigration.

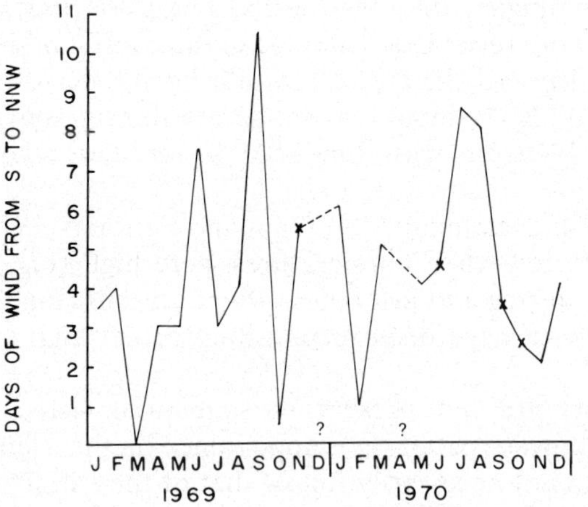

Fig. 92. Number of days on which wind blew from the mainland, thereby improving the probability of aerial transport of arthropods to One-Tree Island. All winds from a southerly direction, swinging westerly through an arc to NNW were considered as blowing from the mainland, see map in Fig. 3. X's indicate data incomplete and dashed line indicates no data available.

Wholesale extinction had probably occurred during the drought. Afterwards there were few species of epigeal arthropods left on the island so absolute extinction rates were of necessity low (Fig. 93). They remained low well into the period of recovery from the drought until with the attainment of a greater pool of species the extinction rate began to rise reaching a peak in January–February 1970. It then declined subsequently to rise again to near peak values in early 1971 (Fig. 93). For both 1969 and 1970 the lowest extinction rates

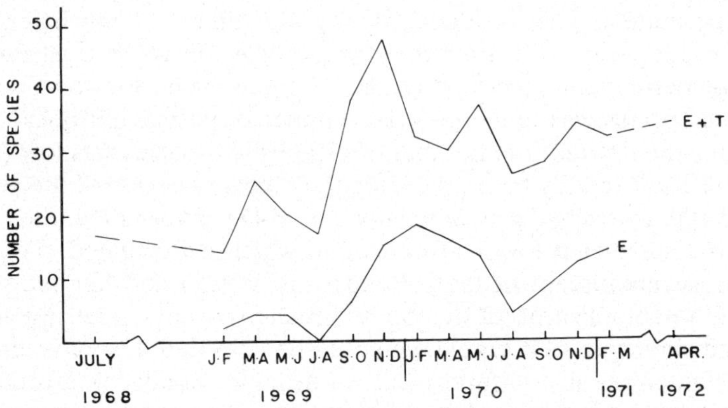

Fig. 93. Disappearance of epigeal arthropods from One-Tree Island. Values are numbers of species that had disappeared since the previous visit (slightly less than a two-month interval). E indicates extinction of previously established species. T indicates disappearance of transients, i.e. species that were recent arrivals at the previous visit. I + T indicates total disappearances.

occurred in July–August, the 1969 period being the only one in which no extinctions at all were recorded. These were the coolest mid-winter months of both years (Fig. 10) and the animals would be subjected to less heat stress than at other periods. Rainfall was moderate during and just prior to that time of year in 1969 but was somewhat lower than average in mid-1970 (Fig. 13).

The periods of highest extinction (January–February 1970 and February–March 1971) occurred when temperatures were high (Fig. 10); rainfall was high or moderate during and just before those periods (Fig. 13) and therefore dryness is unlikely to have been responsible for the high extinction rates at those times.

Since transients are not present in sequential visits, their pattern of disappearance followed that of their appearance on the island (compare Figs. 91 and 93) but did not necessarily follow that of the extinction of established species (Fig. 93). Total disappearances combined the patterns for both groups.

The total number of species present on an island at any one time reflects the interaction of immigration and extinction rates and the temporary presence of transients. Although the transients may be important collectively in their impact on the insular community, it is informative to examine numerical trends separately for those species which are long-term co-inhabitants of the island (i.e. the established species, which include permanents and residents, but exclude transients as defined above). In any event, it is realistic to speak separately of the 'equilibrial number of established species' whether or not one generally chooses to include transients in the concept of equilibrial number.

On One-Tree Island only 26 species of epigeal arthropods were permanently present during the study. However, the total number of established species (permanent plus resident) at any one time was much higher, and except for the period just after the drought was always at least double the number of permanent species (Fig. 94). At one stage, the total number of established species was four times the number of permanent ones.

Just after the drought the number of established species was low (less than 50 species) but it rapidly built up during the recovery period of the vegetation to exceed 100. It reached a peak in November–December 1969, then dropped sharply to a somewhat lower level around which it fluctuated. The form of this curve is reminiscent of that obtained by Wilson and Simberloff (1969) during the re-colonization of defaunated mangrove cays. The two situations are indeed similar in that One-Tree could be considered to have undergone partial defaunation as a natural result of drought. Similar interpretations of the two curves probably apply. After the drought when number of species was low, many subsequent arrivals would represent species not present on the island and the increase in species-number would reflect a combination of re-colonization by species previously driven to island-wide extinction by the drought and colonization by species previously unknown to the island. Just after the drought, when population densities of all species were low, resources would probably be sufficient to accommodate many of the arriving species, even potential competitors. Consequently, extinction rates during recovery would be expected to be relatively low (as indeed was the case, Fig. 93) and many of the species reaching the island could become established, the number perhaps being influenced more by conditions affecting dispersal from the mainland than by ecological conditions on the island. In late 1969 there was an unusually large number of immigrants reaching One-Tree Island. This early phase (Fig. 91) during which re-colonization occurs from a base of low

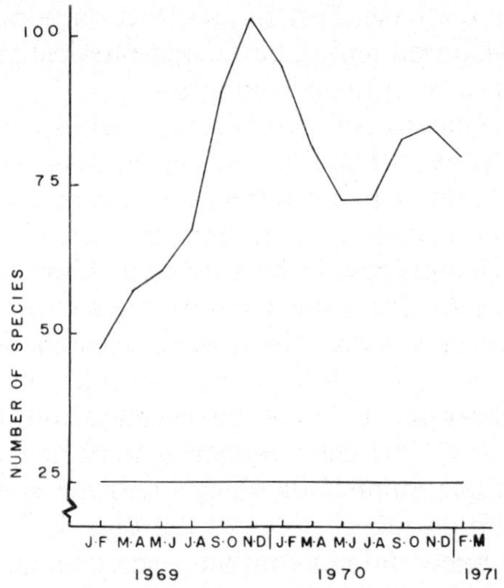

Fig. 94. Temporal change in number of established species of epigeal arthropods on One-Tree Island. Upper line = total number of established species. Lower line = number of species of permanent residents.

species numbers and low population densities was called the non-interactive phase by Simberloff (1969), because of his supposition that immigrating species were probably not influencing survival or population levels of each other to a great extent (see Chapter 22). This resulted in an overshoot of the usual equilibrial number both in Simberloff's study and in the present one[1]. At One-Tree immigration rates were high in comparison to extinction rates during this phase (Table 42) and the I/E ratio averaged 3.39.

During this phase, numbers of individuals were building up (Fig. 49) as well

[1] One difference in the two studies should be pointed out. The drought on One-Tree Island reduced the plant biomass and relative abundances of plant species and may have thus indirectly influenced subsequent establishment of arthropod arrivals. Simberloff's fumigation did not affect the vegetation.

as numbers of species. As population densities increased the likelihood of interspecific competition or other negative biotic interactions resulting in local extinction of species would tend to increase; at the peak in number of individuals in late 1969 (Fig. 49) and subsequent to it, extinction rates increased dramatically (Fig. 93) and few of the arriving species became established (Fig. 91) resulting in a marked decline in number of established species (Fig. 94). The I/E ratio during this period was low (Table 42) and averaged 0.35. Simberloff (1969) called this the interactive phase because of the postulated role of interaction of many species at high densities in contributing to increased extinction and to resistance to establishment of new arrivals. Immediate weather conditions would almost certainly influence the intensity of interactive effects through an effect on population densities and available resources. Consequently, biotic and physical factors may be synergistic in their influence on insular dynamics.

Following the 'overshoot' in number of established species during the non-interactive phase and the decline during the subsequent interactive one, the number of species fluctuated around a somewhat lower value. Simberloff (1969) suggests that this phase represents the enduring equilibrial phase. Gradual long-term changes would be expected, however. As species interactions continue, and involve a spectrum of species that arrive from time to time, one would expect a replacement of species by others that are more compatible with prevailing conditions or with each other and which overlap less in resource utilization. This has been called the assortative phase by Simberloff and Wilson (1970). Such reduction in niche overlap should permit coexistence of a greater number of species and the species-number on the island accordingly should rise slightly. In the One-Tree study, the intensive part of the study probably did not continue long enough after the interactive phase to detect any such possible increase, if indeed it would be detectable in the face of the fluctuations induced by weather changes or stochastic events. If a true equilibrium in species number is reached, the mean I/E ratio should be unity. During the presumed enduring equilibrial phase on One-Tree the ratio showed short-term fluctuations (Table 42) but averaged 1.19.

The question arises as to the applicability of the idea of an equilibrial number on One-Tree Island; and if it is a valid concept what is the correct value for One-Tree? The Equilibrium Theory implies that the number of species on an island remains relatively constant despite change in species composition of the biota. This is clearly not true for One-Tree Island as weather perturbations (such as drought) can have a drastic effect on the number of species. If a single survey of the epigeal arthropods on One-Tree Island had been made with the idea of ascertaining the equilibrial number for relating to island area or distance from mainland, very different results would have been obtained depending on the time that the survey was carried out. The effect would not be negligible and could reach an order of magnitude. If- a survey were made shortly after a drought the usual species number would be underestimated; if made at the peak of the overshoot it would be overestimated.

A further complication is that a single survey would not distinguish established from transient species (and in any event many of the latter would be overlooked on most islands). Including transients as opposed to excluding them on any single survey of One-Tree Island creates a difference as great as 10–30 species, depending on when the survey was carried out. Values could range from 60 to 136 species if all species were included and from 48 to 103 if only established ones were used. Clearly, assessing equilibrial number of species on the basis of single surveys is scarcely possible. Only long-term studies such as the present one can provide a means of confidently assessing it.

Despite the above operational difficulties we believe the concept of equilibrial species number (Ŝ) has significance. We believe it should be restricted to the established species only, or if not it should be clearly stated that transients are included. In the case of One-Tree Island the equilibrial species number could realistically be considered as the mean number around which values fluctuate during the enduring equilibrial phase. Temporary large departures from that general level caused by perturbations such as droughts, and the overshoot during recovery and recolonization could be looked upon as special cases.

If one simply averaged the total species known to be present at the various visits (including transients) during the period of intensive sampling (January 1969 to March 1971), Ŝ would equal 96.5 species (Table 42). The corresponding value for established species only is 76.9 species. The period from March 1970 to March 1971 (the presumed enduring equilibrial phase) provides an Ŝ of 99.3 for all species and one of 79.3 for established ones. We believe the last value to be the most realistic one if some single value is to be accepted.

However, it may be more useful not to consider a single equilibrial value but to think rather in terms of the *tracking* of some hypothetical equilibrial species number. There are at least two sources of variation in species numbers, stochastic events and environmental perturbation and it may be important to distinguish between temporal change in species number arising from environmental fluctuations themselves and those resulting from 'turnover noise' (random immigration and extinction events independent of environmental change) (Diamond and Gilpin 1950). Even if the environment on an island were perfectly constant, which of course it never is, there would still be fluctuations in species numbers arising only from the stochastics of immigration and extinction. Superimposed upon this variation in species number, is that caused by environmental fluctuation. There is always some fluctuation in the environment and each different level of conditions probably has its own characteristic equilibrial number. Since no one level is maintained indefinitely, the equilibrial number also changes; consequently one can think of a floating equilibrium point tracked by an island's biota rather than adhering to the idea of a fixed equilibrium. Future research could profitably be devoted to ascertaining how rapidly species number tracks a shifting equilibrium point, either in terms of short term environmental perturbations or in following systematic long term trends of environmental change. In a theoretical vein, it is pertinent to ask just how much change in species number can be tolerated while still retaining the concept of an equilibrial species number.

22. Community organization

There are two major components of community organization – structure and function. In this chapter various aspects of both are examined from several standpoints not considered in previous chapters. Preliminary consideration is given to some aspects of bioenergetics, the relative abundance of different species is described, and trophic relations are analyzed and cast within the framework of current theory.

Bioenergetics

An important aspect of the organization of a community is the degradation of energy and the cycling of materials such as carbon, water, nutrients or ions through the various trophic links. The present study was directed primarily towards other aspects of community ecology, although weight of matter (biomass) was examined in relation to the vegetation and to a few selected animals. Most of the samples of insects were required for use in identification and neither biomass nor caloric content was determined on a regular basis. One way that the present study could be extended would be to look at energy flow in the One-Tree community, either on a gross level or by paying attention to particular links in the food web. We have made a small step toward such studies by ascertaining the weights of individuals of some species, oxygen consumption of some insects[1] and caloric content of some organic materials in the One-Tree community, using a Parr bomb calorimeter (Appendix VI). These data can serve as a reference resource for future studies.

The caloric contents of different plant parts varied slightly among species. In some plants, roots had the highest energy content, in others it was the stems. Buds and fruits were relatively high in energy. *Canavalia* and *Ipomoea* tended to have a greater caloric content than did the corresponding parts of

[1] Measured in pipettes submerged in water baths. CO_2 absorbed by soda lime. For methods see Muir (1975).

Abutilon and *Melanthera.* It is tempting to speculate that the higher proportionate insect damage to the former two as compared to the latter two (Figs. 62, 63) might relate to selection of the energetically richest foliage by generalized grazing insects.

Dead standing stems (a mixture of all species) had an energy content similar, or slightly below that of live stems of species tested, whereas leaf litter had a lower caloric content than fresh leaves, and the energy in humified organic material was quite low.

In all cases the energy content of One-Tree plants was low in comparison to values presented for their respective families by Cummins and Wuycheck (1971).

Individual weights of various animals and seeds also occur at appropriate places in Chapters 14 and 18.

Equitability

Equitability refers to the distribution of individuals among species. If all species have about the same number of individuals, equitability is high. If a few species have many individuals and others are rare, equitability is low (individuals are unequally distributed among species).

The epigeal arthropods of the herb flat were treated as a community and the distribution of individuals among species examined for each sampling period. The total fauna, including transients, was treated.

Equitability can be expressed in a number of ways. The most informative one is a graphical method in which the species are arranged in order of decreasing relative abundance and their contributions to the total numbers of individuals expressed as a percentage (Fig. 95). Cody (1975) reviewed the use of indices to express equitability and concluded that as a one-parameter description of the distribution, 'the simple dominance measure, $d=$ (number of individuals in the most abundant population)/(total number of individuals), seems as good as any'. The d value is represented by the tallest bar in each of the histograms of Fig. 95. The graphic method thus shows not only the index d but also the way the rest of the individuals are distributed among species. The shape of the frequency distributions reveal much more about community structure than can be conveyed by a single number such as the d index, or other indices of equitability.

In the herb flat, the most abundant species accounted for 13% or more of the total number of individuals at all sampling periods and on three occasions values were high (between 35% and 45%). At 7 of the 14 sampling periods, between 15% and 25% of the individuals were of the most abundant species (Fig. 95).

The second most abundant species often accounted for 1/2 to 3/4 as many of the individuals as did the most abundant one, although the relation between

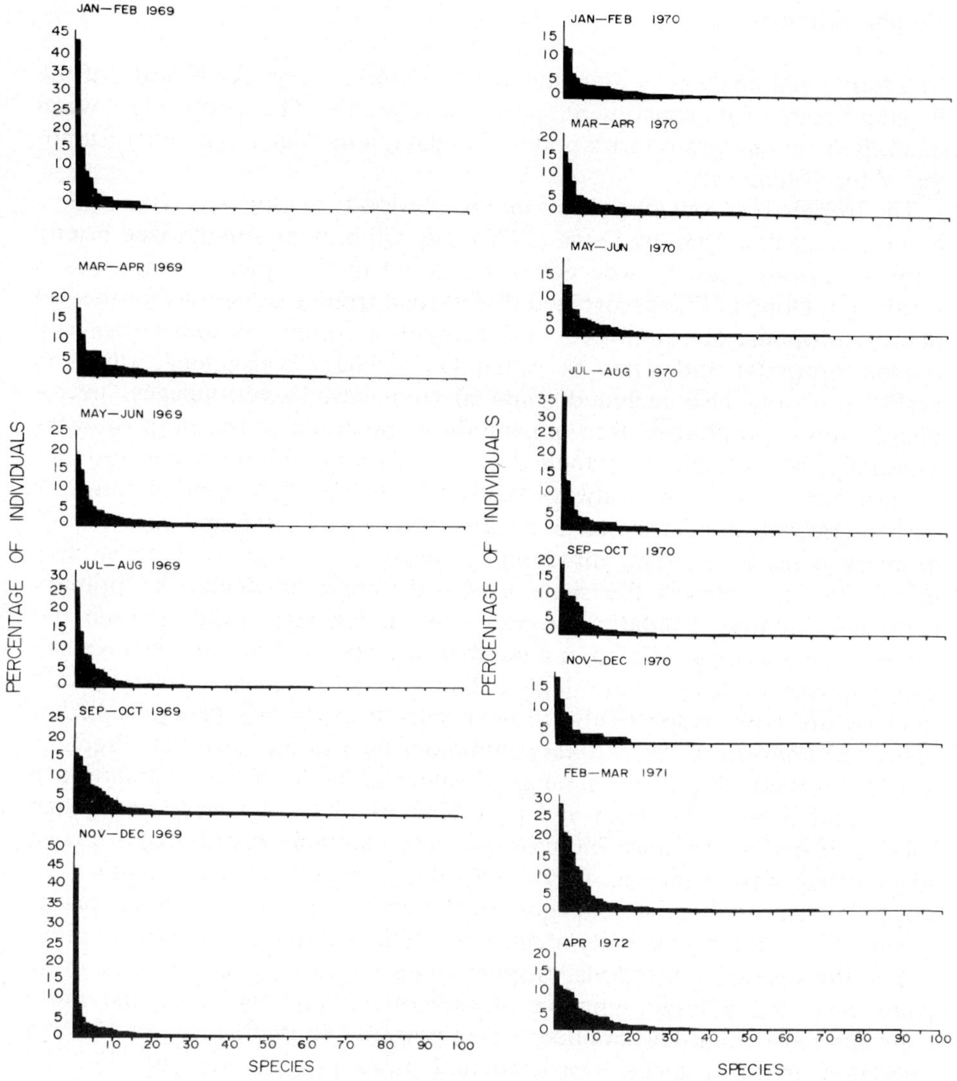

Fig. 95. Equitability of epigeal arthropods in the herb flat. Species are arranged in decreasing order of relative abundance.

the two most abundant species varied widely, ranging from the second being 1/5 to over 9/10 as abundant as the first species. The greatest decline in relative numbers of individuals was usually among the 2–4 most abundant species, after which the distribution of individuals among species declined more gradually. There was usually a rather large number of relatively rare species, many of which were transients.

Trophic relations

In a functional analysis of any animal community, precise knowledge of the feeding habits of its constituent species is desirable. This knowledge, when quantified and integrated with population data, adds much to an understanding of the community.

The trophic relations of the soil microarthropods of One-Tree Island have already been described by Done (1975) and will only be summarized briefly here. Following that, a more detailed account of the epigeal arthropods is presented. Done (1975) recognized three broad trophic categories for the soil microarthropods, (1) omnivores, (2) carnivores (predators upon other soil microarthropods) and what he called (3) 'primary consumers', which in reality probably also included some mycophages, bacteriophages, necrophages and coprophages. Recognizing the imprecision of the third category especially, he was able nevertheless to formulate the following conclusions.

Carnivore biomass was always between 2% and 14% that of 'primary consumers' whereas carnivore abundance was between 5% and 25% that of 'primary consumers'. The direction of change in carnivore biomass was usually but not always the same as the direction of change in 'primary consumer' biomass. Predator and prey size relationships in the herb flat and *Pisonia* communities differed in a number of respects. Prey species diversity and biomass diversity were higher than predator species diversity and biomass diversity, respectively. Prey species diversity was generally higher and more uniform in the *Pisonia* community than in the herb flat. Predator species diversity had a similar range of values in the herb flat, *Pisonia* grove and succulent mat. Predator and prey biomass diversities were lower than species diversities because biomass was less equitably distributed amongst species than were numbers. In the herb flat, predator diversity changes did not parallel prey diversity changes. In the *Pisonia* community both species diversity and biomass diversity of predators followed species diversity of prey.

For the epigeal arthropods, trophic levels could be assigned with more confidence and a larger number of categories were employed and more conventional terminology applied. It is important to note that we use the term 'predator' in its original, restricted and more precise meaning, i.e., an organism that kills animals for food (Hanson 1962, Child and Child 1971).[2] We deplore the practice (whether through ignorance or through deliberate design as a presumed convenience) of using the term 'predator' to refer to granivores (e.g. 'seed predators') or using 'predation' in any context of herbivory.

On each occasion that we observed a trophic interaction in the field, its nature and the species involved were recorded; such observations were very

[2] Also the Shorter Oxford English Dictionary (revised and corrected third edition, 1974) lists under the heading *Predatory:* 'Of an animal: that preys upon other animals; carnivorous.'

incomplete and information on the feeding habits of some species is not available. Certainly, the full range of food species utilized by generalized feeders would not have been recorded. Our observations were supplemented by reference to the literature and by accounts of the general feeding habits of various higher taxa.

Some of the species on One-Tree Island were rather catholic in food habits, i.e., the predators such as spiders and centipedes had a wide range of prey, the grasshoppers, locusts and some other herbivores fed on a number of species of plants, many of the nectarivores visited a variety of flowers, the scavengers fed on almost any species of carrion available, and a number of species of omnivores were present. Other species had narrow food preferences, in some cases being restricted to a single food species. For example, the moth *Utetheisa pulchelloides* fed only on *Argusia*, a gall fly was attracted only to *Melanthera,* an halticine beetle was restricted to *Solanum*, a lygaeid bug only fed on the seeds of *Abutilon*, a pentatomid bug was restricted to *Pisonia* fruits, and it is likely that many of the entomophagous parasites were restricted to particular species, although this was not ascertained in any case. One species of hymenopteran (unidentified) that emerged from *Pandanus* seeds leaving large holes (Fig. 22), was apparently parasitizing a borer feeding upon the seeds. Both the host and parasite were probably restricted in their diet, at least on One-Tree Island.

Trophic structure of species in the herb flat

In Chapter 14 it was indicated that throughout the study very marked fluctuations occurred in numbers of species and in population densities (Figs. 48, 49 and 50), and major alterations in taxonomic composition of the epigeal arthropod fauna took place both at the species level and at those of higher categories (Figs. 48 and 50). A picture emerged of a rapidly changing, highly dynamic situation. Because of a previous suggestion that trophic structure may remain stable under otherwise fluctuating conditions, even during early periods of colonization (Heatwole and Levins 1972b), trophic structure of the One-Tree fauna was analyzed. The herb flat had the highest number of epigeal species and individuals and experienced the most intensive and long-term sampling and was therefore the most suitable site for analysis of trophic structure.

Herbivores constituted the largest trophic group and usually accounted for about 25–35% of the species (Fig. 96). Predators made up the next largest category and usually represented 15–20% of the species. Omnivores followed and were about equal in number to nectarivores. The smallest trophic groups were the parasites (mostly entomophagous hymenopterans) and the scavengers.

Those trophic categories feeding entirely on live plants (chewing and

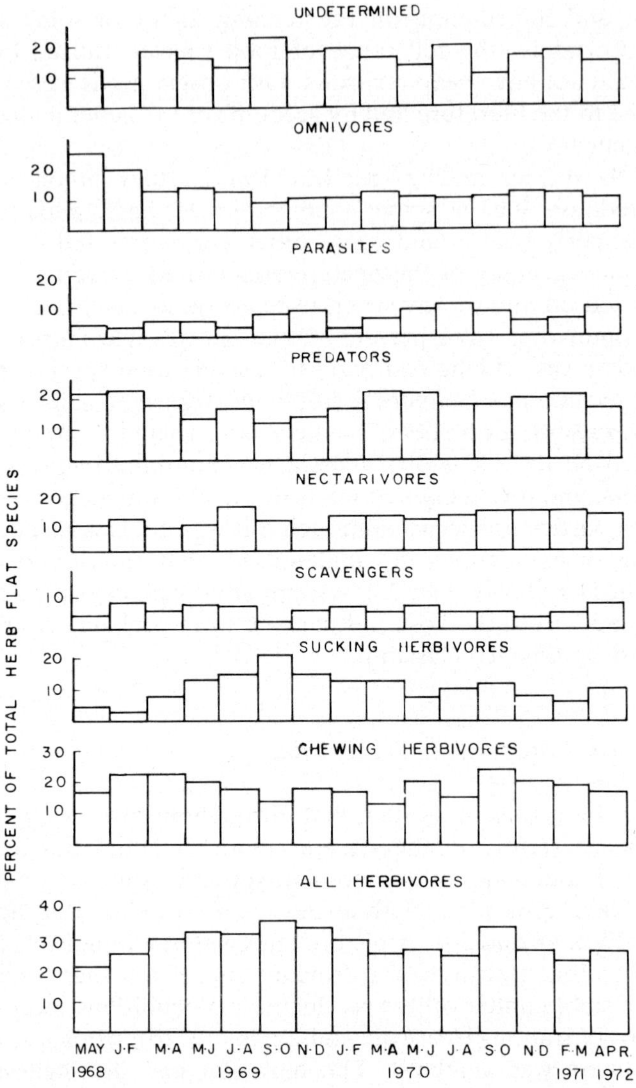

Fig. 96. Temporal change in trophic structure (by species) of the epigeal arthropods of the herb flat.

sucking herbivores plus nectarivores) collectively made up about 35–40% of the species, and the strictly animal eaters (predators, parasites) collectively accounted for about another 25%. The omnivores, or those feeding on both plant and animal matter, accounted for about 15% of the species and the scavengers (including those feeding on dead animals and dead plant material) less than 10%. The number of species of plant feeders thus was larger than the number feeding on animals. On the average about 15% of the species' trophic positions could not be assigned reliably to any particular trophic group.

The most striking feature of trophic structure analyzed by numbers of species, was that it remained remarkably stable throughout the study; with a few minor exceptions the fluctuations was only about 10% of the given trophic level over the entire period from May 1968 to April 1972 (Fig. 96).

The empirical fact that the trophic structure remained nearly stable during the study is subject to several possible interpretations, including: (1) Trophic structure stability was an artifact of sampling methods and/or assignment of trophic status. (2) The One-Tree fauna at any one time represented a random subset of the pool of species which easily dispersed from the mainland, and hence merely reflected the trophic structure of that pool. (3) Community regulation of trophic structure occurred, i.e., biotic interactions affected establishment and/or extinction of species in such a way that a given trophic structure was maintained. These will be examined in turn.

Since a considerable number of species could not be assigned a trophic designation with confidence, there was a residual number of 'undetermined' status at each sampling period. We assumed that the trophic structure of the 'undetermined' species did not vary greatly over time. However, if on some occasions many of these unknowns were in one trophic category but on other occasions were mostly in another, the observed constancy of trophic structure could be an artifact; correct assignment of all species to their proper trophic categories would then produce a variable trophic structure. We believe this unlikely as only about 15% of the total species were of uncertain trophic relations and even considerable bias arising from temporal change in trophic structure among them would not alter the end results greatly (but see discussion of undetermined species of transients, below).

It would seem to stretch credulity to suppose possible errors in assigning trophic status to species in a community with changing trophic structure would abolish those changes and result in the artifactual appearance of stability.

If sampling procedures favour catching certain trophic categories on some occasion and others at other times, and if that bias was such each time that real differences in trophic structure were equalized, an artifact of trophic structure stability would result. In addition to the statistical improbability of such events occurring, the fact that two independent methods of sampling were employed, each of which tended to catch those species escaping from the other method (see Chapter 14) renders this source of bias extremely unlikely.

If there are only a small number of species, each one accounts for a greater percentage of the total than when there are more species in the sample. This effect would tend to promote greater differences in trophic structure when species numbers were low (an extreme case would be a species number of 2 in which for any given trophic category the only possibilities are 0%, 50% or 100%). In the present data, the trophic structure was temporally stable despite widely different numbers of species involved. The above effect would be expected to increase the temporal variability rather than reduce it; it seems unreasonable to postulate that the effect would always be in a direction to produce the artifactual appearance of less temporal change than in fact occurred.

In conclusion, we do not believe that the observed temporal constancy in trophic structure is an artifact of either sampling or statistical error, or bias in assignment of trophic status to species.

The hypothesis that One-Tree Island's arthropod trophic structure merely reflects that of the easily dispersed component of the mainland pool cannot be tested without knowledge of the composition of that pool. However, a comparison may be made between the trophic roles of those arrivals which become established and those which do not. If they are similar it would appear that establishment is random with respect to trophic status; if they are quite different, it would suggest that ability to become established is not independent of trophic status.[3]

A comparison was therefore made between those species which established themselves on the island and those which did not (i.e., transients – see above for definitions of permanence and transience). Fig. 97 indicates that in general the trophic structure of the established immigrants was more constant than that of the transients. (May 1968 was an apparent exception in this regard; however, this preliminary sampling was quite cursory and restricted to the plot method, giving less complete and reliable data than was obtained later. The trophic structure of the May 1968 sample is therefore quite probably artifactual.)

Part of the difference between established and transient species may be attributed to the smaller numbers of transients than established immigrants sampled at any one period (Table 42) and the consequent greater percentage of the total represented by each species in the former category. However, this source of variability would not seem to account for all of the difference, and it is clear that much of the temporal variability of the trophic structure of the entire herb flat epigeal fauna (Fig. 97) arises from the transient component.

[3] The simplest form of the hypothesis, i.e., that composition of the established fauna should mirror that of the arrivals, is the one discussed here. More sophisticated ones could be designed if there were sufficient data to test them. For example, a more complicated and realistic hypothesis would weight the probability of establishment of the arrivals by the number of individuals of each species arriving. Diamond (pers. comm.) has suggested that an alternative test might be based on comparison of the actual fluctuations in trophic composition of the established species with the expected fluctuations based on the number of arrivals. If the number of established individuals is small compared to the number of individuals arriving within a time span that is a significant fraction of the lifetime of the established individuals, than one would expect a high turnover rate of established species if there were no trophic interaction; trophic interactions would tend to reduce turnover rate. Unfortunately, precise quantitative data on the rate of arrival of individuals to One-Tree (as opposed to the standing crop of transients) is not available.

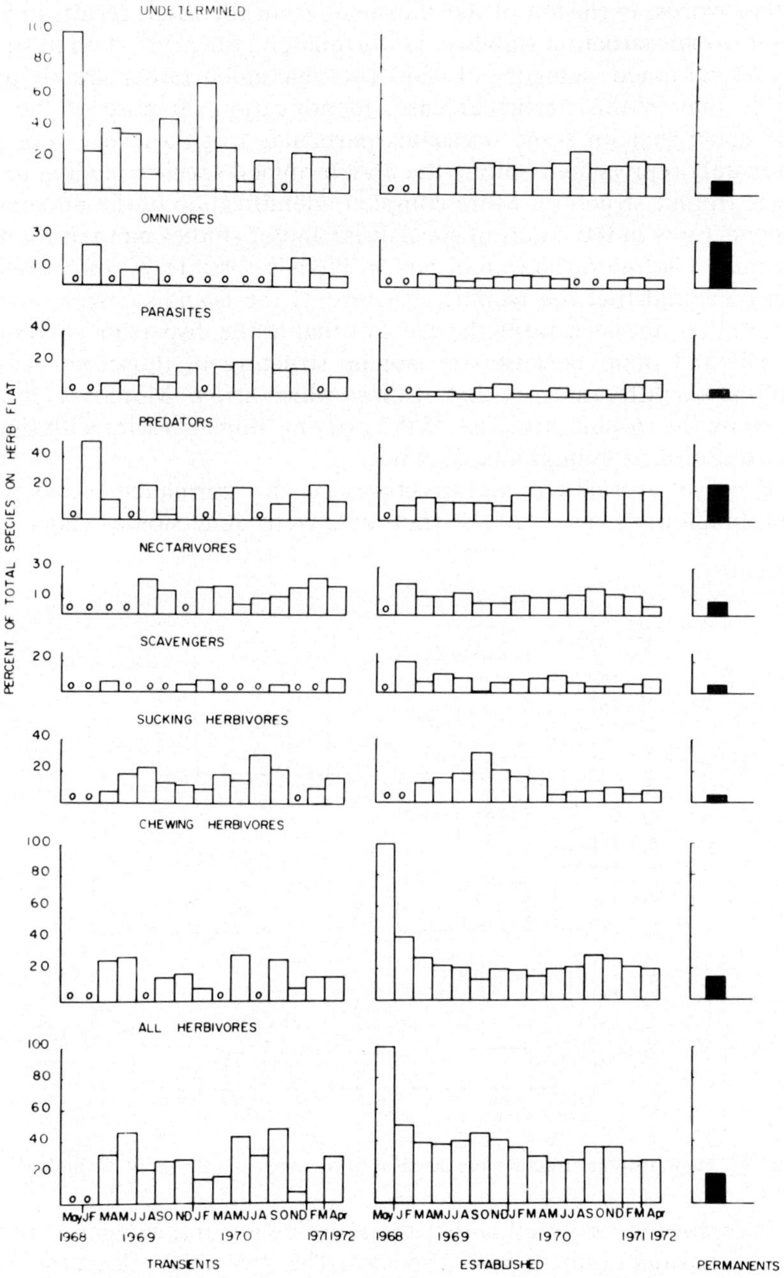

Fig. 97. Temporal change in trophic structure (by species) of transients, non-permanent established species and permanent species of epigeal arthropods in the herb flat.

In other words, exclusion of the transients from the total, results in an even greater trophic structure stability. Unfortunately, the proportion of species in the 'undetermined' category of transients fluctuated rather greatly and it is possible that in this particular case, identification of more of the species would show that on some occasions particular trophic levels were disproportionately represented among the undetermined species, leading to alterations of trophic structure. More complete identification of the undetermined transients (now in the hands of specialists) and/or studies on their food habits are required before resolution of this problem is possible. At the present time it can be stated that the trophic structure of the island's epigeal arthropod community is not necessarily the same as that of the dispersing component of the mainland pool, because the trophic structure of those which become established is different from that of those which arrive. Moreover, it appears that, while the trophic structure of the arriving animals varies with time, that of the established immigrants does not.

The mean overall trophic structures of the permanent residents, the establishing immigrants and the transients were quite similar (Figs. 97, 98).

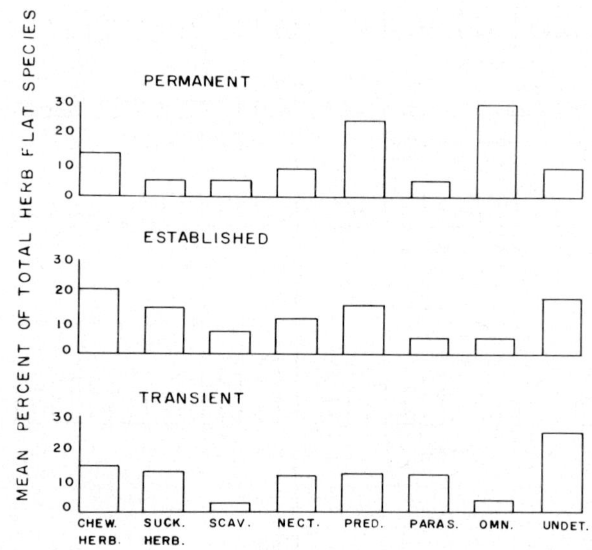

Fig. 98. Mean trophic structure (by species) of the epigeal arthropods of the herb flat.

One difference to be noted is that the more ephemeral categories contain a larger proportion of unassignable species. This probably reflects the fact that common, persistent species are more likely to be described and their habits known, and more likely to be collected in sufficient numbers to permit easy identification. More interesting differences shown by Figs. 97 and 98 are that (1) there were relatively few scavengers among the transients, and (2) there were proportionately more predators and omnivores among the permanent species.

These differences are probably related to differences in mode and frequency of dispersal. Flotsam dispersal to One-Tree seems to be of relatively low frequency compared with aerial transport (Fletcher 1973). The flotsam mode is of particular importance in flightless groups or weak fliers (e.g., ants, earwigs, centipedes and some spiders) a disproportionately high number of which are scavengers, omnivores and predators. Consequently these would arrive infrequently but would accumulate over time, eventually reaching a proportion in the total fauna far above their representation in the transient category which is dominanted by aerially transported species. The transient and more variable component of the omnivores, scavengers and predators consisted mostly of flying forms such as dipterans or of spiders that disperse aerially by gossamer.

The herbivores, nectarivores and parasites are represented largely by strong flying insects or those fliers which are minute and light in weight and consequently easily transported passively by air currents. Many species could reach One-Tree Island and become established. Many others, not able to establish themselves on One-Tree Island would nevertheless reach it and be counted among the transients.

The above trophic structure may be interpreted in the framework of MacArthur and Wilson's (1967) model. Flotsam transported species, biased towards scavengers, omnivores and predators arrive infrequently, become established, and accumulate over time. Once most of the potentially viable colonists have arrived few new species will be added. Many species in these trophic categories are nocturnal, cryptic in habits, or nest underground or otherwise escape the full brunt of the epigeal environment. Both immigration and extinction rates are thus probably low.

By contrast, it may be postulated that turnover rate of the aerially dispersed species is high, or that species of herbivores, nectarivores and parasites which arrive by air, are unsuited for the insular conditions and do not become established. Immigration rates are high, allowing vacant niches to be quickly filled. On the other hand, because most of such species do not seek refuge underground, and are more subject to the vicissitudes of the weather, extinction rates can be high. Interspecific competition among residents or between residents and newcomers might also result in displacement of some species by others.

Two explanations can now be offered in regard to the remarkable stability of the island's trophic structure through time (Fig. 96).

(1) The trophic structure is quite stable over time because there is a large number of species arriving and these become established (or extinct) at rates which are independent of trophic status. The observed trophic structure stability is a reflection of that of the mainland pool with easily-dispersed species dominating and their structure superimposed upon a group of poorly

dispersed but persistent species (especially scavengers and omnivores). The observation that the trophic structure of established species over time is more stable than that of the transients is an artifact which would be eliminated if taxa in the 'undetermined' trophic category were correctly assigned (as discussed above). Environmental perturbations cause a constant proportion of extinction across trophic levels with the result that trophic structure is not altered.

(2) The trophic structure is stable over time because for a given type of community the number of species in one trophic level is determined at least in part by the number in other trophic levels. Given the large number of species arriving on the island any chance niche vacancy can be quickly filled and trophic structure re-established. The constant proportion of different trophic classes indicates an absence of vacant niches, and so new arrivals either fail to establish and thereby join the ranks of the transients, or displace an established species. The stability of trophic structure of the established species reflects these *in situ* interactive processes; the instability among transients results from the vagaries of over-ocean dispersal and a possible diversity of geographic origins. If environmental perturbation causes a change in resident species-number resulting in abnormal representation of certain trophic types, normality is restored by extinction of residents and/or establishment of new arrivals.

We do not favour the first of these two hypotheses because it demands what we consider to be an unacceptable level of coincidence. We do favour the second because it is consistent with much of the other modern information relating to community structure and function (see below). However, internally the One-Tree data could be reconciled with either hypothesis. At this stage we feel that it is parsimonious to accept the second alternative as the more likely of the two. This procedure has the added advantage of leading to further investigation into the role of interactive processes in influencing community structure and to the design of research into mechanisms of community organization. Within the past few years such studies have been fruitful indeed (for a review see Cody and Diamond 1975).

Trophic structure of individuals in the herb flat

When the relative abundance of individuals in each trophic category is plotted, a considerable degree of change through time is indicated (Fig. 99). This change is in contrast to the rather constant proportional representation of numbers of species discussed above. It must first be pointed out that the sampling methods used are likely to estimate numbers of species more accurately than total abundance of individuals or number of individuals in any trophic category. This being the case, more of the variability in the relative abundance spectrum through time would result from sampling error than

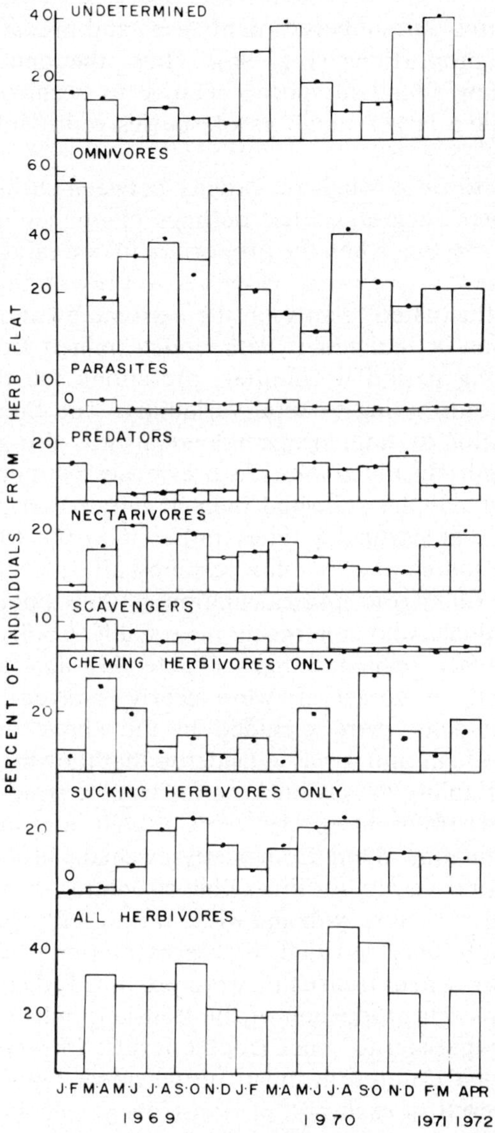

Fig. 99. Temporal change in trophic structure (by individuals) of the epigeal arthropods of the herb flat. Histograms indicate values for the total fauna, dots those after exclusion of transients.

would be true for species numbers. However, for reasons outlined in Chapters 7 and 14, we believe the estimates of population densities to be indicative of actual levels.

The trophic category with the greatest proportion of individuals was that of omnivores (Fig. 99), and thus contrasted with the trophic structure by species in which herbivores outnumbered omnivores, and predators and omnivores were in similar proportions (Fig. 96). Thus, the omnivores tend to be represented by few abundant species relative to herbivores and predators, which tend to have larger numbers of species represented by fewer individuals.

There seemed to be a rough reciprocity between sucking herbivores and chewing herbivores suggesting that perhaps plants could not support high numbers of both and that when the proportion in one category was low that in the other increased or *vice versa*. However, representation of all herbivores collectively also fluctuated greatly, in the extreme by about five-fold.

The proportions of individuals with undetermined trophic status varied from one sampling period to another, presenting the problems discussed previously and contributing to total variability. The fluctuation in proportional representation of omnivorous individuals was also great. This is partly because the category included ants. As previously mentioned, the number of individual ants in samples reflected their activity patterns and the environmental conditions at particular times rather than the total number in the population. This source of variability certainly affected the values obtained. Exclusion of the omnivores and calculation of the trophic structures of the remaining individuals, did not result in a stable trophic structure; rather temporal fluctuation remained high (Table 44) and factors other than temporal difference in ant activity were clearly involved.

Transient individuals were included in the above analyses of trophic structure of individuals and it was considered that they may have contributed to the overall variability as was the case for trophic structure categorized by species. However, transient species were usually low in numbers of individuals and eliminating them from analyses made little difference to the amount of temporal variability in trophic structure of individuals (Fig. 99, Table 45). Trophic structure with and without transients was very similar. The only consistent difference was that the representation of chewing herbivores was always lower when transients were excluded than when they were included. Thus, a high proportion of the transient individuals were chewing herbivores in comparison to other trophic levels. Chewing herbivores were the only group that differed in mean proportional representation by more than 0.8% as a result of exclusion of transients (Table 45).

The peaks of high proportional representation of herbivore individuals did not relate in any consistent way to peaks of standing crop biomass of total plants or of leaves in the herb flat (compare Table 7 and Figs. 25 and 99). However, proportion of herbivorous individuals was high in September–October 1970 (Fig. 99) after a period of high net production of leaves (Table 13) but had dropped sharply by November–December after a period of low or even negative net production of leaves by the major herb flat plants. The high proportion of nectarivorous individuals in May to October 1969 occurred at a period of great flowering by several herb flat plants, but substantial repre-

Table 44. Temporal change in trophic structure of epigeal arthropods in the herb flat, after transients and omnivores have been excluded.

	1969						1970						1971	1972	
Trophic category	J–F	M–A	M–J	J–A	S–O	N–D	J–F	M–A	M–J	J–A	S–O	N–D	F–M	April	$\bar{X}$
Chewing herbivores	9.7	34.8	29.2	9.9	15.1	12.0	12.7	2.6	19.5	5.9	42.7	12.3	6.0	12.8	16.1
Sucking herbivores	0	2.1	12.2	33.6	32.5	33.4	9.3	17.9	22.6	40.6	14.8	14.6	14.8	14.6	18.8
Scavengers	13.1	13.6	10.1	5.8	5.9	1.8	5.7	0.7	6.4	0	1.3	1.0	0	1.4	4.8
Nectarivores	6.0	18.5	32.8	28.1	26.7	26.3	17.2	20.0	12.8	13.0	10.5	7.8	19.9	26.7	19.0
Predators	32.1	8.4	4.1	5.5	5.5	8.4	13.6	9.8	15.0	17.9	15.1	20.0	7.5	7.0	12.1
Parasites	0	5.3	0.8	0	3.7	5.0	3.6	3.4	2.2	6.4	0.06	0.3	0.7	3.3	2.5
Undetermined	39.1	17.5	10.8	17.2	10.4	13.1	37.9	45.6	21.5	16.1	15.5	43.9	51.1	34.2	26.7

Table 45. Mean trophic structure of individuals of epigeal arthropods in the herb flat.

Trophic category	% of total no. of individuals	
	Total fauna	Excluding transients
Chewing herbivores	14.0	11.9
Sucking herbivores	13.1	13.2
Scavengers	3.4	3.3
Nectarivores	13.2	13.6
Predators	8.2	8.4
Parasites	1.9	1.8
Undetermined	19.0	19.8
Omnivores	27.2	28.0

sentation by nectarivores in late 1970 occurred when flowering was slight in the herb flat (compare Figs. 54 and 99).

Trophic structure of foliage faunas

The trophic structure of the epigeal arthropods of the herb flat has been discussed at length. That habitat accounted for 62% of the vegetated area of the island (Table 1), and being the dominant vegetation type was subjected to the most intensive and extensive study. In a sense, that habitat can be considered as representative of the island as a whole.

In order to examine the situation in more restricted communities, the trophic structure of the species of arthropods on the foliage of *Argusia* and *Pisonia* was investigated. As would be expected for a foliage fauna, the largest categories were the herbivorous ones (see Table 46), and were about equally divided between sucking and chewing forms, with nectarivores being represented proportionately less. Predators and omnivores also were represented well with the remaining trophic levels being relatively minor in importance.

Over time the trophic structure was more stable on *Argusia* than on *Pisonia*, but both showed more variation than occurred in the herb flat (Fig. 100).

This difference can be attributed to a variety of factors. Flowers were much more abundant on *Argusia* than on *Pisonia*. In the former, the foliage was in close proximity to inflorescences and nectarivores would probably often rest on the foliage or seek shelter there. The proportion would vary with time of day and/or weather conditions as these factors influenced nectarivores' feeding or resting activities. Thus a source of temporal variation was introduced with selective sampling of only part of the community that would not exist with overall sampling such as carried out in the herb flat.

A lot of species were shared among the various foliage types and between them and the herb flat (Table 22). The numerically much larger population of

Table 46. Mean trophic structure ($\bar{X}$ % of species) of the arthropod faunas from different habitats on One-Tree Island. Established fauna includes permanent and non-permanent established species.

	Pisonia foliage		*Argusia* foliage	
Trophic category	Total fauna	Established fauna	Total fauna	Established fauna
Chewing herbivores	17	19	25	34
Sucking herbivores	19	28	23	32
Nectarivores	15	14	13	14
Scavengers	4	7	0	0
Predators	21	15	19	13
Parasites	7	7	2	1
Omnivores	10	11	11	7
Undetermined	7	0	7	0

the latter habitat (see Chapter 23) might tend to overwhelm the foliage fauna because of large numbers of strays, or because of migration of generalized species to the foliage during periods of high densities in the herb flat. Because the numbers of individuals were much smaller on the arborescent foliage than in the herb flat, the effect of the reverse movement of generalized species would be relatively slight.

The trophic structure of the total fauna on *Pisonia* and *Argusia* foliage differed somewhat from that of the established species alone (Table 46). The more marked differences were higher proportions of herbivores and lower proportions of predators in the established fauna than in the total fauna.

The relative proportions of *individuals* from the various trophic categories changed markedly over time. It is important to relate those changes to other, concurrent phenomena in order to arrive at a possible explanation of the causal influences.

The proportion of individuals in the chewing herbivore category varied temporally and showed a remarkably similar seasonal pattern to that of leaf production in both *Pisonia* (compare Figs. 38, 39 and 101) and *Argusia* Compare Figs. 38, 40 and 101) except that in the latter there was a lag in drop of relative abundance of chewing herbivores behind that of a net decrease in leaf production at the end of 1970.

The relative abundance of individuals of chewing herbivores increased when leaf production was high and new leaves were being formed but decreased when net production was low. In neither plant species did the relative abundance of chewing herbivores relate in any consistent way to temporal changes in leaf biomass (compare Figs. 37, 40 and 101). These results in conjunction with similar but less extensive ones from the herb flat suggest that herbivore numbers may be more closely related to food quality (rapid production of tender, new leaves) than to the total amount of food available (standing crop biomass). This finding is in contrast to that of Evans

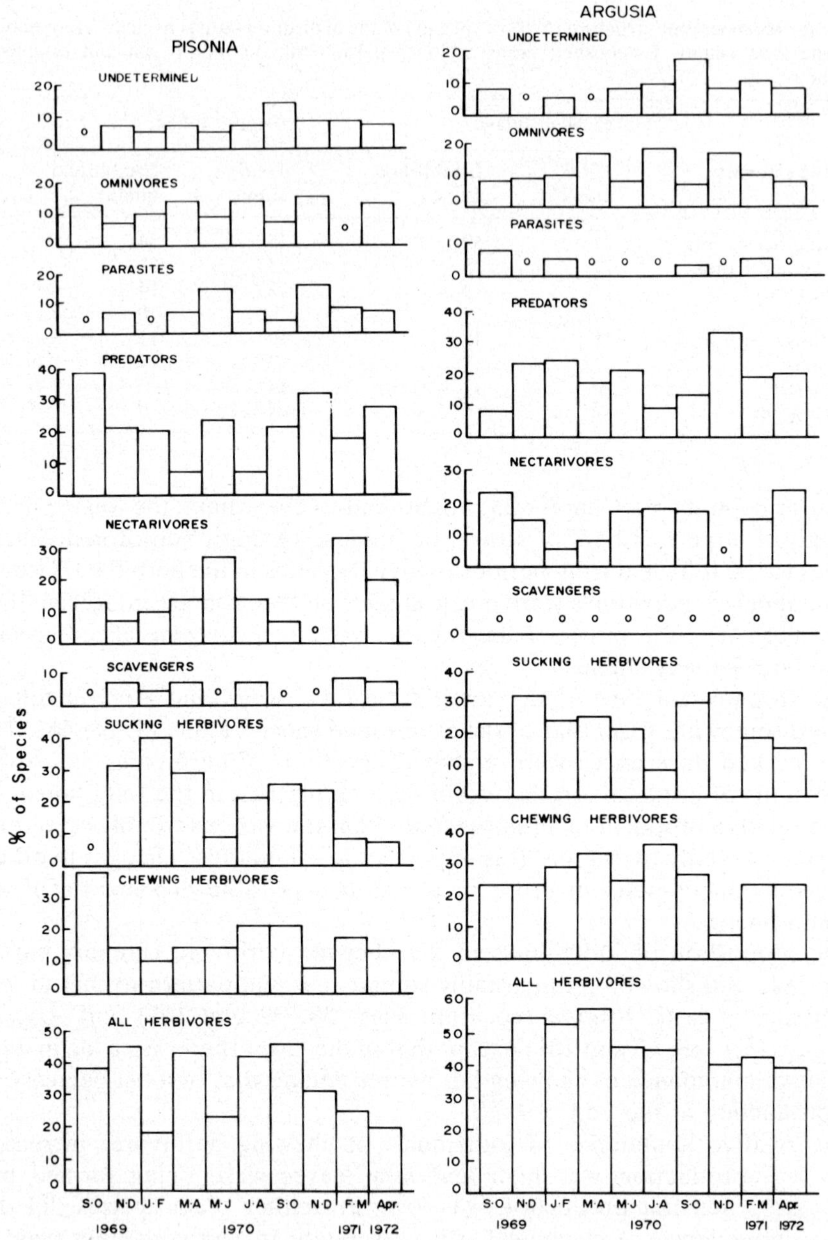

Fig. 100. Temporal change in trophic structure (by species) of the foliage fauna of *Pisonia* trees and *Argusia* shrubs.

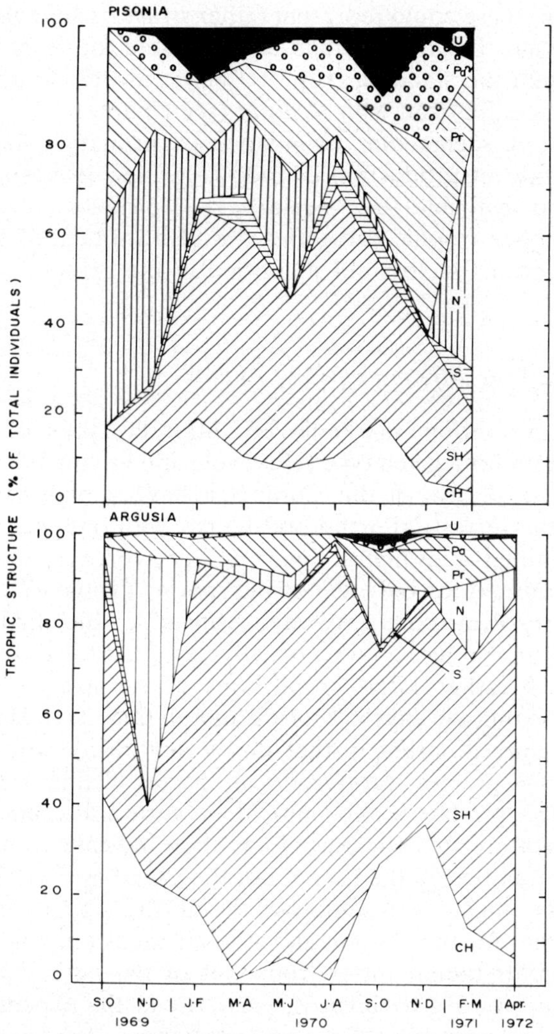

Fig. 101. Temporal change in trophic structure (by individuals) of the foliage fauna of *Pisonia* trees and *Argusia* shrubs. CH = Chewing herbivores. SH = sucking herbivores. S = scavengers. N = nectarivores. Pr = predators. Pa = entomophagous parasites. U = undetermined.

and Murdoch (1968) who noted that abundance of insects followed the same seasonal pattern as plant biomass in a mainland grassland.

The relative abundance of individuals in the nectarivore trophic level on *Argusia* showed peaks in November–December 1969 and September–October 1970, in both cases coinciding with the yearly peaks in flower numbers for that shrub (compare Figs. 35 and 101); this suggests that when large quantities of nectar are available at *Argusia* shrubs, the nectarivore numbers attracted by it become associated with, and included in the foliage fauna. As discussed above, these could represent either strays or animals seeking shelter among the foliage. Insufficient data are available for *Pisonia* flowering for analysis, although the standing crop of *Pisonia* flowers at any one time was always small in comparison with *Argusia* (see Chapter 11).

In summary, it seems that abundance of primary consumers may be influenced by rate of production of food (e.g. new leaves) and perhaps by amount of food available (e.g. nectar). It is probable that predators and parasites and other trophic categories are also influenced by food supply. Undoubtedly various factors not identified in the present study are also of importance.

Trophic structure on different types of islands

Trophic structure differs markedly on different types of islands, but is relatively constant for a given type (Heatwole and Levins 1973). For example, non-vegetated sand cays in the Coral Sea have a high preponderance of scavengers in the terrestrial arthropod fauna, with only ectoparasites of sea birds also attaining values higher than 7%; herbivores (represented solely by transients) accounted for only 2% of the species (Table 47).

A small Puerto Rican sand cay, Cayo Ahogado, which had a small amount of low vegetation also had a fauna heavily weighted towards scavengers, but in addition had rather large proportions of herbivores, omnivores and predators. On mangrove cays in Florida and on One-Tree Island, the trophic structure was quite different in that the herbivores constituted the trophic category with the greatest proportion of species (Table 47). On mangrove cays, scavengers, predators and omnivores were also abundant. The One-Tree Island data are not comparable as they indicate trophic structure of particular communities on the island, not the island as a whole. The herb flat is probably the most representative as all the epigeal macrofauna are included, whereas for the *Pisonia* and *Argusia* areas only the foliage fauna is treated. The latter faunas differ from that of the herb flat in being more heavily weighted toward herbivores, primarily at the expense of scavengers (which tend to be ground-dwelling) and omnivores. If the macrofauna on the trunks and stems and in the litter were also included for *Pisonia* and *Argusia* areas, it is likely that the trophic structure would shift more toward that of the mangrove cays.

Taking all three habitats together, still accounts for only 74% of the vegetated area of One-Tree Island and excludes the fauna of the succulent mat, *Pandanus* grove, *Scaevola* shrubs and the beach. All of the latter habitats are small in comparison to the three for which data were tabulated, except for the beach which represents about 21% of the total island area (Table 1). However, it is sparsely vegetated and supports few arthropod species (Table 16). The fauna of the *Scaevola* shrubs would probably be similar to that of *Argusia*. The succulent mat had few species, most of which

Table 47. Comparison of the overall mean trophic structure (% of total species) of communities of various types.
H = Herbivores (sucking and chewing herbivores, granivores, woodeaters and nectarivores); S = Scavengers, dung eaters, and detritivores; P = Predators; Pa = Entomophagous parasites; O = Ants and other omnivores; E=Ectoparasites of vertebrates; ? = Trophic level unknown. + indicates present in negligibly small proportions. Predominate category underlined.

Island type	H	S	P	Pa	O	E	?	H / (P + Pa)	Source
Bare Sand Cays, Coral Sea (N = 10 isls)	2	63	7	0	0	27	0	0.29	Heatwole 1971; Heatwole and Levins 1973
Motmot Island, New Guinea (Recolonizing volcanic island; years 3 and 4)	12	34	25	0	30	0	0	0.48	Ball and Glucksman 1975
Cayo Ahogado, Puerto Rico. (Sand cay with low vegetation only) (N = 12 visits)	23	34	24	0	17	0	+	0.96	Heatwole and Levins 1973
One-Tree Island, Herb Flat (N = 13 samplings)	45	6	18	11	17	0	2	1.55	
Pisonia foliage (N= 8 samplings)	51	4	21	7	10	0	7	1.82	Present
Argusia foliage (N = 8 samplings)	61	0	19	2	11	7	7	2.91	Study
Mangrove cays, Florida; before defaunation. (N = 6 isls)	39	13	20	7	19	0	+	1.44	Heatwole and Levins 1972b, 1973, based on Simberloff and Wilson 1969

also occurred in the herb flat (Chapter 14). The *Pandanus* grove had few species (Chapter 14) and given the low amount of leaf damage suffered by *Pandanus* plants (Chapter 15) would be expected to have fewer herbivores than the other types of arborescent vegetation. Taking all these things into consideration, a trophic structure for the epigeal macroarthropod fauna for the entire island, including all habitats, would probably reduce the proportion of herbivores on One-Tree Island below the level indicated for specific communities or subcommunities in Table 48, i.e., it would probably bring the figures more in line with those of mangrove cays. Whether or not this prediction would be sustained in the face of additional, more refined data, it is clear that the mangrove cays and One-Tree Island have a much more similar trophic structure to each other than they do to either the bare sand cays or Cayo Ahogado.

Table 48. Estimated total numbers of individuals of arthropods in the herb flat.

Date	Epigeal arthropods		Soil microarthropods	Epigeal arthropods as % of soil microarthropods
	Total	Most abundant species*		
1969				
J–F	590,790	255,221	–	–
M–A	842,852	150,028	–	–
M–J	2,941,132	549,992	–	–
J–A	3,354,929	878,991	29×10^8	0.12
S–O	3,984,779	633,580	46×10^8	0.09
N–D	4,526,743	2,009,874	43×10^8	0.11
1970				
J–F	3,421,454	451,632	29×10^8	0.12
M–A	1,306,085	213,871	76×10^8	0.02
M–J	2,102,552	276,107	37×10^8	0.06
J–A	804,097	108,505	65×10^8	0.01
S–O	2,754,374	513,636	44×10^8	0.06
N–D	837,969	142,815	–	–
1971				
F–M	5,682,690	1,628,659	–	–
1972				
April	3,809,923	356,723	–	–
Mean (July 1969–Oct. 1970)	2,781,877	–	46.1×10^8	0.07

* Not always the same species.

The observed differences in trophic structure among different types of islands suggests that trophic structure does not merely reflect the composition of arriving species. Providing that the arriving composition is independent of the nature of the target island – as would be expected for passive aerial and sea-borne waifs – a stochastic model would predict that all target islands in the same dispersal stream would share the same trophic structure of their established faunas. Differences in trophic structure among islands of different types imply that island resources select established species from the arriving pool.

The type of energy base may be important in determining the trophic composition selected. The trophic structures of islands such as bare sand cays that rely entirely or largely on marine-derived carrion or guano as the energy base for the rest of the community (see Chapter 20) have a high proportion of scavengers among their arthropod faunas but few herbivores. This condition also applies to very early successional stages in the colonization of new islands or recolonization of completely denuded ones. For example, Dammerman (1948) on the basis of his study of the recolonization of Krakatau following its sterilization by hot lava, suggested that the first colonizers of islands are species feeding on animal and vegetable debris washed up by the sea, followed by omnivores, and finally by herbivores, predators and parasites. Similarly, Fridriksson (1975) reports that on Surtsey, an island formed in 1963 off Iceland as a result of volcanic eruption, the arthropods which easily became established (1964–65) were scavengers. Most of the species recorded from the island were aerially dispersed animals but these were unable to become established because they belonged to trophic categories inappropriate for the condition of the island. He stated: 'It must be borne in mind that the greatest number of individuals belonging to the various species of terrestrial invertebrates so far found on Surtsey are only casual visitors that happened to be carried to the island and have little chance of survival. Only a small number have been able to live for any length of time, and only a few of these have been able to breed and may be considered to have become temporary or permanent inhabitants. The main hindrance to a successful colonisation is the lack of available food supply for the animals in this new habitat, as the vegetation has, until recently, not produced any substantial amount of organic matter. The major food source has been the organic matter washed ashore from the sea. Insects that can breed and rear their offspring on such carcasses have therefore had the best chance to colonise.'

Finally, the island of Motmot was formed in 1968 by volcanic eruption. It occurs in a freshwater lake (Lake Wisdom) on Long Island just off northern Papua-New Guinea; it lies only 3 km from the nearest land. Ball and Glucksman (1975) studied recolonization for several years. Plants arrived early. Mosses and sedges were there within a year; by three years 14 species of plants had arrived and by 4 years there were 18 species present. Most were probably brought by ducks although flotation and aerial dispersal were probably also involved.

The first year all the animals observed were predators; there were 2 species of Odonata (strong fliers and probably transients) and one species of lycosid spider which probably subsisted on 'aerial plankton'. After 3 and 4 years respectively there were 8 and 20 species of arthropods recorded. The trophic structure was weighted toward scavengers (3 of the 8 species, or 38%) during the third year after colonization.[4] The fourth year saw a large number of omnivores, predominantly ants (35%), although scavengers were still in high proportions (30%). Herbivores were poorly represented throughout (Table 47) even though a number of plants were established. The mean trophic structure for years 3 and 4 following recolonization (Table 47) was intermediate in some characteristics between that of bare sand cays and the sparsely vegetated Cayo Ahogado.

There are striking parallels between new volcanic islands in the tropics (Krakatau, Motmot) and in the cold temperate region (Surtsey) and non-vegetated sand cays in the tropics (Coral Sea cays). In each case a variety of

[4] Calculated from their tabular data; trophic status determined by reference to Borror and De Long (1955) and C.S.I.R.O. (1970).

arthropods, belonging to many different trophic categories, are available for colonization but only those succeed in doing so which have appropriate trophic relations. The trophic structure which results is heavily weighted toward scavengers.

Insular arthropod communities that rely primarily on endogenous food produced by local autotrophic plants are characterized by a trophic structure heavily weighted toward herbivores (mangrove cays, One-Tree Island). Frith (1979) has also indicated that the vast majority of insect species on Aldabra Atoll were phytophagous.

Intermediate types of islands exist which are small and consequently have a large perimeter to surface area ratio, and possess a very small amount of autotrophic vegetation. These conditions favour a higher input of marine-derived carrion, relative to endogenous autotrophic production than is true for larger, more heavily vegetated islands. Cayo Ahogado (Table 47) is a good example of such an island, and it has a trophic structure intermediate between bare sand cays and more heavily vegetated islands.

The shift of arthropod trophic structures from (a) scavenger-dominated to (b) scavenger-based but increasingly inclusive of other categories to (c) herbivore-dominated, as seen on a gradation of island types probably represents the sequence that occurs on a newly formed island during its development. As a bare island becomes vegetated and the vegetation itself undergoes successional changes one would expect a transition in type of trophic structure from scavenger-based to herbivore-based, with each intermediate stage having a characteristic trophic structure. Islands that never develop beyond a certain stage (such as some of the Coral Sea sand cays) would probably maintain a particular trophic structure indefinitely. On the other hand, disasters which reduced an island to a previous developmental stage (complete devegetation of a small cay by storm waves, or sterilization of volcanic islands by hot lava) would probably result in reversion in trophic structure to levels characteristic of early developmental stages. The few available empirical data support this idea. However, detailed analysis of long-term changes that may have taken place in trophic structure on Krakatau or Surtsey during recolonization would be most instructive.

In essence, we propose that there is a characteristic trophic structure of the macroarthropod fauna for each developmental stage of an island and that that trophic structure gradually changes over long periods of time as an island develops. However, on a short-term basis (or perhaps even indefinitely if no major development of the island occurs), the trophic structure of species remains stable in the face of various environmental perturbations which are not severe enough to alter the basic character of the island. Changes in species dominance patterns in the flora or mild successional set-backs are insufficient to greatly influence trophic structure of the arthropods. Succinctly, we suggest that there is a gradual long-term shift in trophic structure if major island development or succession is occurring, long-term stability if it is not, and short-term stability in either case, unless disasters of sufficient magnitude to alter the type of energy source take place.

It should be pointed out, that the environmental perturbations to which One-Tree Island was subjected during the study (severe droughts, storms), although having great ecological impact, *did not* alter the basic character of the island. It remained a vegetated coral cay containing forested areas and low vegetation and maintained a high degree of autotrophic self-sufficiency influenced by considerable input of marine-derived energy. The aspects of the island influencing the kinds of trophic structure which could be maintained were not altered and as has been mentioned above, trophic structure was remarkably constant throughout these environmental vicissitudes.

The same consideration applies to the defaunation experiments on mangrove cays (Simberloff and Wilson 1969). Fumigation of the islands destroyed only the animals and left the mangroves intact. The cays' food webs maintained their terrestrial autotrophic base and consequently, as on One-Tree, the requisite conditions for re-establishment of the original trophic structure of arthropods were still there. These changes are quite different from the changing of an island from one type to another in terms of the relative importance of endogenous autotrophic input as opposed to allochthonous input from the sea. We suggest the former types of changes have at most a small temporary effect on trophic structure whereas the latter may profoundly affect it.

If the sequence of trophic structures are related to relative degree of endogenous autotrophic production as opposed to allochthonous marine-derived energy in the way proposed above, then mainland terrestrial communities should be shifted further toward herbivory as many such communities would be almost entirely dependent on endogenously produced food, and they would not have any significant input of marine-derived materials.[5] There are few data appropriate to the testing of this idea. Vats and Singh (1978) found that the epigeal insect species of a tropical grassland were 84% herbivores, 11% predators, 3% omnivores, and 0.5% parasites. Inclusion of non-insectan arthropods (e.g. spiders) probably would have increased the proportions of predators at the expense of herbivores. Evans and Murdoch (1968) found that in a Michigan grassland (abandoned field), the trophic structure of the adult insect community[6] was more heavily weighted toward herbivores than that of any of the insular communities listed in Table 47, i.e., over 80% of the species were herbivores, with predators accounting for about 15% and all other categories for less than 5%. Inclusion of all arthropods would probably have modified these proportions somewhat. The trophic

[5] Study of the transfer of food from one terrestrial mainland community to another either as carrion, plant matter or living organisms would be rewarding, but discussion of this topic is beyond the scope of this book.

[6] Data are from their 'short list' which excludes transients; the 'long list' of all species was similar to the short list for adult insects but more heavily weighted toward predators for the larvae.

structure of the larvae showed a herbivore proportion of 59%. This is close to the highest values for insular foliage communities (believed to be artificially inflated in favour of herbivores, see above). Predators (including entomophagous parasites) accounted for 36% of the species and all other categories for 5%. These data suggest a trend toward further reduction in the scavenger category and an increase in relative numbers of species of herbivores (and perhaps also of predators) at the expense of other trophic levels, relative to the condition found in the insular communities. However, knowledge of the trophic relations of the non-insect arthropods in this mainland community would be necessary before firm comparisons with insular communities could be made.

Evans and Murdoch (1968) noted considerable trophic structure stability in their grassland; the ratio of herbivores to entomophagous species did not change greatly over 14 fortnightly samples stretching from mid-April to late October, even though absolute numbers in both categories showed considerable seasonal change. They attributed this stability to interactive effects in the following terms. 'We have been impressed by the evidence that the feeding activities of the organisms impose a pattern on the community which overrides taxonomic composition . . . This diversity of habits, which must result in a complex series of interactions between species, may well be the means by which a stable ratio of herbivores to carnivores is maintained for most of the season.'

We calculated ratios of herbivores to entomophagous species (based on larval food habits) from the data presented by Evans and Murdoch (1968). Values averaged 2.01 with a range of 1.23–2.23 except for one exceptional value of 4.5, during which time number of species of herbivorous larvae ranged from 12–98 and the number of species of entomophagous species ranged from 4–64. Similar data are not available for the insect larvae of One-Tree Island. However, we can compare ratios of herbivorous to entomophagous species (H/P +Pa) based on food habits of adults of the total epigeal arthropods for different habitats on One-Tree Island and on other islands. The mean value for adult insects in Evan's and Murdoch's (1968) study was about 5.3, and that from the data of Vats and Singh (1978) was 7.3, both far higher than any values obtained on One-Tree Island (Table 47), even those from some subhabitats such as *Pisonia* and *Argusia* foliage which would be expected to contain more herbivores than entire habitats encompassing all parts of the vegetation and the ground beneath. Inclusion of non-insectan arthropods in the data from the two mainland grasslands should decrease the ratio to a certain extent, but would be unlikely to bring them down to the One-Tree level, especially to that of the most comparable habitat, the herb flat. These results are consistent with Becker's (1975) observation that the proportion of carnivores to herbivores is higher on islands than would be predicted if their faunas were merely representative samples of mainland species. Bare sand cays, Cayo Ahogado and Motmot had ratios much lower even than One-Tree Island, again emphasizing the relative paucity of herbivores on those particular kinds of islands. The herb flat and the mangrove cays were similar.

The beach habitat, even of large complex islands, has similar conditions of proximity to the sea, substrate type, salinity and sparseness of vegetation as do some of the small, lightly vegetated sand cays. Communities further inland increase in amount of autotrophic vegetation, whereas access to marine carrion and extent of nesting by sea birds decreases. These considerations suggest that progressing inwards from the beach habitat, the relative dependence on endogenously produced food might increase and dependence on marine-derived materials decrease. If trophic structure is influenced by relative importance of these two sources of energy, then there should be a shift from scavenger-dominated communities on the beach toward herbivore-dominated ones inland. We received the qualitative impression of such a shift between the beach habitat and those further inland on One-Tree Island. However, we do not have a rigorous test of this idea as the beach fauna was not adequately sampled in a quantitative way. A study designed specifically to test this idea would be instructive.

Theoretical considerations

Application of new techniques to the study of the structure and function of communities has led to important modifications of theory and has resulted in a realization of the complexity of some of the interactive phenomena involved. Previously unsuspected properties of communities have been revealed. For example, Levins (1975) through the application of loop analysis has shown that the course of natural selection in a particular species may not be comprehensible without reference to the whole matrix of community interactions, especially trophic ones, and that there are effects which are contrary to previous intuitive expectations.

Cody and Diamond (1975) have reviewed many of these developments and have pointed the way toward further research. Much attention has been given to topics which are of relevance to trophic structure stability.

Examining continental situations on an evolutionary time scale, Rosenzweig (1975) suggested that adaptive radiation on continents is likely to be due to a decline in extinction. He suggested that each predator added to a system allows the coexistence of additional prey species, but fewer additional ones than any previous predator addition, and that a steady-state between predator species and prey species is reached. He also suggested the same arguments apply to any trophic level, and concluded: 'If there is a steady-state diversity in each trophic level when diversity in others is imagined to be constant, then one exists when all trophic levels are permitted to vary in diversity'. One may ask whether similar steady-states might not arise on a shorter time scale on islands when species are added by oversea dispersion rather than by speciation. Cody (1975) addressed this topic on a niche-space basis rather than specifically on an insular one. He stated: 'After the local extinction of a species, an immigrant arrives whose identity is statistically predictable from a probability distribution of likely immigrants; this probability distribution is generated from the amounts, proximities, and similarities of the habitats nearby that supply potential colonists. Thus, we might be able to generate the most likely arrival sequence and schedule. Potential immigrants arrive early, and often if the depauperate patch of habitat is not isolated, and these immigrants are appropriate consumers for the unused resources if the habitat in which the extinction occurred is well represented in the surrounding environment. The more appropriate and rapid the colonization, the less likely that these unused resources will be accounted for by density compensation or habitat-breadth compensation by the neighbouring species in niche space. The populations of colonists with low efficiencies on the unused resources would grow slowly; if the immigration schedule is rapid, they will soon be replaced by more appropriate consumers, but if it is slow they might possibly withstand later invasions by better adapted species, as in the interim their populations will have built up.' He reviewed literature including that dealing with insular situations, demonstrating that after extinction in some part of niche space, expansion by species in neighbouring niches is slow in the food dimension but relatively rapid in the habitat one. He likened niches to contour ellipses in two dimensions which are more expansible in one dimension (habitat) than in the other (food). Hespenheide (1975) also supported this view. He pointed out that as species are added to communities, diets remain relatively unchanged in comparison to the restriction of the foraging range in which the food is sought (habitat specialization). If the above viewpoints are correct, then a given arrival at an island might be able to fit into any one of several habitat niches, but only into one specific food niche. Thus, its establishment would be more likely to be determined by its trophic position than its usual habitat preferences. In this sense, establishment is seen to be deterministic and based primarily on trophic position. A stochastic element is introduced, however, in that it is largely chance that determines which of various potentially suitable occupants of a given food niche arrives at a time when that niche happens to be vacant (or occupied by a readily displaceable species). Stated in another way, the trophic structure of the community is deterministic and interactive, but taxonomic structure of the community is stochastic and consequently variable in time. Such a view accords with the data from One-Tree Island.

One of the most thought provoking approaches relating to the structure of insular communities to have appeared in recent years is that of Diamond (1975, 1979a, b) in which the empirically based technique of incidence analysis was applied. He constructed curves relating J (the incidence of occurrence of a particular avian species on New-Guinean islands with certain species-richness) to S, the total number of bird species. Species were classified into 6 categories; High-S species, confined to the most species-rich islands; A-tramps, B-tramps, C-tramps, and D-tramps, all present on the most species-rich islands but also to a respectively greater extent on islands of progressively lower species-numbers; supertramps, absent from species-rich islands but present on species-poor ones. Different species had incidence functions of different shapes and the faunas of the islands were non-random subsets of the total species pool. Detailed examination of four guilds revealed that there were some potentially possible combinations of species that did occur (permissible combinations) but others that did not (forbidden combinations). Permissible combinations resisted invaders that would transfer them to a forbidden combination. Diamond identified recurrent characteristics that he called 'assembly rules'. They were:

> 'A combination that is stable on a large or species-rich island may be unstable on a small or species-poor island.
>
> On a small or species-poor island, a combination may resist invaders that would be incorporated on a larger or more species-rich island.
>
> Some pairs of species never coexist, either by themselves or as part of a larger combination.
>
> Some pairs of species that form an unstable combination by themselves may form part of a stable larger combination.
>
> Conversely, some combinations that are composed entirely of stable subcombinations are themselves unstable.'

He explained assembly rules as arising from competition for resources, with harvesting of resources by permitted combinations minimizing the unutilized resource availability to potential invaders. Much of this competition was probably of the diffuse type, i.e., resulting from the sum of competitive effects exerted by a *combination* of species with the presence or absence of a particular given species not being explained entirely in terms of a correlated distribution of any other *single* species. Diamond stated: 'Communities are assembled through selection of colonists, adjustment of their abundances, and compression of their niches, in part so as to match the combined resource consumption curve of all the colonists to the resource production curve of the island. Members of permitted combinations must also be 'companions in starvation' – i.e., must be similar in their tendencies to overexploit and in their tolerances for lowered resource levels, thereby starving less tolerant species off the island. Thus, consumer species form hierarchies with respect to exploitive strategy.' Permissible combinations were also considered to be those combinations in which component species were relatively immune to displacement by the invading species.

Assembly was also related to dispersal abilities and to 'transition probabilities', i.e., ease of assembling of particular combinations in one or a few steps from a pre-existing combination.

In the former case it was pointed out that the high-S strategy of overexploitation is dangerous on a very small island because of its effect in reducing population size and hence probability of random extinction. Thus, species with high dispersal rates and large populations (supertramps) may occupy small islands for a greater proportion of the time than those with superior competitive ability or small populations.

While many authors have found Diamond's ideas useful in interpreting biogeographic data, statistical testing of the assembly rules has cast some doubt on their validity. Connor and Simberloff (1979) contend that 'every assembly rule is either a tautological consequence of the definitions employed, a trivial logical deduction from the stated circumstances, or a pattern which would largely be expected were species distributed randomly on the islands [with some specified constraints] . . .' While accepting Diamond's premise that competitive exclusion may be of importance in shaping community structure, they reject that its operation may be demonstrated by analysis of distributional data alone.

In the current study, we propose that species interactions are at some times more influential to community structure than stochastic events and at other times the contrary is true.[7] Our time series data are quite suggestive of the explanations offered by Diamond, but like Diamond, we lack definitive autecological or experimental basis for our interpretation. Our interactive interpretation is based on the theory proposed by Heatwole and Levins (1972b), although as Simberloff (1976b) has pointed out, the theory has not been proven even for a specific case and cannot be considered a valid generalization for all insular communities. Simberloff (1978) and Connor and Simberloff (1979) favour more stochastic explanations for plant, insect, bird and bat distributional data. Simberloff apparently felt that it is nearly impossible to test the Heatwole and Levins theory as he states (1978): 'If patterns of energy flow *do* constrain colonization, these patterns must first be worked out in detail for at least most of the species before a proof can be given.' We feel that this is perhaps an overstatement of the case but do agree that detailed studies are required for rigorous testing of the idea.

[7] For example, Sale (1978) reviews convincing evidence that the composition of assemblages of some reef fishes on coral patches is determined by stochastic events. The various species successfully defend territories against other individuals of the assemblage, either of the same or different species and regardless of size (no species is competitively superior to another); all available sites are usually occupied and living space is at a premium. Since the various species produce a large number of pelagic larvae that are distributed by ocean currents it is a matter of chance which species happens to arrive at a newly available site. This system depends on the combination of successful interspecific territoriality, high fecundity and passive dispersal and may apply to other groups with similar characteristics, but cannot be considered as generally applicable in all situations. This combination of characteristics does not apply, for example, to the One-Tree arthropods, especially the component of interspecific territoriality.

Diamond (1975) cautioned that diffuse competition can be used to explain almost anything, and if it *is* important, its operation is likely to be so complicated that its existence becomes difficult to establish and impossible to refute; nevertheless in a few cases he was able to relate species distributions to diffuse competitive effects from specific combinations of related species. We suggest that experimental studies designed specifically to test among alternative theories are essential. The present study goes some way towards Simberloff's ideal of looking at each individual species but still does not permit unqualified decision among alternative theories regarding the mechanism of trophic structure stability. Both Heatwole's and Levins' deterministic theory and Simberloff's stochastic one remain theories requiring further research. Neither has been proven; neither has been falsified. We feel that some of the data relating to these theories have been squeezed rather dry and re-interpretations have been re-interpreted. There seems little point in prolonging this process as we feel those data have little more to offer. Also, we are confident that models could be proposed supporting either theory (in a sense a model is a formalized statement of a theory). In order to provide additional insight into this problem we feel that modelling will have to have a firmer empirical and experimental basis than is currently available. It is important to recognize that deciding among alternative theories is going to be exceedingly difficult, will require extensive empirical data, and will almost certainly require experimental approaches. However, neither theory should be dismissed in a facile manner merely because of difficulty of proof (reality does not cease to exist because it is ignored). Rather, the difficulty should be a challenge.

The current fund of relevant empirical data is so small and the present state of knowledge in this field so rudimentary that various possibilities must still be taken seriously. The important task is to widen the base of empirical data and to carry out specific studies designed to decide among various alternatives, all of which might be consistent with available information and none of which can be invalidated by presently known facts.

McNaughton (1977) in an important review of ideas regarding species diversity and stability in communities found that the less vigorous traditional verbal model was considerably more robust in application than more recent and more rigorous mathematical models, and he expressed the view that 'the marked instability of attitudes . . . arises from a low diversity of empirical tests of the hypothesis'. Such a situation is likely to occur whenever theoretical constructs outstrip their empirical base. Resolution of the various ideas regarding trophic structure is a case in point; the urgent need is primarily for empirical tests.

For many decades, biotic interaction, especially interspecific competition was used uncritically as an explanation of distributional patterns, community composition and structure, spacing, habitat selection and interspecific differences in morphology and behaviour. The theory was accepted and was consistent with empirical observations. Further empirical studies were consistent with the traditional theory and were cited as 'supportive' without critical assessment. Although competition was often invoked it was seldom proven. In recent years this situation has been increasingly recognized and alternative theories are being explored including the role of predation. This is a healthy and long overdue development. However, within it lies the danger of a new fashion, that of dismissing competition without critical evaluation. Although the concept may not apply as widely as formerly supposed, trading one dogma for another is not an advance. Certainly some of the more thoughtful modern studies of community structure, such as those of Cody (1975) and Diamond (1975) have implicated competitive interactions as we have in the interpretation of our One-Tree data. We feel that Kroes (1977) has made a realistic assessment of competitive interactions. He indicated that the competitive exclusion principle should be replaced by the competitive exclusion *possibility*, and that it was up to ecologists to 'find out case for case where this possibility is realized and where it is not'.

Connell (1975) has drawn attention to the roles of predation and of physical environment as determinants of community structure. He found two relationships of some importance (1) the change in vulnerability of prey as its size varied in relation to that of the predator and (2) the change in intensity of predation as physical conditions varied. He noted that predation seemed to be more intense under more benign physical conditions and suggested that competition consequently should be prevented more regularly under such regimes (predation keeping populations at sufficiently low densities that competition among species is eliminated), whereas escapes from predation leading to competitive interaction would be more prevalent in harsher environments. Again, assessment of relative importance of various factors must be made for different situations individually. There is probably no one blanket statement applicable to all communities.

We may speculate about the ecological causes of a rather constant spectrum of relative species numbers in trophic categories on one hand, and on the other an unstable spectrum of relative abundances of individuals in those same categories. One possibility is that the two represent phenomena with very different functional bases. Trophic structure by species may indicate the number of *types* of niches that are filled whereas the number of individuals in particular trophic positions may reflect the *magnitude* or *availability* of resources within a particular niche. If the former remains constant but the latter varies, then stability of trophic structure based on species but variability of trophic structure based on individuals would be expected.

Arnold (1972) suggested in relation to his study of snakes, 'where there are more species of prey, more species of predators can coexist because there are more ways to avoid competition for food.' Reduction in prey species would have the reverse effect. Diamond (1975) examined species-coupling in relation to resource levels, and presented a model based on New-Guinean insular birds. When resources dropped below a certain level either because of unfavourable environmental conditions or because of establishment of a high-S species he suggested that supertramps become locally extinct because the other species overharvest the resources to levels they can tolerate but the supertramps cannot. In highly productive years or when high-S species temporarily disappear, supertramps can become established. Such substitutions of supertramps by high-S species or *vice versa* depending on resource level did not markedly alter species-numbers in the guild or greatly affect trophic structure. However, the direct effect of changes in productivity *per se* could influence both to some extent. On One-Tree Island the relatively constant trophic level suggests that the latter effect was slight, although it is possible that part of the mechanism of change in species number during the drought in comparison to later more favourable times could have arisen from species-coupling. MacArthur and Wilson (1967) and Cody (1975) provided postulates on the interdependence of population sizes and species richness which pertain to trophic structure stability. They suggested that each species on an island has a threshold of population size below which it will become extinct. Should the resource or resources which limit one species become superabundant, other species formerly excluded from the island by deficiencies of that resource may establish populations (unless the original species can monopolize the resource quickly or otherwise competitively exclude potential newcomers). Should a subsequent decline in the resource cause a population to fall below its threshold, it will become extinct.

Our data on epigeal arthropods may be understood partly in terms of these concepts. Here we reconsider briefly a major resource of this community – the vegetation itself which is the base of the arthropod food web and provides shelter and substrate. We have already reported on two aspects of the dynamics of the herb flat vegetation. Firstly, the standing crop of plant species individually and *in toto* fluctuated dramatically during the study, and devastation by drought and subsequent recovery occurred. Secondly, despite these dramatic fluctuations, the species composition of the flora remained constant. In other words, the variety of food sources available to primary consumers remained constant, the structural elements of the habitat remained constant, but the amount of food, shelter and substrate varied greatly through the study. The intercorrelated fluctuations of numbers of species and individuals may be a direct result of the fluctuations in levels of available food and shelter taking species above and below population thresholds as described above. The constancy of the distribution of these species among trophic categories may reflect the constancy of types of plants, and hence of food for herbivores, and of shelter and substrate.

Variability in relative abundance in the trophic categories may be a function not only of the temporal variability of food and shelter, but also of the morphological and taxonomic diversity of the arthropods themselves. Within trophic categories, arthropods ranged in body length from one to several millimetres and as is usually true (Schoener and Janzen 1968) smaller species were generally more abundant in samples than larger ones. Thus, the size spectrum of species present (whether determined by the nature of the food, or whether largely fortuitous) can have a significant effect on the abundance in particular trophic categories, and therefore on numerical stability over time. Given considerations such as these, there is little reason to expect a constant relationship between numbers of individuals in different trophic groups, even were one category (such as predators) exclusively limited by another (such as herbivores).

There are numerous reasons why there should not be a simple, constant relationship between the abundance (or biomass or productivity) of a consumer and that of its resource. In the predator–prey case they include: time lags in predator population change in response to changes in prey levels; prey escape mechanisms (in time, by adopting activity cycles asynchronous with that of its predators; in space, by adopting various aggregative and/or hiding behaviours); physiological and morphological defence mechanisms in prey (i.e. obnoxious taste, spines, etc.); relative prey specificity or generality of predators and its effect on degree of competition among them. All of these aspects of consumption would tend to complicate the numerical relationship between food and consumer which might be expected solely on the grounds of energy requirements of the consumer category, and its efficiency of extracting it from its food source.

The predator–prey trophic interaction has been used in the above discussion as an example. Similar considerations would apply to other trophic levels.

Interpretation of the changes in community structure

At this point we present a speculative account of the temporal changes of community structure which took place during the study. Although based on the herb flat plant-arthropod community, the account provides a general overview of the way we see the whole island system fluctuating through time. The account has parallels with the Simberloff and Wilson (1969) scheme for the recolonization of defaunated islands, but a difference in emphasis is made of the possible role of interspecific competition.

Before the drought, a certain trophic structure of species was established, including, we hypothesize, a balance between the number of species of herbivores and the numbers of species of plants. The drought did not cause extinction of any plant species but did cause a massive mortality among the arthropods and caused numerous extinctions. At all trophic levels, these extinctions were caused directly (through heat and water imbalances) and indirectly (through destruction of species' habitats and food sources, i.e., food components of the vegetation, prey organisms). With the return of favourable weather and reestablishment of pre-drought vegetation biomass and diversity, arriving herbivorous species with suitable niche characteristics could establish populations which could themselves support additional and larger populations of predators. We suggest that interspecific competition was not necessarily a major determinant of community structure in the early recovery stage, but can envisage specific situations in which it may have been important (see below).

The overshoot (1972) which was accompanied by a rather constant proportional trophic structure may be explained as follows. During the post-drought recovery, equivalent herb flat niches were in some cases filled in different parts of the island by different epigeal species, the precise establishment location of each being determined by chance. The wide area coverage of our sampling methods would result in some instances in both (or all) of these ecologically equivalent (but spatially separated) species being sampled, thus contributing to the overshoot. Later, as the populations expanded and caused their spatial ranges to overlap, some of the ecologically equivalent species would be eliminated (competitive exclusion); other ecologically equivalent groupings would persist perhaps at population levels kept low enough by predation or other factors which prevent competitive exclusion. Notwithstanding the details, the loss of *some* of these ecological equivalents could result in the decline which followed the overshoot. The relative constancy of trophic structure is, we suggest, a result of the Eltonion interdependence between trophic levels being of greater importance than stochastic events, i.e., the number of species at one trophic level is determined to a large degree by the number of species in the level below.

Finally, we suggest the epigeal community entered an 'enduring equilibrial phase' during which the establishment of new species required either (a) the occupancy of a previously existing but unutilized niche on the island, (b) the creation by normal temporal fluctuations of vegetation and dependent fauna, of new niches which can be filled and in some cases, competed for by one or more immigrating species, or (c) the displacement of entrenched species from an existing niche. The last alternative may be facilitated by the immigrant species becoming established during a period when the entrenched species is at a low density within the bounds of its usual population fluctuations.

Regarding this replacement process, Kroes (1977) has remarked that where competition with one resident species determines success or failure of an 'entering' species in establishing a niche two factors are important: the degree of similarity of both species ecological potentials and the structure of the ecosystem concerned.

The above processes act in a time scale encompassing seasonal variability in the physical environment and quantitative and qualitative changes in the trophic base (i.e. the vegetation). Non-catastrophic levels of species extinctions and a continuing but variable proportion of establishment of immigrants might be expected. The stochastic element in each of the processes might tend to displace the community from any constancy of trophic structure, and indeed, absolute numbers of species in trophic levels did vary considerably through time. Our observation that a relatively constant proportional trophic structure was maintained through time in spite of ongoing environmental and compositional change again suggests an Eltonion interdependence between trophic levels, and that our predominantly deterministic explanation is more plausible than a predominantly stochastic one.

This phase is referred to as an equilibrium because unlike the early recolonization, overshoot and recovery stages, there is no directional change in species number and/or trophic structure. It endures through a time span which on average would encompass a period of years and would be terminated by some catastrophic event, such as another serious drought.

Competitive interactions may be of prime importance during the decline following the overshoot and during the equilibrial phase. As mentioned earlier such interactions are not necessary for explaining trophic structure stability during early colonization. However, theoretically, competitive interactions could be quite important even in early phases of colonization. If a species became established early in colonization and was restricted to a particular segment of the resource spectrum, and soon after another species, similarly restricted, also arrived, competition between them could be intense, regardless of the resources available in other niches at that time. Similarly, the population size is not necessarily a prime feature. A very small, recently established resource population could represent an inadequate supply for two nearly simultaneously established exploitative species. Such a phenomenon could easily occur during recolonization (following drought or defaunation) of an island characterized by high immigration rates. Only in the case of the resource species becoming established and building up a large population well in advance of the arrival of exploiting species might there be a long period in which several potentially competitive species would be freed from competition merely because resources were overly abundant. Failure to recognize the above possibilities may lead to an undue emphasis on the 'non-interactive' quality of the colonization phase (see Chapter 21). We disagree with Simberloff (1976b) in that we contend that competition could occur in very early stages of colonization and have an effect on trophic structure. We also recognize, however, that it is not a necessary condition for maintenance of trophic structure stability during that phase. It is likely that in nature a mixture of both phenomena (competition and increase in kinds of exploitable resources) are involved. They are not mutually exclusive phenomena.

The above hypothesis accounts for the observed stability in trophic structure of species in the face of changing numbers of individuals and numbers of species and during a recolonization characterized by a recovery phase, an overshoot, and a decline to an equilibrial phase. The main determinants of trophic structure stability are postulated to be a restitution of the number of kinds of resources and kinds of species exploiting them, and competitive exclusion once the upper levels of kinds of resources are reached. We believe this to be the simplest explanation of our empirical data when all aspects of the study are considered.

In making this statement, it is perhaps necessary to point out a philosophical difference in viewpoint regarding the law of parsimony that exists between a strict adherence to stochasticism and strict adherence to interactive determinism. Traditionally, it is accepted that events are considered to be random unless proven otherwise and this viewpoint is the basis on which statistical tests have been developed, and rightly so. Applied to the problem at hand, the null hypothesis could be set up that community structure arises from stochastic processes, the hypothesis not falsified and the stochastic viewpoint accepted. As mentioned above, adherents of this viewpoint have indicated that even if community structure does have an interactive basis, it is exceedingly difficult to demonstrate without complete autecological studies of most species in the community, a nearly impossible task. That is, the stochastic theory cannot easily be falsified regardless of whether it is true or not, and it becomes 'true' merely by definition; a certain circularity is interjected.

Another important difficulty with the stochastic theory has been pointed out by Grant and Abbott (1980). They showed that the analysis of character displacement by Strong *et al.* (1979) was seriously flawed by a lack of realism and by bias that tended to favour acceptance of the null hypothesis. In general terms they concluded that 'stochastic models may be a useful tool in hypothesis testing in biogeography, but until they can be made realistic their usefulness will be severely limited' and suggested that tests permitting a clear rejection, as opposed to merely non-acceptance of an hypothesis were needed. They further expressed the opinion that detailed ecological studies, especially if directed towards specific questions of competition and dispersal, may be more promising than refinement of stochastic models.

An alternative interpretation is that in such highly complex phenomena as community structure, formal testing of null hypotheses have limited usefulness for the reason mentioned above. Since interactions among species have been clearly demonstrated in a number of specific situations,[8] it does not seem parsimonious to dismiss them entirely and to suggest that a phenomenon known to operate at the level of two, or a few, species plays no role at the community level. The important approach is to maintain open-mindedness in regards to both possibilities and continue the difficult, but necessary, empirical evaluation of the relative importance of stochastic and interactive effects in particular situations. Comparison of situations where

[8] The fact that competitive exclusion often has been uncritically used as a blanket explanation for many distributional patterns and ecological differences among species does not nullify those studies in which it has been convincingly shown to operate.

stochasticism seems to be paramount with those where interactive effects apparently predominate may extend current theory in valuable ways.

Finally, there is an additional consideration that should be recognized in adopting an operational hypothesis for further research on structure of insular communities. Accepting a purely stochastic explanation at present is a dead end. Once one has accepted that community structure is a purely stochastic phenomenon, the topic becomes exceedingly uninteresting and there are no further questions to ask. However, if one operationally accepts an interactive one (recognizing that it may eventually be necessary to discard it in favour of stochasticism) mechanistic questions can be asked, and the theory tested by experimental studies of species interactions. In this sense we admit to the bias toward a theory that stimulates research and maintains our fascination. We are prepared, however, to accept a duller alternative and direct our research interest elsewhere, if and when the empirical data demonstrates such to be the most reasonable course of action.

23. An overview

This book has been devoted to a study of the community ecology of One-Tree Island; details of the ecology of various species have been elaborated and a synthesis has been attempted in terms of the overall community. In the present chapter two additional topics are treated. The first is an account of factors influencing the vegetation. Then follows a review of the absolute numbers of individuals in various taxa or ecological categories, and the importance of total population sizes is discussed in the context of insularity.

Finally, a brief summary of all the major findings of the One-Tree Island study closes the book.

The ecology of plants of coral cays

Islands have been looked upon as ecosystems in a general way (Fosberg 1961, 1963a, b, 1965) and various general accounts of the ecology of vegetation of coral cays have been presented (Fosberg 1948, 1949, 1957, 1974, Heatwole 1976). Spicer and Newberry (1979) and Newberry and Spicer (1979) have related vegetation to environmental gradients on a cay from an Indian Ocean atoll. In this section, several of the most important influences are evaluated in the context of One-Tree Island. The ecological characteristics and habitat preferences of specific One-Tree plants have already been outlined in Chapter 8.

A number of environmental factors have important influences on the vegetation of strands, beaches and small islands. Scarcity of water, high salinity and the effects of sea birds are three which have been cited often and it is important to consider them in relation to the One-Tree flora.

Water

Factors influencing the availability of moisture to strand plants are not completely understood. The water-holding capacity of most beach and dune sand is relatively low (Salisbury 1952, Oosting 1954, Van der Valk 1974), and it has been calculated that the amount available should sustain plants for only 1–4 days before wilting would occur (Chapman 1976). However, plants in sandy areas grow for much longer periods, even up to 6 weeks, without rain and their substrate often is moist only a short distance below the surface (Chapman 1976, De Jong 1979).

The origins of such water was long a mystery (Ranwell 1972). Several theories were advanced to account for it. One theory was that it resulted from 'internal dew formation', i.e. the formation of dew from either atmospheric water vapour (Salisbury 1952) or from water vapour moving up from deeper in the soil (Olsson-Seffer 1909, Willis *et al.* 1959). Another was that it moved upward via capillarity or as liquid along a water-potential gradient (De Jong 1979).

De Jong (1979) working on Californian beaches found that the sand at 100 cm depth remained relatively moist throughout the year but during rainless periods dried out at shallow depths. On the basis of experiments utilizing plastic barriers buried in the sand, he found that water moved upward from the deeper areas and became available to the plants. He thus was able to discard the theory of internal formation of dew from atmospheric vapour, but on the basis of his data could not decide among the various mechanisms postulated for upward movement of water from below. Nor was he able to assess the amounts of water which could be moved by the postulated mechanisms. He did stress, however, that because of the very low water holding capacity of sand, only very small amounts of water per unit volume of sand would be required to maintain water potentials between −10 and −15 bar.

One-Tree Island is not sandy and therefore differs from most strands, beaches and other cays. Much of the island is made up of loose coral shingle, and upward movement of water from the water table would probably be restricted to whatever amounts result from condensation of water vapour from below (at those times of day when the temperatures of surface layers drop below those of deeper layers). Drainage is good in coral shingle and little capillary water would be held. All these factors converge to render One-Tree Island a drier habitat than a sandy cay in a similar climate, and it may account for the very sparse development of vegetation on the One-Tree beach even by strand standards. Few species occur on the beach and many of those represented in the One-Tree flora that are typically beach plants, are found instead further inland in the herb flat.

Ground water is not the only factor involved in the moisture relations of strand vegetation; the plants' physiological attributes such as transpiration rates, stomatal responses, etc., are also involved. De Jong (1979) has pointed out that different species on a Californian beach have different modes of adaptation to their common habitat. Species differ in variability of dawn xylem-sap tension, degree of maximum mid-day tension and rate of recovery from it, all of which are in turn related in part to rooting pattern, especially rooting depth.

One-Tree plants also seem to differ in their responses to moisture. Allaway, Pitman, Storey and Tyerman (pers. comm.) have compared sap flow and stomatal responses in *Argusia, Melanthera* and *Scaevola* on One-Tree Island during drought conditions in 1977 and at about the same time of year but under favourable conditions in 1978. Under favourable conditions when moisture can be easily replaced by the plant, sap flow (and hence transpiration) increases after sunrise, reaching a peak in the hotter parts of the day and drops off again in the evening, remaining low during the night. Variations of this pattern are related to light intensity changes. The stomata tended to be open wider during sunny periods than cloudy ones, thereby facilitating uptake of CO_2 at times when light was favourable for photosynthesis.

Marked changes in the stomatal response occur during drought in *Scaevola*. The stomatal conductance of this species remained relatively low throughout the day (stomata closed), only elevating slightly in late afternoon. Thus, water was conserved during droughts but photosynthesis was retarded. By contrast, *Melanthera*, even during drought, followed the usual pattern of high stomatal conductance at mid-day. *Argusia* was somewhat intermediate. During drought, it opened the stomata earlier in the morning but then closed them beginning about 0800 hrs and continuing progressively until about 1400 hrs when conductance levels reached low values and remained there the rest of the day.

These differences seem to be related to the relative abilities of these plants to take up water. *Melanthera* was able to develop much lower water potentials than *Argusia* or *Scaevola*; in fact *Melanthera* could develop values so low it would have been able to extract water from seawater. One would expect *Melanthera* to be the most drought-resistant of the three, an enigma in view of the facts that *Argusia* and *Scaevola* are often considered as pioneer strand species whereas *Melanthera* is not. Perhaps *Melanthera* is less easily dispersed. Also a paradox is the fact that *Melanthera* suffered more severely during the 1968 drought than any other species. However, comparable information is not available for species such as *Ipomoea tuba* which replaced *Melanthera* in dominance as a result of the drought (see Chapter 9). The type of root system would influence susceptibility to drought and a comparative study of the roots of *Melanthera, Argusia* and *Scaevola* may resolve the apparent paradoxical inversion of physiological capabilities and ecological susceptibilities.

Clegg, Koranda and Robison (1979) also studied *Argusia* (in the Marshall Islands). They found that stomatal resistance in the canopy increased (stomata closing) by 3–4 times of full sunlit values when shaded. They suggest that the heliophilic requirement of *Argusia* is an adaptation to a sub-climax ecology.

Salinity

The ground water of small sand cays is influenced by the sea. However, islands above a minimum size of 350 feet (107 m) in smallest diameter, can hold a freshwater lens in the substrate which is available to plants (Whitehead and Jones 1969). Consequently, ground water can be quite low in salt, and aside from the problems of definition of halophytism (see Chapter 8), plants which are seemingly halophytic on the basis of proximity to the sea may in fact not be subject to very high salt loads via ground water. Indeed, De Jong (1979) indicates that the notion that beach taxa are generally either halophytic, xerophytic or both appears 'to be naive and partially incorrect'.

One of the important factors is that salinity values in the soil can drop very sharply over very small distances landward from the beach (Barbour 1970a). De Jong (1979) found that the salinity of the water in the water table of a Californian beach was always less than 3% sea water throughout the year. Calculated soil osmotic potential was between 0 and −10 bar at 100 cm depth and substantially less at shallower ones.

The loose texture of the One-Tree beach may permit greater lateral penetration of sea water, and greater mixing of sea water with rain water, at least in the upper beach; however, this topic remains to be explored. It is possible that the seaward extent of the shrub ring may be limited by ground-water salinity; flooding of a part of the eastern edge of the island by a storm in 1976 resulted in the death of the *Argusia* in the affected area (Fig. 6). At the same time, roots of *Argusia* were exposed on the southern edge of the island by beach erosion.

Salt spray can have a direct effect and certainly reaches all parts of One-Tree Island at least occasionally. Its level and its effect on the vegetation have yet to be assessed. It is possible that the soil moisture conditions resulting from substrate texture interact with salinity in being responsible for the zonation of vegetation on the island. The pattern of the concentric ridges of storm rubble suggests that the higher, forested areas such as the *Pandanus* clump and the *Pisonia* groves are the oldest parts of the island and have probably been vegetated the longest. (It is probable that the various groves of *Pisonia* and *Pandanus* indicate the sites of previously separate small cays which were subsequently joined into one island by the accumulation of storm rubble between them.) They have the best developed soil with the greatest infilling of small particles such as sand and organic matter (Fig. 15). Consequently they are best suited to the movement upward of deeper soil water, for the retention of rain water, and consequently can support a more mesic type of vegetation. In turn, the plant cover ameliorates the hydric and thermal microclimate at the surface of the soil and would reduce surface evaporation.

The area now covered by the herb flat with its loose surface rubble but some sand infill would more closely resemble strand or beach conditions and be vegetated accordingly. The herb flat seems to be similar to dry beaches in terms of its moisture characteristics but because of its greater distance from the sea may be less affected by salt. It consequently has a mixture of typical beach species and those relatively drought resistant ones unable to tolerate the salinity of the beach. The loose rubble beach of One-Tree is reserved for only the salt-spray tolerant species that in addition can persist in areas with a substrate containing such large interstices that upward movement of deeper water is minimal.

The succulent area, influenced as it is by the brackish water of the pond contains only true halophytes.

Sea birds

Sea birds have been noted to affect the vegetation in areas in which they nest. It has been assumed frequently that the influence is mediated primarily via changes in soil nitrates or phosphates resulting from faecal deposition. Gillham (1961) found that trampling and guano deposition by nesting sea birds can cause the destruction of the vegetation at nesting sites (even for those that nest in shrubs) and may lead to abandonment of the site by birds. Upon regeneration of the vegetation the cycle may begin again. Sobey and Kenworthy (1979) have implicated a variety of mechanical factors including treading, the digging of scrapes and collecting of nest materials, and boundary clashes associated with maintenance of territories around the nest. The overall effect was that the plant community was species-poor with annual and ruderal species predominating to a much greater degree than in areas unaffected by gulls.

Experimental, artificial mechanical disturbance to the vegetation produced some of the same effects. However, mechanical disturbance is not necessarily the sole factor as the vegetation characteristics of non-gull areas can be destroyed by intense defaecation, and conversely, one species of plant seemed to require guano-rich soil for establishment.

In some areas, the occupancy by gulls resulted in increased erosion and an indirect effect on the vegetation, including marked seasonal changes.

On One-Tree Island, the species of birds that nested in the vegetation undoubtedly caused some mechanical damage and certainly added nutrients to the soil. The high proportion of ruderals in the herb flat flora (see Chapter 8) may be in part bird-induced or at least bird-influenced. Hnatiuk and Mertons (1979a) suggest that some parts of Aldabra 'may have been modified from shrub to herb cover by the activities of the birds'. It is even possible that the nesting of birds on One-Tree is partly responsible for the herb flat not proceeding to later seral stages. However, there were no detectable vegetation differences between areas in which birds nested and those in which they did not, nor between years when birds nested in a given area and those when they did not. The birds changed nesting sites from year to year (Chapter 17) and it is possible that because of such movement, their effect was averaged in time over the entire herb flat, obscuring specific local effects.

Generalized nutritional effects are almost certainly profound, but do not seem to be greatly localized, a view supported by the study of Davis *et al.* (1938). The role of birds as dispersal agents has already been discussed as has the relationship between sea birds and the *Pisonia* tree (see Chapters 8 and 18).

Stress and succession

The type of vegetation on One-Tree Island is probably a reflection not only of immediate conditions prevailing at any one time, but of the types and frequencies of various stresses imposed on the system. Storms do not seem to have very long-lasting effects and merely cause breakage or leaf loss in particular species, or localized mortality, but in the intensities experienced during the present study do not seriously affect community organization.

Droughts were a much more important source of stress and did sporadically influence the community. Connell and Slatyer (1977) have pointed out that in many communities major disturbances occur frequently enough that succession will be set back and begun again at some retrograde stage. Grime (1977) noted that competitive characteristics such as rapid growth of leaf area and root absorptive area may be inhibited by stress or damage and that the role of stress changes according to its severity. He suggested that low intensity stress, characteristic of productive habitats, functions as a modifier of competition, whereas severe stress exercised a dominant and more immediate impact upon species competition and vegetation structure through its direct effects on survival and reproduction. He identified three major strategies among plants: (1) the competitive strategy (maximization of capture of resources) prevailing in productive, relatively undisturbed vegetation, (2) the stress-tolerant strategy associated with continuously unproductive conditions, and (3) the ruderal strategy characteristic of severely disturbed but potentially productive habitats. He found that perennial herbs and ferns included the widest range of strategies, that annuals were restricted mainly to severely disturbed habitats while biennials were prominent among the competitive ruderals and stress-tolerant ruderals. Trees and shrubs may be competitors, stress-tolerant competitors or stress tolerators.

These ideas apply to One-Tree Island. The *Pisonia* tree is often regarded as a climax species, although Gillham (1963) pointed out that it may be replaced by taller species under some circumstances. Despite this it would seem to be characteristic of later stages of succession and to be a 'competitor'; it occupies the higher, older sites along with *Pandanus*. The newer parts of the island are covered either by the herb flat or shrubs. Both *Scaevola* and *Argusia* are known as colonizers of new habitats, appearing soon after the very early pioneer species and are resistant to high salinity (see Chapter 8). Accordingly they should probably be considered stress tolerators. As noted previously, they tolerate the strand habitat and are largely restricted to it by their mode of dispersal (although they grow better in more sheltered habitats) (Guppy 1891).

The herb flat species contain a mixture of those usually considered as pioneers of unstable substrates (beaches, dunes) and those normally found further inland under more stable conditions. This mixture probably indicates a transitional successional stage, which is moved toward later stages by favourable weather conditions, but reverts toward earlier ones by the stresses imposed by drought. Mechanical disturbance by sea birds could also favour the earlier successional stages. The transitional character of the herb flat and the periodic effect of drought probably maintain a higher number of species in the herb flat than would otherwise occur as the droughts maintain pioneer species that would probably disappear from the herb flat if succession proceeded unaltered by moisture stress. As mentioned previously, the annual habit is associated with disturbed habitats and perennials have a wide range of strategies. These are the two major categories of plants in the herb flat whereas biennials which are usually competitive or stress-tolerant ruderals are poorly represented (Table 5).

The beach has only pioneer, stress-tolerant species. The succulent mat is under the influence of the brackish water of the pond and water shortages would be uncommon for species tolerant of the salinity. They are thus stress-tolerant.

A summary of the local distribution and succession of the One-Tree vegetation, based on the above considerations and those described elsewhere in the book is as follows. The major stresses on One-Tree Island are drought (generally) and salinity (succulent mat and perhaps beach); there may be minor mechanical disturbance imposed by the nesting of sea birds. Species tolerant to salinity occupy the areas around the brackish pond (succulent mat), those tolerant to both drought and salinity occupy the beach and the ring of shrubs (*Scaevola* and *Argusia*) around the upper beach; these two groups represent pioneer stages, the beach plants appearing first but followed quickly by shrubs. All of these are stress-tolerators. The herb flat contains a mixture of species, predominantly stress-tolerators also found in pioneer communities and perennial and annual ruderals. Most of the latter categories would be stress-tolerators (particularly to drought) but to varying degrees and some could possibly be regarded more as competitors.

The herb flat is in a transitional successional stage, drought temporarily favouring the pioneer species, and periods of greater rainfall favouring the others.

The most advanced successional stage on the island is represented by the *Pisonia* groves and the *Pisonia-Pandanus* grove. These occur in the older parts of the island which have the greatest accumulation of organic matter and retain water the best. The shade-intolerant species characteristic of the other habitats are absent.

The succulent mat is a special case, probably being influenced largely by edaphic conditions (soil salinity and frequent water-logging of soil) and containing only a few stress-tolerators (to salinity).

In terms of Grime's (1977) strategies, the plants of the beach vegetation, the shrub zone, and the succulent mat are stress-tolerators, those of the herb flat a mixture of primarily stress-tolerators and ruderals and those of the groves are competitors. These represent progressive successional stages.

The herb flat has a far greater species richness than any of the other habitats, a fact consistent with the general tendency of intermediate successional stages to be more diverse than either pioneer or climax ones (Odum 1969).[1]

Connell (1978) has presented two different views of communities. In one, a community is regarded as stable, returning to its original state following perturbation. The other is that disturbances interrupt the competitive hierarchy on a short-term scale, preventing communities from reaching equilibrium and a high species diversity is maintained. Without disturbance, competition would eliminate some species and reduce diversity. He points out that these two views are not mutually exclusive. We suggest that the vegetation of the herb flat conforms more to the second hypothesis. The *Pisonia* groves are also of that type but have gone to completion in the sense that they are less subject to the effect of perturbation (drought) because they occupy the older, more favourable sites and they have become reduced in numbers of species.

Connell was referring to stability in terms of species diversity and taxonomic composition. In that sense, in the present case stability of the plant community of the herb flat was maintained. Except for *Solanum* none of the temporary species occurred in the herb flat, nor were species of plants lost from that habitat.

In terms of the animal component of the herb flat community, stability in Connell's sense was not maintained. Numbers of species (or any other indicators of species diversity) varied widely and never reached equilibrium. However, trophic structure stability was maintained throughout these perturbations and in this characteristic seemed to resemble a community described by his first hypothesis.

Human disturbance

The role of human disturbance in the ecology of One-Tree Island has been discussed already in a brief way, and an outline of the procedures we followed

[1] This is not always the case and Odum indicates a bimodal pattern in some cases; also see Tramer (1975) for a contrary example.

in order to minimize our impact on the island during the study is presented in Appendix I. Here we wish to point out an effect we had on the vegetation, which although not operative immediately, and not affecting the areas we subsequently sampled or our results, nevertheless had a long-term effect on the heterogeneity of the herb flat vegetation.

The removal of all the vegetation in the 10 sample plots at each visit left small patches of bare ground. These subsequently became re-vegetated by natural means. The resulting vegetation differed markedly among the bare patches left at different sampling times.

Not all species of plants were fruiting at the same time, and peaks in fruiting time varies markedly among species (Figs. 33, 34). Thus, bare patches produced by sampling at one time would tend to be seeded by one species of plant and those produced at another time, by a different species, depending on which species was fruiting at the time sampling took place. As a result, the sampled areas became local patches of nearly monospecific vegetation, the particular species involved varying from one sampling area to another.

These areas were never resampled subsequently and consequently, all the vegetation samples relate to the original condition of the herb flat vegetation. However, by the end of the study, instead of the herb flat continuing to be rather homogeneous throughout (see Chapters 9 and 10), it had become a mosaic of monospecific patches of varying types set in a matrix of the original homogeneous mixture of species. The herb flat had become highly heterogeneous because of the sampling procedures used. This effect is likely to persist for a long period of time. It is important to recognize the origin of this heterogeneity so that subsequent investigators will not misinterpret the local distribution of the vegetation, and will not receive a false impression of the degree of homogeneity of the herb flat during the present study.

The human-induced heterogeneity may have (or have had) repercussion in the insect fauna. Wilson (1977) indicated that on Costa Rican beaches, the numbers of insect taxa differed greatly between pure stands of two plant species (*Ipomoea* and *Canavalia*), especially among the more important herbivore groups such as Hemiptera and Coleoptera. He attributed these differences to differences between the plant species in toxic chemical composition. Thus, the heterogeneity of the One-Tree herb flat vegetation arising from production of a mosaic of nearly monospecific localized stands may result in an equally patchy distribution of some groups of insects. Sampling of these various patches of vegetation and comparing the results would be a fruitful approach to the study of plant-insect interactions.

Absolute sizes of populations

The previous parts of the book have treated the densities of animals primarily on a unit areal basis. Such notation is appropriate for most types of ecological interpretation. However, in insular situations the absolute sizes of populations are important as they relate to the probability of extinction. In general, very small populations would be more prone to random extinction than would larger ones.

Quantitative studies in the *Argusia* areas dealt only with the foliage fauna and nectarivores, only with the foliage fauna and soil microarthropods in the *Pisonia* groves, only with the soil arthropods in the succulent mat, and no overall quantitative treatment of the *Pandanus* grove and beach areas were attempted. The herb flat was the only habitat on One-Tree Island for which an attempt was made to quantitatively sample the complete arthropod fauna, including the epigeal and soil components. That area can thus be examined in most detail.

The estimated total number of epigeal arthropods (all species) in the herb flat ranged from slightly over half a million just after the drought to over five and a half million in February–March 1971 with the average number being just over 2¾ million (Table 48). The total number of individual species varied widely. As indicated in the discussion on equitability (Chapter 22), the most abundant single species was 45% of the total numbers or less, and usually much less. Species varied in their ranking in terms of relative abundance, a different species being the most abundant on different occasions. In absolute numbers, the most abundant species was represented by about 150 thousand to slightly over 2 million individuals (Table 48). Others were represented by smaller total population sizes, probably down to a very few or even one individual in the case of some of the transients.

Absolute population size was quite different for the soil microarthropods of the herb flat, the estimate for the period of lowest density being almost 3 billion, and the largest population estimated to exceed 7½ billion. The average was over 4½ billion (Table 48). Even partitioning these large numbers into the approximately 130 species of which they were composed, and allowing for differences in relative abundance, even the 'rarer' species would have very large absolute population sizes in comparison to levels of the epigeal species. The total epigeal arthropods were only about 0.01% to 0.12% (mean 0.07%, see Table 48) as numerous as were the total soil microarthropods.

The above condition was not restricted to the herb flat. The *Pisonia* groves and succulent area, although much smaller in total area than the herb flat, still had very large total populations of soil microarthropods, averaging over 1 billion and about 300 million respectively (Table 49). Combining all three areas from which soil microarthropods were sampled gave a combined average total population size of over 6 billion, with the largest value for any sampling period exceeding 10 billion. Inclusion of the *Pandanus* grove, *Argusia* areas, and the beach would, of course, substantially increase the number. When it is remembered that some of the species are found widely on the island and occur in several habitats (Chapter 16), one realises that the

Table 49. Estimated total numbers of soil microarthropods in different habitats on One-Tree Island at different sampling dates.

Date	Herb flat	Pisonia	Succulents	Total no.
1969				
J–A	29×10^8	9.2×10^8	2.8×10^8	41.0×10^8
S–O	46×10^8	5.2×10^8	1.4×10^8	52.6×10^8
N–D	43×10^8	7.0×10^8	1.5×10^8	51.5×10^8
1970				
J–F	29×10^8	9.6×10^8	3.2×10^8	41.8×10^8
M–A	76×10^8	21.0×10^8	4.5×10^8	101.5×10^8
M–J	37×10^8	8.6×10^8	2.3×10^8	47.9×10^8
J–A	65×10^8	15.0×10^8	2.0×10^8	82.0×10^8
S–O	44×10^8	17.0×10^8	6.5×10^8	67.5×10^8
Mean	46.1×10^8	11.6×10^8	3.0×10^8	60.7×10^8

total numbers of individuals of the more common species of soil microarthropods on One-Tree Island reach truly staggering proportions.

Insularity is probably a somewhat relative matter, depending on the type of organism involved. Species-turnover was a common phenomenon among the epigeal arthropods, but was not among the soil microarthropods. Many of the latter were common pantropical or pantemperate species (Chapter 16) and probably were hardy species able to survive harsh conditions either during dispersal to the island or on the island once there. Furthermore, many occupied a variety of habitats on the island and survived in the most favourable refugia during generally harsh conditions and could re-colonize the most severely affected areas upon the return of favourable conditions. For these reasons extinction rates of established species might be expected to be low. However, another important reason for low extinction rates is probably related to the large total population sizes of many of the species. Small absolute populations would be subject to random extinction whereas large ones would not. In this respect, to a soil arthropod with large total population size, One-Tree Island must be more like a continent than an island. Such would not be the case for many of the epigeal species, to whom the insular character of One-Tree Island would be clearly expressed.

Another factor in the low species-turnover of the soil microarthropods is their low potential for dispersal, probably requiring floating objects such as logs with considerable amounts of soil in crevices, birds with mud on their feet, or perhaps unusual winds for dispersal. Transport by such means would be much less frequent than by the means available to epigeal flying arthropods.

In summary, we suggest that species-turnover is low in soil arthropods because of low immigration rates and low extinction rates, the latter being greatly reduced by the large total population sizes of many of the species.

By contrast, the observed high turnover rate of epigeal species (see

Chapter 21) probably results from generally greater ease of dispersal and higher extinction rates, the latter resulting in part from the greater susceptibility of small populations to random extinction.

Although the total faunas of areas other than the herb flat were not sampled quantitatively, certain components were. For example, the numbers of herbivores in the herb flat, and on *Argusia* and *Pisonia* foliage were sampled. The chewing herbivores had the greatest total population sizes in the herb flat, on the average accounting for over 99% of the total numbers of individual herbivores for the 3 areas combined (Table 50).

Table 50. Estimated total numbers of chewing herbivores on One-Tree Island at different sampling periods.

Date	Herb flat	Argusia	Pisonia	Total
1969				
J–F	41,807	–	–	–
M–A	239,856	–	–	–
M–J	595,062	–	–	–
J–A	271,592	–	–	–
S–O	444,618	239	0	444,857
N–D	282,273	4,061	5,202	291,536
1970				
J–F	377,788	2,150	2,081	382,019
M–A	76,290	478	1,040	77,808
M–J	429,970	1,195	0	431,165
J–A	35,093	0	1,387	36,480
S–O	899,307	956	2,081	902,344
N–D	123,590	2,150	694	126,434
1971				
F–M	376,873	5,973	0	382,846
1972				
April	667,690	2,628	0	670,318
Mean (Sept. 1969– Apr. 1972)	371,349	1,983	1,249	374,581
Percentage	99.1	0.5	0.3	–

The total population size ranged from just over 35,000 to almost 900,000 for the herb flat, but only from practically nil to almost 6,000 for the *Argusia* foliage and from nearly nil to slightly more than 5,000 for the *Pisonia* foliage.

The more abundant species, especially those which occur in all three areas would have reasonably sized populations but many species must be of rather small absolute size, and of course a number of them were merely transients. Species restricted to only *Pisonia* or *Argusia*, especially, might be represented by relatively few total numbers of individuals. Clearly, a number of species, particularly the more restricted ones, occur in such small populations that they would be endangered by random extinction. It is not surprising that turnover rates are high among the epigeal arthropods.

Generalized herbivores have a reservoir of individuals in the herb flat which raises total population size. From this reservoir individuals could move to the other types of vegetation. It is probable that generalized feeders accordingly are less subject to random extinction than are those more specialized ones restricted to a vegetation type covering only a small area. Conversely, specialized feeders may be able to exploit a given resource more effectively and maintain larger populations on that resource than would more

Table 51. Estimated total numbers of sucking herbivores on One-Tree Island at different sampling periods.

Date	Herb flat	Argusia	Pisonia	Total
1969				
J–F	0	–	–	–
M–A	14,329	–	–	–
M–J	244,114	–	–	–
J–A	704,535	–	–	–
S–O	964,317	24,176	0	988,493
N–D	688,065	16,541	18,855	723,461
1970				
J–F	249,766	42,944	18,436	311,146
M–A	202,443	44,852	11,732	259,027
M–J	439,433	29,265	8,380	477,078
J–A	195,396	30,856	14,246	240,498
S–O	311,244	19,722	10,894	341,860
N–D	103,908	15,905	7,542	127,355
1971				
F–M	664,875	48,669	12,570	726,114
1972				
April	430,521	76,980	–	–
Mean (Sept. 1969– March 1971)	424,383	30,326	11,406	466,115
Percentage	91.1	6.5	2.5	–

generalized species. Which occurs may vary from taxon to taxon. We do not have sufficiently detailed information on the feeding habits of enough of the One-Tree species, especially those which were established for only a short time and/or in low densities, to resolve this question. A project designed to compare turnover rates and absolute population sizes between generalized and specialized users of a particular resource would be most instructive.

The above discussion of the chewing herbivores would also apply to the sucking herbivores as the results obtained on both groups were similar. Over 90% of the total individuals of sucking herbivores were in the herb flat

(Table 51). Numbers there ranged up to over 960,000 and averaged over 424,000 whereas means for the *Argusia* and *Pisonia* foliage were just over 30,000 and 11,000 respectively.

The nectarivores in the herb flat, *Argusia* shrubs and *Pisonia* groves were also compared and the results obtained were generally similar to those from herbivores. The herb flat with its large area, had far greater total numbers of

Table 52. Estimated total numbers of nectarivores on One-Tree Island at different sampling periods.

Date	Herb flat	Argusia*	Pisonia	Total
1968				
July	...	4,161	–	–
1969				
J–F	15,361	1,000	–	–
M–A	126,428	12,699	–	–
M–J	655,872	22,309	–	–
J–A	587,113	22,920	–	–
S–O	788,986	46,522	4,609	840,117
N–D	543,209	86,632	73,744	703,585
1970				
J–F	465,318	23,123	3,771	492,212
M–A	225,953	15.992	4,190	246,135
M–J	248,101	5,491	6,285	259,877
J–A	62,720	28,237	1,257	92,214
S–O	220,350	59,695	2,095	282,140
N–D	55,306	140,678	0	195,984
1971				
F–M	897,865	115,073	37,291	1,050,229
1972				
April	788,654	–	–	–
Mean (Sept. 1969– March 1971)	389,756	41,752	14,805	446,313
Percentage	87.3	9.4	3.3	–

* Based on flower plus foliage samples.

nectarivores than either the *Pisonia* or *Argusia* areas. On the average, the herb flat had 87% of the total individuals of nectarivores, *Argusia* shrubs about 9% and *Pisonia* about 3% (Table 52).

The highest total numbers for all three areas ranged from about 90,000 individuals to over 1,000,000 individuals, with the average being slightly less than 450,000.

The relative proportions of the total numbers of individuals found in the herb flat varied somewhat with trophic level. Chewing herbivores are more strongly represented in the herb flat relative to other areas (99.1%) than are sucking herbivores (91.1%) or nectarivores (87.3%). The highest relative representation achieved by the *Argusia* shrubs was in the nectarivores. The *Argusia* shrubs, when flowering heavily, seemed to have a much higher proportion of flowers to foliage than is usually observed in the herb flat.

The total number of individuals of herbivores and nectarivores in the epigeal arthropod fauna of the three areas combined are probably close to the totals for the entire island. The habitats not included (beach, *Pandanus*, succulents) all represent relatively small areas of the island, except the beach which was sparsely vegetated and consequently contained few arthropods. The *Pandanus* suffered little damage by chewing herbivores (Chapter 15) and flowered infrequently and then only briefly (Chapter 11) and consequently probably contained few herbivores or nectarivores. Densities of all epigeal arthropods in the succulent mat and *Pandanus* grove, although unquantified, were quite low (Chapter 14).

Summary

1. The study was conducted to fill the need for a long-term, detailed study of as many aspects of community ecology of an island as possible, and involving as many taxa as feasible.
2. Early scholars seldom thought about island biology, or if they did, it was couched in an unrealistic theological framework. Insular biogeography and ecology was placed on a modern theoretical footing by Charles Darwin and Alfred Wallace. They can be regarded as the founders of this field; the concept of evolution was critical to its development.
3. The equilibrium theory of MacArthur and Wilson provided renewed interest in the field, stimulated research along new lines, and fostered an experimental approach to the subject.
4. One-Tree Island is a coral rubble island on the most seaward and exposed reef of the Bunker-Capricorn Reef, southern Great Barrier Reef, Australia. The vegetation consists of beach vegetation, herb flat, shrub ring, groves of *Pisonia* trees and of *Pandanus* trees, and a succulent mat surrounding a brackish pond. The climate is maritime subtropical with equable, mild temperatures showing regular seasonal fluctuation. There are periodic droughts and frequent tropical storms. The island had suffered little disturbance by humans prior to the study.
5. The flora included 21 permanent residents, 4 temporarily established species and various transients, including human introductions.
6. The established flora was characterized by being dispersed predominantly by ocean currents and by seabirds. The transient plants were mostly transported by humans and wind.
7. The flora consisted mainly of strand plants and ruderals, most of which were perennial and many of which were halophytic; there were high

proportions of chamaephytes and therophytes. A variety of germination strategies were employed.

8. All established species of plants except one were either hermaphroditic or monoecious, suggesting a high potential for autogamy; however, outcrossing is probably frequent as most are cross-pollinated by insects.
9. Plant biomass in the herb flat was high early in the study, was reduced drastically by a drought in late 1969, from which it took 10 months for the original levels of biomass to recover. During recovery, different species recovered at different rates and there was a shift in dominance from *Melanthera biflora* to *Ipomoea macrantha* followed by a relative increase in *Abutilon indicum* and *Canavalia rosea*. Eventually *Melanthera* seemed to be approaching its original status of dominance.
10. Plant height decreased as a result of the drought, and percentage cover was greatly reduced.
11. Different species of plants had different period of flowering and fruiting. Their seasons were staggered and flowering by several species was in progress at all times of year. Numbers of flowers produced varied from year to year.
12. New leaf production in *Pisonia* and *Argusia* showed a decline in early 1970 and resulted in a net loss of leaves, but there was a peak of production in late 1970 resulting in a net gain in leaf biomass. *Pandanus* showed a net loss of foliage during much of the study period; leaves of this species are long-lived (several years) and persist on the stem long after death. In all species leaf production varied more than leaf fall.
13. Annual production of *Argusia* twigs was 378 kg dry weight for the island as a whole.
14. Plants of the herb flat killed by the drought remained as dead standing stems for a long time and only gradually became incorporated into the litter.
15. Many seeds wash up on the One-Tree beach. Most are of species not otherwise represented on the island and few germinate. The flotsam is largely woody and not usable by the local organisms. Jetsam occurs on the beaches but plays a minor role in the terrestrial community.
16. The *Pisonia* and *Pandanus-Pisonia* groves occupy the older parts of the island and probably represent the climax vegetation for the cay. The succulent mat is made up of salt-tolerant species, the beach of pioneer species that tolerate salt and drought, and the herb flat is characterized by a mixture of pioneer and later stage species. The effect of drought periodically favouring the earlier stage species, alternating with wetter periods favouring the later stage ones probably maintains the herb flat in its intermediate condition. Sea birds through guano deposition and/or mechanical disturbance may also affect the nature of the herb flat vegetation.
17. Sampling the vegetation of the herb flat left bare patches that were

seeded by different species at different times, depending on which species were in seed at that time. This has increased the spatial heterogeneity of the herb flat vegetation.

18. The One-Tree epigeal arthropod fauna was imbalanced in comparison with the Australian mainland in that the former had disproportionately larger numbers of species of Hymenoptera and Diptera and unusually low numbers of species of Lepidoptera and Coleoptera.
19. Total numbers of species of epigeal arthropods in the herb flat fluctuated with time. Following the drought, species numbers were low but built up to a peak by late 1969. Then there was a decline to a lower number which persisted throughout most of the rest of the study.
20. Total numbers of individuals of epigeal arthropods in the herb flat were also low just following the drought, but subsequently recovered and peaked in late 1969 before dropping to lower, but variable levels; there was a second peak in early 1971. Numbers of individuals of different taxa departed to varying degrees from the pattern of total numbers.
21. The succulent mat had few individuals and species of epigeal arthropods and most seemed to be strays from other habitats. The *Pandanus* grove was poor in foliage-inhabiting arthropods.
22. Numbers of individuals of epigeal arthropods associated with *Argusia* flowers were low just after the drought but subsequently reached a peak in late 1969. A second peak occurred in late 1970; in both years the insect peak coincided with, or immediately followed the flowering peak of *Argusia*. On the average, 89% of the species associated with *Argusia* flowers were nectarivores. Intensity of usage of flowers varied among taxa and over time.
23. Ant activity at *Argusia* was most intense at night and on cloudy days. Some of the more common visitors at *Argusia* flowers (two species of fly and a bee) were diurnal.
24. The numbers of species and individuals of arthropods on *Pisonia* and *Argusia* foliage varied with time; peaks on *Argusia* coincided with times of highest leaf production but no such consistent correlation occurred with respect to *Pisonia*. The relative abundance of the various taxa of arthropods varied temporally.
25. There was considerable habitat specificity among the epigeal arthropods, and although each habitat on the island had its distinct fauna, there was broad overlap among the more highly vegetated habitats, especially between *Argusia* and *Pisonia*. The beach had few species and shared a low proportion of them with other habitats. The herb flat had the greatest number of species.
26. The small vegetation-free rubble cays on the edge of the reef had a few species of terrestrial arthropods.
27. Before the drought a large amount of leaf biomass had been consumed by insects; after the drought it was low but increased during recovery, and

varied thereafter, reflecting the level of arthropod densities in general, but not always in detail. Patterns of insect damage to leaves were influenced by temporal changes in leaf production and leaf fall in ways that were difficult to assess. *Argusia* and *Pisonia* differed in their temporal patterns of leaf damage. *Pandanus* sustained very low grazing levels.

28. A variety of species studied in some detail immigrated to the island during the study and/or underwent local extinction (in some cases more than once); others were characterized by very marked fluctuation in population density.
29. Frequency of new arrivals of epigeal arthropods to the island was positively related to frequency of winds blowing from the mainland; there were exceptions, however. In 1969 there was a short period of intense movement of arthropods to the island whereas in 1970 there was a less intensive but more prolonged period of high incidence of arrivals.
30. There was probably wholesale extinction of epigeal arthropods during the drought. Afterward, extinction rates were low and remained low well into the recovery period, then rose. Extinction rates were lowest during the coolest months, and highest during the hottest ones.
31. The number of established species was low just after the drought, rose to a high 'overshoot' after recovery and then dropped to fluctuate around a somewhat lower level. The mean number of total species during the last phase is probably the best estimate of the equilibrial number, $\hat{S}$. The total number of established species reached as high as 4 times the number of permanent ones.
32. Despite large fluctuations in population densities and numbers of species and great variation in taxonomic structure, the trophic structure (by species) of the epigeal arthropods in the herb flat was remarkably constant, even during periods of rather large environmental perturbations. We suggest that community regulation mediated via species interactions is the simplest explanation of this phenomenon. Trophic structure was quite constant among established species but varied more among transients.
33. Trophic structure of individuals of the epigeal arthropods of the herb flat was variable. We suggest that trophic structure by species reflects the number of kinds of niches available and that trophic structure by individuals reflects the resource levels in those niches. The former remains relatively constant, the latter varies.
34. The trophic structure of the foliage fauna of *Argusia* and *Pisonia* varied more than in the herb flat.
35. The epigeal arthropod fauna of the herb flat was relatively inequitable, the most abundant species usually accounting for 15–25% of the total numbers of individuals but with some values reaching to levels between 35% and 45%.

36. Over 130 species of soil microarthropods were recorded from One-Tree Island. Estimates of arthropod numbers and biomass varied over time. There was broad overlap of species among habitats. The fauna of the succulent mat was more heterogeneous spatially than those of the *Pisonia* groves or herb flat. The herb flat is subject to greater temporal variation in microclimate than the *Pisonia* groves and the fauna of the former is more labile than that of the latter.
37. The soil arthropods occur in much larger absolute population sizes than the epigeal species, and consequently have less likelihood of random extinction. They also are less easily transported to the island and have a lower immigration rate. The result is a hardy soil fauna with low turnover rates in comparison to the epigeal species.
38. There were 84 species of birds recorded from One-Tree Island, many of which were transients. Most of the established species were seasonally resident seabirds. There were three species of permanently resident land birds, a rail, a silvereye and a fantail-warbler.
39. The seabirds selected nest sites which varied interspecifically according to vegetation and substrate characteristics. Nesting sites for a given species were often changed to new locations in different years.
40. Seasonal fluctuation in bird numbers was about ten-fold. Total numbers during the nesting season did not change greatly from year to year.
41. Birds are important as predators on insects, in transferring marine-derived energy and organic materials to the island, and perhaps through mechanical disturbance to the low vegetation.
42. Loggerhead turtles occasionally nested on One-Tree Island; nesting attempts were frequently unsuccessful probably because of the nature of the substrate.
43. A brackish pond near the centre of the island periodically filled and dried. Populations of algae and snails fluctuated markedly.
44. There is a considerable flow of material from the sea to the island directly as wash-up of marine carrion, and indirectly through the agency of transfer organisms such as sea birds that deposit guano and their carrion. The large population of beach earwigs is supported mainly by marine-derived carrion.
45. Different kinds of islands each have their own characteristic trophic structure, reflecting their main source of energy. Pioneer stages of recolonization of islands, or of small vegetation-free islands are weighted heavily toward scavengers with no or few herbivores. Herbivores increase in proportion to the extent that autotrophic energy resources increase relative to the marine-derived ones.

Appendix 1. Special procedures in order to minimize the effects of the researchers on the biota, during the investigation of One-Tree Island

As a guide to future investigators of island communities we now describe our procedures and their effectiveness in some detail. A covered plastic rubbish bin containing sea water covered with a layer of motor oil several cm thick was used as a receptacle for all food scraps and empty food tins. A second such container served as a latrine. These containers were removed from the island following each trip. During most of the study, only tinned food was consumed while on the island, and hence fresh food was not available to serve as a nutrient source for local insects. Left-overs were not kept for later meals but were discarded in the bin with motor oil. After meals, utensils and dishes were washed and the water discarded in the bin. Toward the end of the study these procedures were relaxed to the extent that some fresh food was taken to the island but was kept in air-tight containers. We feel that these precautions were observed with sufficient care that our maintenance activities had a negligible effect on local populations; insects were not usually attracted to the campsite. There were several exceptional occasions. Mosquitoes were not always deterred by insect repellent and we occasionally contributed unwillingly to their energy supply. At times we also, and much less reluctantly, became direct agents of their mortality. As mosquitoes were not common except for two trips, and then only in the *Pandanus* clump, we feel our impact on the energetics and demography of this group had only slight repercussions throughout the food web of the community.

Accidents resulted in minor effects. Once some food was inadvertently dropped and attracted two muscid flies; on another occasion a moth and a fruit fly were attracted to a glass of fruit juice and fell in and drowned. The most serious accidental effect resulted from the attraction of moths to light. A gas lamp was used one evening outside the tent and attracted a number of *Utetheisa* moths, five of which flew into the lamp and were killed (about 2½% of the total adult population on the island at that time). Subsequently we used flashlights outside and when a gas lamp was used inside the tent, the zippers of the tent were secured, making it insect-proof. When persons entered or left when the light was on, the zipper was opened as briefly as possible and only far enough to allow the person to pass. A subsequent check for insects was made and any which had entered were evicted. The light shining through the mosquito-net windows may have caused a concentration of moths in the vicinity of the tent, however.

The trampling of vegetation which occurred as a necessary consequence of our scientific activities did, of course, have some influence. We kept this to a minimum by walking on the beach when possible or by using previously trampled paths. Care was exercised when loading gear to be taken to the island and also when unpacking it once there, to prevent introducing any individuals from either the mainland or from Heron Island. Effect of the sampling programme on the fauna and flora is discussed in the text.

Our main expeditions took place every two months over slightly more than two years and involved three to six persons each time. In contrast, other scientific expeditions occurred only once-yearly but included a much larger number of people; for general accounts see Talbot (1967) and Eschmeyer and Eschmeyer (1971). The latter expeditions altered the island by constructing a grill for cooking, and by laying down a cement slab upon which to pitch tents. Both were on the beach and no vegetation was cleared. Boards used for cooking fuel were transported to the island and some remained there until subsequent expeditions. After we began our study, other expeditions cooperated in taking precautions with food so as not to influence the terrestrial energy budget of the island; as on earlier expeditions their empty food tins were burned. The rigid procedures we instigated for ourselves, however, were not observed by other expeditions and they may have had some influence on local populations. For example, a pig taken along to serve as food for a marine expedition, was kept in a cage on the beach and must have had some effect on insect populations; once, it escaped and remained at large for one day and may have influenced vegetation during that time. The same expedition believed that they probably introduced a cockroach (*Periplaneta australasiae*) to the island in their equipment. An individual was later found beneath a box stored on the island between trips by the expedition; this species was not found during our study and it probably failed to become established then. Roaches subsequently became established after erection of the marine station. Other probable introductions by other expeditions are discussed in the section on species-turnover.

The sea bird colonies were sometimes disturbed by our sampling procedures and our presence; other expeditions undoubtedly influenced them as well.

The establishment of the marine station on One-Tree in 1971 increased the effect of humans on the local biota. Some of these influences are known and are discussed at appropriate places in the text.

Appendix 2. The One-Tree Island arthropods that have been identified at least to genus

Subphylum Chilopoda (centipedes)
Scolopendridae
Scolopendra sp.

Subphylum Arachnida

Order PSEUDOSCORPIONIDA
Garypus sp.
Neocheiridium sp.

Order ARANEIDA (spiders)
Clubionidae
Chiracanthium sp.
Araneus sp.
Lycosidae
Lycosa sp.
Artoria sp.
Salticidae
Bianor maculatus
Segestriidae
Ariadna sp.
Theridiidae
Dipoena sp. nov.
Theridion sp.
Lithyphantes cingulatus
Achaearanea sp.
Steatoda limitata
Thomisidae
Diaea haematodactyla L. Koch
Diaea variabilis
Diaea ?multipunctata
Uloboridae
Uloborus barbipes

Subphylum Hexapoda

Class Insecta

Order ODONATA
Libellulidae
Pantala flavescens Fabr.

Order DERMAPTERA
Labiduridae
Anisolabis maritima

Order ORTHOPTERA
Tettigoniidae
Conocephalus sp.
Oecanthus sp.
Gryllidae
Ornebius sp.
Teleogryllus oceanicus (Le Guill.)
Acrididae
Valanga irregularis (Walk.)
Valanga sp.
Locusta migratoria (Linné)

Order PSOCOPTERA
Ectopscocidae
Ectopsocus pilosoides Smithers
Ectopsocus baliosus Thornton and Wong

Order HEMIPTERA
Lygaeidae
Oxycarenus sp.?
Scutellaridae
Tectocoris diophthalmus?

Order NEUROPTERA
Hemerobiidae
Micromus tasmaniae Walker
Chrysopidae
Chrysopa sp.

Order COLEOPTERA
- Carabidae
  - *Abacetus* sp.?
- Coccinellidae
  - *Scymnodes lividigaster*
  - *Coelophorus inaequalis*
- Tenebrionidae
  - *Gonocephalum torridum*?
- Curcullionidae
  - *Notiosomus* sp.

Order DIPTERA
- Culicidae
  - *Anopheles amictus* Wood and Lee
  - *Aedes vigilax*? (Skuse)
  - *Aedes kochi* (Dönitz)
  - *Aedes* sp.
- Chironomidae
  - *Chironomus* sp.
- Sciaridae
  - *Sciara* sp. A
  - *Sciara* sp. B
  - *Sciara* sp. C
  - *Sciara* sp. D
  - *Sciara* sp. E
- Mycetophilidae
  - *Clastobasis tryoni* Sk.
- Dolichopodidae
  - *Sciapus* sp.
- Phoridae
  - *Megaselia* sp. A
  - *Megaselia* sp. B
- Platystomatidae
  - *Rivellia* sp.
- Tephritidae
  - Gen. nov.
  - *Trypanea* sp.
  - *Rhabdochaeta* sp.
  - *Spathulina* sp.
- Lauxaniidae
  - *Homoneura indecisa* Mall.
  - *Homoneura* sp. nov.
  - *Sapromyza* sp.
  - *Steganopsis* sp.
- Lonchaeidae
  - *Lamprolonchaea metatarsata* (Kert.)
  - *Lonchaea* sp. nov.
- Agromyzidae
  - *Liriomyza caulophaga* (Kleinschmidt)
  - *Phytomyza syngenesiae* Griff.
  - *Phytomyza* sp.
- Ephydridae
  - *Placopsidella cyanocephala* Kert.
- Drosophilidae
  - *Drosophila serrata* Mall.
  - *Drosophila* sp.
- Milichiidae
  - *Desmometopa* sp.
  - *Leptometopa albipennis* Mall.
- Chloropidae
  - *Cadrema* sp. nov.
  - *Tricimba* sp.
  - *Lioscinella* sp. A
  - *Lioscinella* sp. B
  - *Conioscinella* sp.
  - *Botanobia* sp.
- Anthomyiidae
  - *Anthomyia vicariens* Schin.
- Muscidae
  - *Atherigona* sp. nov.
  - *Atherigona* sp. A
  - *Atherigona* sp. B
  - *Coenosia* sp.
- Tachinidae
  - *Actia* sp.
- Hippoboscidae
  - *Icosta* sp.

Order LEPIDOPTERA
- Butterflies
- Nymphalidae
  - *Danaus plexippus* L.
  - *Hypolimnas bolina*
- Lycaenidae
  - *Syntarucus plinius pseudocassius* (Murray)

Moths
- Arctiidae
  - *Utetheisa pulchelloides*

Order HYMENOPTERA
- Braconidae
  - *Apanteles opacus* Ashmead
  - *Apanteles* sp. (2 spp.)
  - *Yelicones* sp.
  - *Myropotes creon* Nixon
  - *Iphiaulax* sp.
  - *Agathiella* (2 spp.)
  - *Bracon* sp. nov.
  - *Heterospilus* sp.
- Formicidae
  - *Pheidole megacephala* (Fabricius)
  - *Cardiocondyla nuda* (Mayr)
  - *Tapinoma minutum* (Mayr)
  - *Iridomyrmex glaber* (Mayr)
  - *Iridomyrmex* sp.
  - *Camponotus* sp.

Appendix 3. List of soil mites and collembolans from the herb flat, a *Pisonia* grove and the succulent mat on One-Tree Island.

SUBCLASS ACARI	
Order Parasitiformes	
Suborder Mesostigmata	
Supercohort Monogynaspida	
Cohort Sejina	
S.F. Sejoidea	
F. Segiidae	*Epicroseius porosus*
Cohort Uropodina	
S.F. Uropodoidea	
F. Uropodidae	Uropodid I
	Uropodid II
	Uropodid III
Cohort Gamasina	
S.F. Parasitoidea	Parasitoid I
F. Rhodocaridae	*Rhodacarus marksae*
	Gamasiphis setosus
	Rhodacaroides new sp.
	Gamasellus discutatus
	(in complex, new sp.)
F. Ascidae	
Sub F. Platyseiinae	*Chieroseius cassiteridum*
Sub F. Ascinae	*Asca major*
	Protogamasellus primitivus
	Protogamasellus massula
	Gamasellodes (in *bicolor* group)
F. Dermanyssidae	*Hypoaspis* sp. I
	Hypoaspis sp. II
F. Digamasellidae	*Digamasellus australis* n. sp.
F. Macrochelidae	*Genolaspis* sp.
F. Parholaspidae	*Holaspulus* near *tweediei*
Order Acariformes	
Suborder Prostigmata	

Supercohort Heterostigmata	
Cohort Tarsonemina	
S.F. Tarsonemoidea	Tarasoneminid
F. Tarsonemidae	Tarsonemid I
	Tarsonemid II
Supercohort Endeostigmata	
Cohort Pachygnathina	
S.F. Pachygnathoidea	
F. Pachygnathidae	*Bimichaelia stellaris*
	Pachygnathid II
	Pachygnathid III
F. Alicorhagiidae	Alicorhagiid
F. Nanorchestidae	Nanorchestid
F. Terpnacaridae	*Sebaia palmata*
F. Pediculochelidae	Pediculochelid
Supercohort Promata	
Cohort Eupodostigmata	
S. F. Eupodoidea	
F. Eupodidae	*Eupodes/Protereunutes*
F. Rhagidiidae	*Rhagidiid*
S.F. Tydeoidae	
F. Tydeidae	*Tydeus* sp.
	Coccetydeus sp.
	Tydeid III
F. Paratydeidae	*Paratydeus* sp.
S.F. Bdelloidea	
F. Bdellidae	*Spinibdella depressa*
	Cyta latirostris
F. Cunaxidae	*Cunaxoides* sp.
Cohort Eleutherogonia	
S.F. Caeculoidea	
F. Caeculidae	Caeculid
S.F. Raphignathoidea	
F. Raphignathidae	*Ledermuelleria* sp.
	Mediolata sp.
	Stigmaeid IV
	Stigmaeid V
	Stigmaeid VI
	Stigmaeid VII
	Stigmaeid VIII
F. Cryptognathidae	Cryptognathid
S.F. Cheyletoidea	
F. Cheyletidae	Cheyletid
F. Anystoidea	
F. Teneriffiidae	Teneriffiid
Supercohort Parasitengonia	
S.F. Erythraeoidea	Erythraeid I
F. Erythraeidae	Erythraeid II
S.F. Trombidioidea	
F. Trombidiidae	Trombidiid

Suborder Astigmata	
Supercohort Acaridia	
S.F. Anoetoidea	Anoetid
F. Anoetidae	Anoetid
	Anoetid (hypopus)
S.F. Acaroidea	
F. Acaridae	Acarid (hypopus)
F. Nanacaridae	Nanacarid
F. Glycyphagidae	Acarid glycyphagid
Suborder Cryptostigmata	
Supercohort Palaeacari	
S.F. Palaeacaroidea	
F. Ctenacaridae	*Ctenacarus* sp.
Supercohort Oribatei Inferiores	
S.F. Hypochthonoidea	
F. Brachychthoniidae	*Liochthonius* sp.
	Synchthonius sp.
F. Haplochthoniidae	*Haplochthonius* sp.
F. Cosmochthoniidae	*Cosmochthonius* sp.
F. Sphaerochthoniidae	*Sphaerochthonius* sp.
S.F. Phthiracaroidea	
F. Phthiracaridae	*Holophorella* sp.
F. Euphthiracaridae	*Austrotritia* sp.
S.F. Perlohmannioidea	
F. Perlohmanniidae	*Apolohmannia* sp.
F. Epilohmanniidae	*Epilohmannia* sp.
F. Lohmanniidae	*Haplacarus* sp.
S.F. Nothroidea	Nothrid I
Supercohort Oribatei Superiores	
Cohort Pycnonoticina	
S.F. Hermannielloidea	
F. Hermanniellidae	*Hermanobates* sp.
S.F. Liodoidea	
F. Liodidae	*Liodes* sp.
	Liodoid II
S.F. Eremaeoidea	
F. Eremobelbidae	*Eremelus* sp.
	Fosseremus sp.
S.F. Carabodoidea	
F. Carabodidae	*Carabodes* sp. 1
	Carabodes sp. 2
S.F. Oppiodea	
F. Oppiidae	*Oppia* sp.
F. Suctobelbidae	*Suctobelba* sp. 1
	Suctobelba sp. 2
F. Astegiatidae	*Cultroribula* sp.

Cohort Poronoticina	
S.F. Oribatelloidea	*Oribatelloid*
S.F. Microzetoidea	
F. Microzetidae	*Microzetes* sp.
S.F. Galumnoidea	Galumnoid II
F. Galumnidae	*Galuma* sp. 1
S.F. Orbatuloidea	Oribatuloid I
	Oribatuloid II
	Oribatuloid III
F. Oribatulidae	*Zygoribatula* sp.
Subphylum Hexapoda	
Class Collembola	
Suborder Arthropleona	
Section Entomobryomorpha	Arthropleona indet
Section Poduromorpha	
F. Poduridae	*Brachystomella* new sp.
	Xenylla cf. *manusiensis*
F. Onychiuridae	*Tullbergia krausbaueri*
F. Isotomidae	*Folsomides exiguus*
	Folsomina yossii
	Proisotoma sp.
Suborder Symphypleona	
F. Sminthuridae	*Sminthurinus muscophilus*
	Megalothroax rubidus

Appendix 4. List of taxa of soil animals, other than mites and collembolans from the herb flat, a *Pisonia* grove and the succulent mat on One-Tree Island

Subclass Araneida		
Order ARANEOMORPHA	present but not in samples	
Subclass Pseudoscorpionida		
Order CHELONETHIDA		
Suborder *Monosphyronida*	Family Chieridiidae	1 sp.
Suborder *Diplosphyronida*	Family Neobisiidae	1 sp.
Class Insecta		
Order PROTURA	Family Eosentomidae	1 sp.
Order DIPLURA	Family Japygidae	1 sp.
Order THYSANURA	Family Lepismatidae	1 sp.
Order PSOCOPTERA		
Suborder *Troctomorpha*	Family Liposcelidae	1 sp.
Order COLEOPTERA-LARVAE*		
Section Symphiogastra		
Subsection Dascilliformia		2 sp.
Order DIPTERA-LARVAE*		
Suborder *Brachycerca*		
Section Orthopyga	Family Stratiomyidae	1 sp.
Order HYMENOPTERA	Superfamily Formicoidea	3 sp.
Myriapodous Arthropods		
Class Symphyla		
Order CEPHALOSTIGMATA		1 sp.
Class Chilopoda		
Order GEOPHILOMORPHA		1 sp.
Class Diplopoda		
Subclass Pselathognatha		
Order PENCILLATA	Family Polyxenidae	1 sp.
Class Crustacea		
Order ISOPODA		
Suborder Oniscoidea		1 sp.
Phylum Mollusca		
Class Gastropoda		
Order PULMONATA		
Suborder Stylommatomorpha		2 sp.
Phylum Annelida		
Class Oligochaeta		
Order PROSOPORA	Family Enchytraeidae	1 sp.
Phylum Protozoa	1 testate species	

* Adults present but not in samples; infrequent unidentified larvae also present in samples.

Appendix 5. Species of birds recorded on One-Tree Island, and their status

Species		Status	
Sooty albatross	*Phoebetria fusca*	A	
Wandering albatross	*Diomedea exulans*	A	
Wedge-tailed shearwater	*Puffinus pacificus*	S	
Dove Prion	*Pachyptila desolata*	A	
Australian pelican	*Pelecanus conspicillatus*	A	
Masked booby	*Sula dactylatra*	A	
Brown booby	*Sula leucogaster*	R	
Black cormorant	*Phalacrocorax carbo*	A*	
Little black cormorant	*Phalacrocorax sulcirostris*	A*	
Pied cormorant	*Phalacrocorax varius*	A	
Little pied cormorant	*Phalacrocorax melanoleucos*	R	
Lesser frigatebird	*Fregata ariel*	A	
Greater frigatebird	*Fregata minor*	A	
Red-tailed tropicbird	*Phaethon rubricauda*	A	
White-tailed tropicbird	*Phaethon lepturus*	A	
Reef heron	*Egretta sacra*	RB	
White egret	*Egretta alba*	A	
White-faced heron	*Ardea novaehollandiae*	A → W	
Mangrove heron	*Butorides striatus*	A	
Black swan	*Cygnus atratus*	A	
Grey (white) goshawk	*Accipiter novaehollandiae*	A	
White-breasted sea-eagle	*Haliaeetus leucogaster*	?R	
Peregrin falcon	*Falco peregrinus*	A	
Banded landrail	*Rallus philippensis*	RB	
Royal spoonbill	*Platalea regia*	A	
Pied oystercatcher	*Haematopus ostralegus*	A	
Sooty oystercatcher	*Haematopus fuliginosus*	RB	
Red-capped dotterel	*Charadrius alexandrinus*	A	
Mongolian sand-dotterel	*Charadrius mongolus*	S	T–E mig.
Large sand-dotterel	*Charadrius leschnaultii*	A	T–E mig.
Eastern golden plover	*Pluvialis dominica*	S	T–E mig.
Turnstone	*Arenaria interpres*	S	T–E mig.
Whimbrel	*Numenius phaeopus*	W → R	T–E mig.

Species		Status	
Eastern curlew	*Numenius madagascariensis*	S	T–E mig.
Common sandpiper	*Tringa hypoleucos*	A	T–E mig.
Grey-tailed tattler	*Tringa brevipes*	S → R	T–E mig.
Wandering tattler	*Tringa incana*	A	T–E mig.
Knot	*Calidris canutus*	A → S	T–E mig.
Great knot	*Calidris tenuirostris*	A	T–E mig.
Sharp-tailed sandpiper	*Calidris acuminata*	A	T–E mig.
Red-necked stint	*Calidris ruficollis*	S	T–E mig.
Sanderling	*Calidris alba*	A	T–E mig.
Black-tailed godwit	*Limosa limosa*	A	T–E mig.
Bar-tailed godwit	*Limosa lapponica*	S	T–E mig.
Silver gull	*Larus novaehollandiae*	RB	
Roseate tern	*Sterna dougalli*	SB	
Black-naped tern	*Sterna sumatrana*	RB	
Bridled tern	*Sterna anaetheta*	SB	
Little tern	*Sterna albifrons*	S → R	
Crested tern	*Sterna bergii*	RB	
Lesser crested tern	*Sterna bengalensis*	RB	
White-capped noddy	*Anous minutus*	R	
Pallid cuckoo	*Cuculus pallidus*	A	
Bronze cuckoo	*Chrysococcyx lucidus lucidus*	A	I–A mig.
(2 subspecies)	*Chrysococcyx lucidus plagosus*	A	
Black-eared cuckoo	*Chrysococcyx osculans*	A	
Koel	*Eudynamys scolopaceax*	A	I–A mig.
Barn Owl	*Tyto alba*	A	
Spine-tailed swift	*Hirundapus caudacutus*	A	T–E mig.
Fork-tailed swift	*Apus pacificus*	A	
Sacred kingfisher	*Halcyon sancta*	R	I–A mig.
Forest kingfisher	*Halcyon macleayi*	A	I–A mig.
Dollar-bird	*Eurystomus orientalis*	A	I–A mig.
Welcome swallow	*Hirundo neoxena*	A	I–A mig.
Cicada bird	*Coracina tenuirostris*	A	I–A mig.
Black-faced cuckoo-shrike	*Coracina novaehollandiae*	A	I–A mig.
Little cuckoo-shrike	*Coracina papuensis*	A	
White-winged triller	*Lalage sueurii*	A	I–A mig.
Golden-headed fantail-warbler	*Cisticola exilis*	RB	
Grey fantail	*Rhipidura fuliginosa*	A	
Rufous fantail	*Rhipidura rufifrons*	A	I–A mig.
Leaden flycatcher	*Myiagra rubecula*	A	I–A mig.
Satin flycatcher	*Myiagra cyanoleuca*	A	I–A mig.
Black-faced flycatcher	*Monarcha frater*	A	I–A mig.
White-eared flycatcher	*Monarcha leucotis*	A	
Spectacled flycatcher	*Monarcha trivirgata*	A	I–A mig.
Rufous whistler	*Pachycephala rufiventris*	A	I–A mig.
Grey-breasted silvereye	*Zosterops lateralis*	RB	I–A mig.
Mangrove honeyeater	*Meliphaga fasciogularis*	A	
Pied honeyeater	*Certhionyx variegatus*	A	
House sparrow	*Passer domesticus*	A	

Species		Status	
Starling	*Sturnus vulgaris*	A	
Spangled drongo	*Dicrurus bracteatus*	A	I–A mig.
Magpie-lark (Peewit)	*Grallina cyanoleuca*	A	I–A mig.
Domestic pigeon	*Columba livia*	A	

Data from the present study, Domm and Recher (1973), Hulsman (1973) and J. Chilvers (1975, pers. comm.).

A Uncommon and/or irregular visitor
A* Uncommon visitor earlier, but resident by 1978
R Resident
S Summer visitor
W Winter visitor
B Breeding
I–A Intra-Australasian migrant
T–E Trans-equatorial migrant

Appendix 6. Some bioenergetic values for some of the organisms on One-Tree Island

Taxon	Biomass (Mean weight per individual)	
	Mean	Range
Birds		
*Diomedea exulans**	8.16 kg	5.9–11.3 kg
*Puffinus pacificus**	—	284–425 kg
*Sula leucogaster**	—	992–1247 kg
Phalacrocorax melanoleucos		
♀	—	567–879 g
♂	—	680–964 g
*Egretta sacra**	—	420–510 g
*Larus novaehollandiae**	—	284–340 g
*Sterna sumatrana**	99 g	—
*Sterna anaetheta**	128 g	—
*Sterna bergii**	—	326–383 g
Anous minutus		
adult	99 g*	—
39 mm chick	81 g (18.8 g)	—
Insects		
Hemiptera		
Pentatomidae (unident.) (N = 10)	(57 mg)	—
Diptera		
Stratiomyidae (unident.) (N = 12)	(2.25 mg)	—
Lonchaeidae		
Lamprolonchaea metatarsata (N = 11)	(3.25 mg)	
Plants		
Argusia		
inflorescence in bud (N = 3)	9.7 g (2.29 g)	8.9–10.9 g (2.19–2.42 g)
inflorescence in fruit (N = 5)	36.2 g (10.94 g)	27.1–45.8 g (8.1–14.9 g)
Trachymene sp.		
Whole plant		
Diameter:		
3 cm	(0.080 g)	—
4 cm (N = 3)	(0.079 g)	(0.047–0.109 g)
4.5 cm	(0.087 g)	—
5.5 cm	(0.132 g)	—

Taxon	Biomass (Mean weight per individual)	
	Mean	Range
6.5 cm	(0.275 g)	—
11.0 cm (N = 2)	(1.055 g)	(1.031–1.078 g)
12 cm	(1.146 g)	—
28 cm	(5.153 g)	—
Ripe umbels (N = 4)	(0.377 g)	—
Mericarps (N = 644)	(2.34 mg)	—
Abutilon indicum (82 cm high)		
Stems and twigs	22.6 g (11.65 g)	—
Leaves	13.9 g (6.17 g)	—

Taxon	Mean oxygen consumption (ml O_2/g dry wt/hr) at 26° ± 1° C
Insects	
Hemiptera	
Pentatomidae (unident.) (N = 10)	0.23
Diptera	
Stratiomyidae (unident.) (N = 12)	1.73
Lonchaeidae	
Lamprolonchaea metatarsata (N = 11)	1.65

Taxon	Calories per gram dry weight
Melanthera	
Buds	4,196–4,200
Leaves	3,905–3,936
Stems	3,896–4,127
Roots	3,731
Ipomoea	
Stems	4,129–4,138
Canavalia	
Leaves	4,015–4,025
Stems	4,122–4,132
Roots	4,120–4,132
Fruits	4,127–4,145
Abutilon	
Leaves	3,752–3,777
Stems	3,863–3,876
Roots	3,873–3,880
Dead standing stems	4,109–4,113
Leaf litter	3,382–3,403
Humus	1,367–1,379

Biomass values are fresh weight when not in parentheses and dry weight when in parentheses.
* = Data from Serventy *et al.* (1971).

Literature cited

Abbott, I. 1980. The floras of 37 south-western Australian islands. West. Aust. Herbarium Res. Notes No. 3 (1980): 19–36.

Airy Shaw, H. K. 1952. On the distribution of *Pisonia grandis* B. Br. (Nyctaginaceae), with special reference to Malaysia. Kew Bull. 1952: 87–97.

Alexander, H. G. L. 1979. A preliminary assessment of the rôle of the terrestrial decapod crustaceans in the Aldabran ecosystem. Phil. Trans. Roy. Soc. London B 286: 241–246.

Allard, R. W. 1965. Genetic systems associated with colonizing ability in predominantly self-pollinated species. Chapter (pp. 49–76) in: The Genetics of Colonizing Species (editors: H. G. Baker and G. L. Stebbins), Academic Press, New York.

Arber, A. 1934. The Gramineae, a Study of Cereal, Bamboo, and Grass. J. Cramer, Weinheim.

Aristotle. ΠΕΡΙ ΖΩΙΩΝ ΓΕΝΕΣΕΩΣ. English translation (1963): Generation of Animals, by A. L. Peck, William Heineman Ltd., London, 608 pp.

Arnold, S. J. 1972. Species densities of predators and their prey. Amer. Nat. 106: 220–236.

Backer, C. A. 1929. The Problem of Krakatao as seen by a Botanist. Martinus Nijhoff, The Hague, 229 pp.

Backer, C. A. 1949. Amaranthaceae. pp. 67–98 in: Flora Malesiana, Series I, vol. 4 (ed. C. G. G. J. van Steenis). Noordhoff-Kolff, N.V., Djakarta, 631 pp.

Backer, C. A. 1951. Aizoaceae. pp. 267–275 in: Flora Malesiana. Series I, Vol. 4 (ed. C. G. G. J. van Steenis). Noordhoff-Kolff, N.V., Djakarta, 631 pp.

Bailey, F. M. 1899. The Queensland Flora, H. J. Diddams, Brisbane, 6 vols.

Baker, H. G. 1953. Dimorphism and monomorphism in the Plumbaginaceae. III. Correlation of geographical distribution patterns with dimorphism and monomorphism in *Limonium.* Annals of Botany 17: 615–627.

Baker, H. G. 1955. Self-comparibility and establishment after "long-distance" dispersal. Evolution 9: 347–349.

Baker, H. G. 1967. Support for Baker's law – as a rule. Evolution 21: 853–856.

Baker, H. G. and G. L. Stebbins (editors). 1965. The Genetics of Colonizing Species. Academic Press, New York, 588 pp.

Ball, E. and J. Glucksman. 1975. Biological colonization of Motmot, a recently-created tropical island. Proc. Roy. Soc. London B. 190: 421–442.

Barbour, M. G. 1970a. Germination and early growth of the strand plant *Cakile maritima*. Bull. Torrey Bot. Club 97: 13–22.

Barbour, M. G. 1970b. Seedling ecology of *Cakile maritima* along the California coast. Bull. Torrey Bot. Club 97: 280–289.

Barbour, M. G. and J. E. Rodman. 1970. Saga of the west coast sea-rockets: *Cakile edentula* ssp. *californica* and *C. maritima.* Rhodora 72: 370–386.

Barbour, T. 1914. A contribution to the zoögeography of the West Indies, with special reference to amphibians and reptiles. Mem. Mus. Comp. Zool. Harvard 44: 209–359.

Barbour, T. and G. K. Noble. 1915. A revision of the lizards of the genus *Ameiva*. Bull. Mus. Comp. Zool. Harvard 49: 417–479.
Bates, M. and P. S. Humphrey (editors). 1956. The Darwin Reader. Charles Scribner's Sons, New York, 470 pp.
Beadle, N. C. W., O. D. Evans and R. C. Carolin. 1962. Handbook of the Vascular Plants of the Sydney District and Blue Mountains. Brown Gem. Print, Armidale, 597 pp.
Beaver, R. A. 1977. Non-equilibrium "island" communities: Diptera breeding in dead snails. J. Anim. Ecol. 46: 783–798.
Beaver, R. A. 1979. Non-equilibrium "island" communities: A guild of tropical bark beetles. J. Anim. Ecol. 48: 987–1002.
Becker, P. 1975. Island colonization by carnivorous and herbivorous Coleoptera. J. Anim. Ecol. 44: 893–906.
Blanchard, D. C. 1968. Surface active, organic material on airborne salt particles. Proc. International Conference on Cloud Physics, August 26–30, 1968, Toronto: 25–29.
Booth, J. 1970. Birds of Fairfax Island (Bunker Group) and Wreck Island (Capricorn Group), Great Barrier Reef. Sunbird 1: 85–92.
Borror, D. J. and D. M. Delong. 1955. An Introduction to the Study of Insects. Rinehart and Co., New York, 1030 pp.
Brackbill, H. 1976. Yellow-crowned night herons defecate, disgorge pellets on shore. Wilson Bull. 88: 670–671.
Britton, E. B. 1970. Coleoptera (beetles). Chapter 30 (pp. 495–621) in: The Insects of Australia (CSIRO), Melbourne Univ. Press, Melbourne.
Brown, J. H. 1978. The theory of insular biogeography and the distribution of boreal birds and mammals. Great Basin Naturalist Memoirs 2: 209–227.
Brown, W. H., E. D. Merrill and H. S. Yates. 1917. The revegetation of Volcano Island, Luzon, Philippine Islands, since the eruption of Taal Volcano in 1911. Philippine J. Sci. 12: 177–243.
Brown, W. L. Jr. 1958. A review of the ants of New Zealand. Acta Hymenopterologica 1: 1–50.
Bullock, S. H. and R. B. Primack. 1977. Comparative experimental study of seed dispersal on animals. Ecology 58: 681–686.
Burr, E. J. 1968. Cluster sorting with mixed character types. I. Standardization of character values. Aust. Comput. J. 1: 97–99.
Bustard, H. R. 1979. Population dynamics of sea turtles. Chapter 24 (pp. 523–540) in: Turtles Perspectives and Research (eds. M. Harless and H. Morlock), John Wiley and Sons, New York, 695 pp.
Bustard, H. R. and P. Greenham. 1968. Physical and chemical factors affecting hatching in the green sea turtle, *Chelonia mydas*. Ecology 49: 269–276.
Bustard, H. R., P. Greenham and C. Limpus. 1975. Nesting behaviour of loggerhead and flatback turtles in Queensland, Australia. Proc. Koninkl. Nederl. Akad. van Wetenschappen Series C 78: 111–122.
Bustard, H. R. and C. Limpus. 1971. Loggerhead turtle movements. Brit. J. Herpetology 4: 228–230.
Buwalda, P. 1949. Umbelliferae. pp. 113–140 in: Flora Malesiana, Series I, Vol. 4. (Ed. C. G. G. J. van Steenis), Noordhoff-Kolff, N.V., Djakarta, 631 pp.
Byrnes, N. B., S. L. Everist, S. T. Reynolds, A. Specht and R. L. Specht. 1977. The vegetation of Lizard Island, North Queensland. Proc. Roy. Soc. Qld. 88: 1–15.
Cain, S. A. 1950. Life-forms and phytoclimate. Bot. Rev. 16: 1–32.
Caldwell, D. K. 1959. The Atlantic loggerhead sea turtle, *Caretta caretta caretta* (L.), in America. III. The loggerhead turtles of Cape Romain, South Carolina. Bull. Fla. State Mus. (Biol. Sci.) 4: 319–348.
Caldwell, D. K., A. Carr and L. H. Ogren. 1959a. The Atlantic loggerhead sea turtle, *Caretta caretta caretta* (L.), in America. I. Nesting and migration of the Atlantic loggerhead turtle. Bull. Fla. State Mus. (Biol. Sci.) 4: 295–308.
Caldwell, D. K., F. H. Berry, A. Carr and R. A. Ragotzkie. 1959b. The Atlantic loggerhead sea

turtle, *Caretta caretta caretta* (L.), in America. II. Multiple and group nesting by the Atlantic loggerhead turtle. Bull. Fla State Mus. (Biol. Sci.) 4: 309–318.
Caldwell, M. M. 1974. Physiology of desert halophytes. Chapter (pp. 355–378) in: Ecology of Halophytes (eds. R. J. Reimold and W. H. Queen), Academic Press, New York, 605 pp.
Campbell, A. J. and S. A. White. 1910. Birds identified on the Capricorn Group, during expedition of R.A.O.U., 8th to 17th October, 1910. The Emu 10: 195–204.
Carlquist, S. 1966. The biota of long-distance dispersal. IV. Genetic systems in the floras of oceanic islands. Evolution 20: 433–455.
Carlquist, S. 1974. Island Biology. Columbia Univ. Press, New York, 660 pp.
Carr, A. 1967. So Excellent a Fishe. Natural History Press, Garden City, N.Y.
Cayley, N. W. 1946. What Bird is That? Angus and Robertson Ltd., Sydney, 319 pp.
Chadwick, C. E. 1962. Some insects and terrestrial arthropods from Heron Island, Queensland. Proc. Linn. Soc. N.S.W. 87: 196–199.
Chapman, V. J. 1976. Coastal Vegetation (second edition). Pergamon Press, Sydney, 292 pp.
Chew, R. M. 1974. Consumers as regulators of ecosystems: an alternate to energetics. Ohio J. Sci. 74: 359–370.
Child, J. and P. Child. 1971. Australian Dictionary of Biology. Lansdowne Press, Melbourne, 112 pp.
Chilvers, J. 1975. A barn owl, a sooty albatross, and a dove prion from One-Tree Island. Sunbird 6: 48.
Clegg, B. R., J. J. Koranda and W. L. Robison. 1979. The teleoclimate of *Messerschmidia argentea*. Bull. Ecol. Soc. America 60: 140 (Abstract).
Cody, M. L. 1975. Towards a theory of continental species diversities. Chapter 10 (pp. 214–257) in: Ecology and Evolution of Communities (eds. M. L. Cody and J. M. Diamond), Belknap Press of Harvard University, Cambridge.
Cody, M. L. and J. M. Diamond (editors). 1975. Ecology and Evolution of Communities. Belknap Press of Harvard University, Cambridge, 545 pp.
Colless, D. H. and D. K. McAlpine. 1970. Diptera (flies). Chapter 34 (pp. 656–740) in: The Insects of Australia (CSIRO), Melbourne Univ. Press, Melbourne.
Common, I. F. B. 1970. Lepidoptera (moths and butterflies). Chapter 36 (pp. 765–866) in: The Insects of Australia (CSIRO), Melbourne Univ. Press, Melbourne.
Connell, J. H. 1975. Some mechanisms producing structure in natural communities. Chapter 16 (pp. 460–490) in: Ecology and Evolution of Communities (eds. M. L. Cody and J. M. Diamond), Belknap Press of Harvard University, Cambridge.
Connell, J. H. 1978. Diversity in tropical rainforests and coral reefs. Science 199: 1302–1310.
Connell, J. H. and R. O. Slatyer. 1977. Mechanisms of succession in natural communities and their role in community stability and organization. Amer. Nat. 111: 1119–1144.
Connor, E. F. and D. Simberloff. 1978. Species number and compositional similarity of the Galapagos flora and avifauna. Ecological Monographs 48: 219–248.
Connor, E. F. and D. Simberloff. 1979. The assembly of species communities: chance or competition? Ecology 60: 1132–1140.
Cooper, R. P. 1948. Birds of the Capricorns – Great Barrier Reef. The Emu 48: 107–126.
Coquillat, M. 1951. Sur les plantes les plus communes a la surface du globe. Bull. Mensual Soc. Linn. Lyon 20: 165–170.
Cribb, A. B. 1969. The Pisonia. Qld. Nat. 19: 110–114.
Cribb, A. B. 1972. Vegetation of Hoskyn I. and reef. Qld. Nat. 20: 92–100.
Crowe, T. M. 1979. Lots of weeds: insular phytogeography of vacant urban lots. J. Biogeogr. 6: 169–181.
Crowell, K. L. 1968. Rates of competitive exclusion by the Argentine ant in Bermuda. Ecology 49: 551–555.
C.S.I.R.O. 1970. The Insects of Australia. Melbourne University Press, Melbourne, 1029 pp.
Culver, D. C. 1970. Analysis of simple cave communities. I. Caves as islands. Evolution 24: 463–474.

Cummins, K. W. and J. C. Wuycheck. 1971. Caloric equivalents for investigations in ecological energetics. Mitt. Internat. Verein. Limnol. 18: 1–158.

Dammerman, K. W. 1948. The fauna of Krakatau 1883–1933. Verhand. Koninkl. Nederl. Akad. van Wetenschappen, afd. Natuurkunde 44: 1–594.

Darwin, C. 1859. On the Origin of Species by Means of Natural Selection, or the Preservation of Favoured Races in the Struggle for Life. 8 vo. London.

Darwin, C. 1872. On the Origin of Species by Means of Natural Selection, or the Preservation of Favoured Races in the Struggle for Life. Sixth edition, with additions and corrections to 1872, 8 vo. London, 1882 (Dated Jan. 1872).

Darwin, C. and A. Wallace. 1859 (read 1 July 1858). On the tendency of species to form varieties: and on the perpetuation of varieties and species by natural means of selection. J. Linn. Soc. 3: 45.

Davis, C., M. F. Day and D. F. Waterhouse. 1938. Notes on the terrestrial ecology of the Five Islands. I. Proc. Linn. Soc. N.S.W. 63: 357–388, and plates XV–XIX.

Davis, J. H. Jr. 1942. The ecology of the vegetation and topography of the sand keys of Florida. Papers from Tortugas Lab. 33: 114–195 + 7 plates.

De Jong, T. M. 1979. Water and salinity relations of Californian beach species. J. Ecol. 67: 647–663.

Department of Science. 1975. Climatic averages Queensland. Aust. Gov't Publ. Service, Canberra, 76 pp.

de Vlaming, V. L. 1967. Regurgitation by killdeer as a possible means of dispersal of seeds and aquatic organisms. Wilson Bull. 79: 449–450.

Diamond, A. W. 1979. Dynamic ecology of Aldabran seabird communities. Phil. Trans. Roy. Soc. London B. 286: 231–240.

Diamond, J. M. 1975. Assembly of species communities. Chapter 14 (pp. 342–444) in: Ecology and Evolution of Communities (eds. M. L. Cody and J. M. Diamond), The Belknap Press of Harvard University, Cambridge.

Diamond, J. M. 1979a. Population dynamics and interspecific competition in bird communities. Fortsch. Zool. 25: 389–402.

Diamond, J. M. 1979b. Community structure: is it random, or is it shaped by species differences and competition? Chapter 8 (pp. 165–181) in: Population Dynamics (editors: R. M. Anderson, B. D. Turner and L. R. Taylor), Blackwell Scient. Publ., Oxford.

Diamond, J. M. Reconstitution of bird community structure on Long Island, New Guinea, after a volcanic explosion. National Geographic Research Reports, in press.

Diamond, J. M. and M. E. Gilpin. 1980. Turnover noise: contribution to variance in species number and prediction from immigration and extinction curves. Amer. Nat. 115: 884–889.

Diamond, J. M. and R. M. May. 1977. Species turnover rates on islands: dependence on census interval. Science 197: 266–270.

Dobzhansky, T. 1937–1951. Genetics and the Origin of Species. Columbia University Press, New York, 364 pp.

Dockrill, A. W. 1969. Australian Indigenous Orchids. Society for Growing Australian Plants, Vol. 1, 825 pp.

Docters van Leeuwen, W. 1919. The flora and the fauna of the islands of the Krakatau-group in 1919. Annales du Jardin Botanique Buitenzorg 31: 103–139.

Docters van Leeuwen, W. 1936. Krakatau 1883–1933. A. Botany. Annales du Jardin Botanique de Buitenzorg 56–57: 1–505 + 36 plates.

Domm, S. B. 1971. The uninhabited cays of the Capricorn Group, Great Barrier Reef, Australia. Atoll Res. Bull. 142: 1–27.

Domm, S. and H. F. Recher. 1973. The birds of One Tree Island with notes on their yearly cycle and feeding ecology. Sunbird 4: 63–86.

Donaldson, A. and B. A. Whitton. 1977a. Chemistry of freshwater pools on Aldabra. Atoll Res. Bull. No. 213: 1–28.

Donaldson, A. and B. A. Whitton. 1977b. Algal flora of freshwater habitats on Aldabra. Atoll Res. Bull. No. 215: 1–26.
Done, T. J. 1975. The ecology of soil-arthropod communities on One Tree Island. Unpublished Ph.D. thesis, University of New England, Armidale.
Duncan, W. H. 1974. Vascular halophytes of the Atlantic and Gulf coasts of North America north of Mexico. Chapter (pp. 23–50) in: Ecology of Halophytes (eds. R. J. Reimold and W. H. Queen), Academic Press, New York, 605 pp.
Edgell, J. A. 1927. Bird-life in the Capricorn Group. The Emu 26: 292–296.
Ehrendorfer, F. 1965. Dispersal mechanisms, genetic systems, and colonizing abilities in some flowering plant families. Chapter (pp. 331–352) in: The Genetics of Colonizing Species (eds. H. G. Baker and G. L. Stebbins), Academic Press, New York.
Endean, R., R. Kenny and W. Stephenson. 1956. The ecology and distribution of intertidal organisms on the rocky shores of the Queensland mainland. Aust. J. Marine and Freshwater Res. 7: 317–342.
Eschmeyer, W. and L. Eschmeyer. 1971. One Tree Island. Pacific Discovery 24: 12–22.
Evans, F. C. and W. W. Murdoch. 1968. Taxonomic composition, trophic structure and seasonal occurrence in a grassland insect community. J. Anim. Ecol. 37: 259–273.
Fairbridge, R. W. 1950. Recent and Pleistocene coral reefs of Australia. J. Geol. 58: 330–401.
Fien, I. 1971. A note on the avifauna of Lady Elliot Island. Sunbird 2: 63–67.
Fletcher, B. S. 1973. Observations on a movement of insects at Heron Island, Queensland. J. Aust. Ent. Soc. 12: 157–160.
Fosberg, F. R. 1948. Salinity and atoll vegetation. Hawaiian Academy of Science, Proceedings 23: 8 (Abstract).
Fosberg, F. R. 1949. Atoll vegetation and salinity. Pacific Science 3: 89–92.
Fosberg, F. R. 1954. Soils of the northern Marshall Atolls, with special reference to the Jemo series. Soil Science 78: 99–107.
Fosberg, F. R. 1955. Pacific forms of *Lepturus* R. Br. (Gramineae). Occ. Pap. Bernice P. Bishop Museum 21: 285–294.
Fosberg, F. R. 1957. Soils, vegetation and agriculture on coral atolls. Proc. 8th Pac. Sci. Congr. of the Pac. Sci. Assoc. 3a: 1037–1047.
Fosberg, F. R. 1961. Qualitative description of the coral atoll ecosystem. Atoll Research Bulletin No. 81: 1–11.
Fosberg, F. R. 1963a. Disturbance in island ecosystems. Chapter (pp. 557–561) in: Pacific Basin Biogeography. Bishop Mus. Press, Honolulu.
Fosberg, F. R. 1963b. The Island Ecosystem. Chapter (pp. 1–6) in: Man's Place in the Island Ecosystem. Bishop Mus. Press, Honolulu.
Fosberg, F. R. (editor). 1965. Man's Place in the Island Ecosystem, a symposium. Bishop Mus. Press, Honolulu, 264 pp.
Fosberg, F. R. 1974. Phytogeography of atolls and other coral islands. Proc. Second International Coral Reef Symp. 1: 389–396.
Freeland, W. J. 1979. Primate social groups as biological islands. Ecology 60: 719–728.
Fridriksson, S. 1975. Surtsey, Evolution of Life on a Volcanic Island. Butterworths, London, 198 pp.
Frith, D. W. 1979. A twelve month study of insect abundance and composition at various localities on Aldabra Atoll. Phil. Trans. R. Soc. London B. 286: 119–126.
Gannon, G. R. 1936. Plants spread by the Silvereye. The Emu 35: 314–316.
Geesink, R. 1971. Portulacaceae. Pp. 121–133 in: Flora Malesiana (ed. C. G. G. J. van Steenis), Series I., Vol. 7., Wolters-Noordhoff, Groningen., 1023 pp.
Gilbert, P. A. 1925–27. The biology of North-west Islet, Capricorn Group. (B.) Birds. Aust. Zoologist 4: 210–226.
Gillham, M. E. 1961. Alteration of the breeding habitat by sea-birds and seals in Western Australia. J. Ecology 49: 289–300.

Gillham, M. E. 1963. Coral cay vegetation Heron Island, Great Barrier Reef. Proc. Roy. Soc. Qld. 73: 79–92.

Gilpin, M. E. and J. M. Diamond. 1976. Calculation of immigration and extinction curves from the species–area–distance relation. Proc. Natl. Acad. Sci. USA 73: 4130–4134.

Grant, P. R. and I. Abbott. 1980. Interspecific competition, island biogeography and null hypotheses. Evolution 34: 332–341.

Great Barrier Reef Committee. 1978. A Coral Reef Handbook. GBRC Handbook Series No. 1, GBRC, Brisbane, 92 pp.

Green, J. 1968. The Biology of Estuarine Animals. Sidgwick & Jackson, London, 401 pp.

Gressitt, J. L. and C. M. Yoshimoto. 1963a. Dispersal of animals in the Pacific. pp. 283–292 in: Pacific Basin Biogeography, a symposium. Bishop Mus. Press, Honolulu.

Gressitt, J. L. and C. M. Yoshimoto. 1963b. Aerial transport of insects in polar areas, particularly in the Antarctic. Proc. 3rd International Biometeorological Congress, p. 911.

Grime, J. P. 1977. Evidence for the existence of three primary strategies in plants and its relevance to ecological and evolutionary theory. Amer. Nat. 111: 1169–1194.

Gunn, C. R. and J. V. Dennis. 1976. World Guide to Tropical Drift Seeds and Fruits. Demeter Press, New York, 240 pp.

Guppy, H. B. 1891. The dispersal of plants as illustrated by the flora of the Keeling or Cocos Islands. Trans. Victoria Instit. London 24: 267–306.

Guppy, H. B. 1906. Observations of a Naturalist in the Pacific between 1896 and 1899. II. Plant Dispersal. MacMillan, London, 627 pp.

Guppy, H. B. 1917. Plants, seeds and currents in the West Indies and Azores. Williams and Norgate, London, 531 pp.

Hahn, D. E. 1962. A list of the designated type specimens in the Macleay Museum. The Macleay Museum, Univ. of Sydney, 184 pp., mimeogr.

Hanson, H. C. 1962. Dictionary of Ecology. Philosophical Library, New York, 382 pp.

Hanson, W. R. 1968. Estimating the number of animals: A rapid method for unidentified individuals. Science 162: 675–676.

Harper, J. L. 1965. Establishment, aggression, and cohabitation in weedy species. Chapter (pp. 243–268) in: The Genetics of Colonizing Species (eds. H. G. Baker and G. L. Stebbins), Academic Press, New York.

Hart, R. 1977. Why are biennials so few? Amer. Nat. 111: 792–799.

Haskins, C. P. and E. F. Haskins. 1965. *Pheidole megacephala* and *Iridomyrmex humilis* in Bermuda – Equilibrium or Slow Replacement? Ecology 46: 736–740.

Haukioja, E. and P. Niemalä. 1979. Birch leaves as a resource for herbivores: seasonal occurrence of increased resistance in foliage after mechanical damage of adjacent leaves. Oecologia 39: 151–159.

Hawken, M. 1972. Notes on the birds of Hoskyn Island, December–January, 1972. Qld. Nat. 20: 120–122.

Heatwole, H. 1971. Marine-dependent terrestrial biotic communities on some cays in the Coral Sea. Ecology 52: 363–366.

Heatwole, H. 1975. Biogeography of reptiles on some of the islands and cays of eastern Papua-New Guinea. Atoll Res. Bull. No. 180, 32 pp.

Heatwole, H. 1976. The ecology and biogeography of coral cays. Chapter 11 (pp. 369–387) in: Biology and Geology of Coral Reefs, vol. III, Biology 2, (eds. O. A. Jones and R. Endean), Academic Press, New York.

Heatwole, H. A Coral Island. Collins Press Ltd., Syndney, in press.

Heatwole, H. and R. Levins. 1972a. Biogeography of the Puerto Rican Bank: Flotsam transport of terrestrial animals. Ecology 53: 112–117.

Heatwole, H. and R. Levins. 1972b. Trophic structure stability and faunal change during recolonization. Ecology 53: 531–534.

Heatwole, H. and R. Levins. 1973. Biogeography of the Puerto Rican Bank: Species-turnover on a small cay, Cayo Ahogado. Ecology 54: 1042–1055.

Heatwole, H. and F. MacKenzie. 1967. Herpetogeography of Puerto Rico. IV. Paleogeography, faunal similarity and endemism. Evolution 21: 429–438.

Hedley, C. 1925. Coral shingle as a beach formation. Trans. Roy. Geogr. Soc. Australasia (Qld). 1: 66.

Heine, H. 1976. *Argusia* Boehmer. Pp. 108–112 in: Flore de la Nouvelle - Calédonie et dependances. Part 7 Acanthaceés Bignoniaceés Boraginaceés Solanaceés.

Hemsley, W. B. 1892. A drift-seed (*Ipomoea tuberosa*, L.). Annals of Botany 6: 369–373.

Henderson, R. J. F. 1974. *Solanum nigrum* L. (Solanaceae) and related species in Australia. Contrib. Qld. Herb. No. 16: 1–78.

Hespenheide, H. A. 1975. Prey characteristics and predator niche width. Chapter 7 (pp. 158–180) in: Ecology and Evolution of Communities (eds. M. L. Cody and J. M. Diamond), The Belknap Press of Harvard University, Cambridge.

Hespenheide, H. A. 1979. Are there fewer parasitoids in the tropics? Amer. Nat. 113: 766–769.

Hindwood, K. A., K. Keith and D. L. Serventy. 1963. Birds of the southwest Coral Sea. C.S.I.R.O. Div. Wildlife Res. Tech Paper No. 3: 1–44.

Hnatiuk, R. J. and L. F. H. Merton. 1979a. Vegetation of Aldabra, a reassessment. Atoll Res. Bull. no. 239: 1–22.

Hnatiuk, R. J. and L. F. H. Merton. 1979b. A perspective of the vegetation of Aldabra. Phil. Trans. Roy. Soc. London B. 286: 79–84.

Hnatiuk, S. H. 1979. A survey of germination of seeds from some vascular plants found on Aldabra Atoll. J. Biogeogr. 6: 105–114.

Holm, L. G., D. L. Plucknett, J. V. Pancho and J. P. Herberger. 1977. The World's Worst Weeds, Distribution and Biology. Univ. Press of Hawaii, Honolulu, 609 pp.

Hou, D. and C. G. G. J. van Steenis. 1958. Rhizophoraceae, pp. 429–493 in: Flora Malesiana, Series I, Vol. 5 (ed. C. G. G. J. van Steenis) Noordhoff-Kolff N.V., Djarkarta, 598 pp.

Howard, R. A. 1950. Vegetation of the Bimini Island Group Bahamas, B.W.I. Ecological Monographs 20: 317–349.

Hughes, G. R. 1974a. The sea turtles of south-east Africa. I. Status, morphology and distributions. Oceanogr. Res. Instit. Investigational Report No. 35: 1–144.

Hughes, G. R. 1974b. The sea turtles of south-east Africa. II. The biology of the Tongaland loggerhead turtle *Caretta caretta* L. with comments on the leatherback *Dermochelys coriacea* L. and the green turtle *Chelonia mydas* L. in the study region. Oceanogr. Res. Instit. Investigational Report No. 36: 1–96.

Hunt, D. R. 1967–68. Nyctaginaceae. In: Tropical African Plants: XXVIII. Kew Bulletin 21: 251.

Hulsman, K. 1973. Addendum. Sunbird 4: 86.

Hulsman, K. 1974a. Notes on the behaviour of terns at One Tree Island. Sunbird 5: 44–49.

Hulsman, K. 1974b. Silver gulls hawking insects. Sunbird 5: 52.

Hulsman, K. 1975. The skimming behaviour of terns. Sunbird 6: 41–43.

Hulsman, K. 1976. The robbing behaviour of terns and gulls. The Emu 76: 143–149.

Hulsman, K. 1977a. Breeding success and mortality of terns at One-Tree Island, Great Barrier Reef. The Emu 77: 49–60.

Hulsman, K. 1977b. Daily attendance of terns and gulls at One Tree Island. Sunbird 8: 9–19.

Huxley, J. (editor). 1940. The New Systematics. Oxford University Press, London, 583 pp.

Ingleton, G. C. 1944. Charting a continent. Angus and Robertson Ltd., Sydney, 146 pp.

Ivens, G. W. 1967. East African Weeds and their Control. Oxford Univ. Press, Nairobi, 244 pp.

Janzen, D. H. 1968. Host plants as islands in evolutionary and contemporary time. Amer. Nat. 102: 592–595.

Johnson, N. K. 1975. Controls of number of bird species on montane islands in the Great Basin. Evolution 29: 545–563.

Jones, W. T. 1971. The field identification and distribution of mangroves in eastern Australia. Qld. Nat. 20: 35–51.

Jukes, J. B. 1847. Narrative of the Surveying Voyage of H.M.S. Fly, commanded by Captain F. P. Blackwood, R. N. in Torres Strait, New Guinea, and other islands of the eastern archipelago, during the years 1842–1846; together with an excursion into the interior of the eastern part of Java. 2 vols. T. and W. Boone, London.

Keast, A. (editor). 1921. Biogeography and Ecology in Australia (new edition). Dr. W. Junk Publishers, The Hague, 2182 pp.

Keddy, P. A. 1976. Lakes as islands: The distributional ecology of two aquatic plants, *Lemna minor* L. and *L. trisulca* L. Ecology 57: 353–359.

Kenneally, K. 1972. Tropical seeds and fruits washed up on the south-west coast of Western Australia. W.A. Nat. 12: 73–80.

Key, K. H. L. 1970. Orthoptera (grasshoppers, locusts, crickets). Chapter 21 (pp. 323–347) in: The Insects of Australia. Melbourne Univ. Press, Melbourne.

Kikkawa, J. 1970. Birds recorded at Heron Island. Sunbird 1: 34–48.

Kikkawa, J. 1973. The status of silvereyes *Zosterops* on the islands of the Great Barrier Reef. Sunbird 4: 30–37.

Kikkawa, J. 1976. Seabird islands, Heron Island, Queensland. The Australian Bird Bander 14: 3–6.

Kleinschmidt, H. E. and R. W. Johnson. 1977. Weeds of Queensland. S. R. Hampson, Gov't. Printer, Brisbane, 469 pp.

Koranda, J. J., W. L. Robison, S. Thompson and M. Stuart. 1979. The distribution of biomass, nutrients, and radionuclides in *Messerschmidia argentea* (Boraginaceae) at Enewetak Atoll, Marshall Islands. Bull. Ecol. Soc. America 60: 139 (Abstract).

Kozlowski, T. T. 1972. Seed Biology, Vol. III. Insects, and Seed Collection, Storage, Testing, and Certification. Academic Press, New York, 422 pp.

Kroes, H. W. 1977. The niche structure of ecosystems. J. Theoretical Biology 65: 317–326.

Kugler, H. 1973. Zur Bestäubung von *Scaevola plumieri* (L.) Vahl und *Ipomoea pes-caprae* sweet, zwei tropischen Strandpflanzen. Flora 162: 381–391.

Kuris, A. M., A. R. Blaustein, B. O'Connor, W. Dritschilo and D. Natus. 1977. Ecto-parasitic mites on rodents: application of the island biogeographic theory? Science 195: 596–598.

Lance, G. N. and W. T. Williams. 1967. Mixed-data classificatory programmes. I. Agglomerative systems. Aust. Comput. J. 1: 15–20.

Lassen, H. H. 1975. The diversity of freshwater snails in view of the equilibrium theory of island biogeography. Oecologia 19: 1–8.

Laughlin, R. 1976. Counting the flowers in the forest: combining two population estimates. Aust. J. Ecol. 1: 97–101.

Lavery, H. J. and R. J. Grimes. 1971. Sea-birds of the Great Barrier Reef. Qld. Agric. J. 97: 106–113.

Lea, A. M. and J. T. Gray. 1935a. The food of Australian birds. An analysis of the stomach contents. The Emu 34: 275–292.

Lea, A. M. and J. T. Gray. 1935b. The food of Australian birds. An analysis of the stomach contents. Part II. The Emu 35: 63–98.

Lea, A. M. and J. T. Gray. 1935c. The food of Australian birds. An analysis of the stomach contents. Part III. The Emu 35: 145–178.

Lea, A. M. and J. T. Gray. 1935d. The food of Australian birds. An analysis of the stomach contents. Part IV. The Emu 35: 251–280.

Le Buff, C. L. Jr. 1974. Unusual nesting relocation in the loggerhead turtle, *Caretta caretta*. Herpetologica 30: 29–31.

Leenhouts, P. W. 1957. Goodeniaceae. pp 335–344 in: Flora Malesiana, Series I, Vol. 5, (ed. C. G. G. J. van Steenis), Noordhoff-Kolff, N.V., Djakarta, 595 pp.

Lesko, G. L. and R. B. Walker. 1969. Effect of sea water on seed germination in two Pacific atoll beach species. Ecology 50: 730–734.

Levins, R. 1975. Evolution in communities near equilibrium. Chapter 1 (pp. 16–50) in: Ecology

and Evolution of Communities (eds. M. L. Cody and J. M. Diamond), The Belknap Press of Harvard University, Cambridge.
Levins, R. and H. Heatwole. 1973. Biogeography of the Puerto Rican Bank: Introduction of species onto Palominitos Island. Ecology 54: 1056–1064.
Levins, R., M. L. Pressick and H. Heatwole. 1973. Coexistence patterns in insular ants. Amer. Scient. 61: 463–472.
Limpus, C. 1978. The reef, uncertain land of plenty. Chapter (pp. 186–222) in: Exploration North (ed. H. J. Lavery). Richmond Hill Press, Richmond, 250 pp.
Linsley, E. G., C. M. Rick and S. G. Stephens. 1966. Observations on the floral relationships of the Galápagos Carpenter Bee. The Pan-Pacific Entomologist 42: 1–18.
Longman, H. A. 1914. The plants of Mast-head Island. Proc. Roy. Soc. Qld. 25: 17–23.
Longhurst, A. R. 1955. Evolution in the Notostraca. Evolution 9: 84–86.
Lourensz, R. S. 1977. Tropical Cyclones in the Australian Region July 1909 to June 1975. Aust. Gov't. Publ. Service, 111 pp.
Luer, C. A. 1975. The Native Orchids of the United States and Canada, excluding Florida, New York Botanical Garden, N.Y., 361 pp.
Lynch, J. F. and N. K. Johnson. 1974. Turnover and equilibria in insular avifaunas, with special reference to the California Channel Islands. The Condor 76: 370–384.
MacArthur, R. H. and E. O. Wilson. 1963. An equilibrium theory of insular zoogeography. Evolution 17: 373–387.
MacArthur, R. H. and E. O. Wilson. 1967. The Theory of Island Biogeography. Princeton University Press, Princeton, 203 pp.
Macfadyen, A. 1962. Control of humidity in three funnel-type extractors for soil arthropods, pp. 158–168 in: Progress in Soil Zoology (ed. P. W. Murphy), London, 398 pp.
MacGillivray, W. 1926. Birds of the Capricorn Islands. The Emu 25: 229–238.
MacGillivray, W. 1928. Bird-life of the Bunker and Capricorn Islands. The Emu 27: 230–249.
MacGillivray, W. D. K. 1931. A May visit to the Capricorn Islands. The Emu 30: 270–276.
MacGillivray, W. D. K. and F. A. Rodway. 1931. Plants on islands of the Bunker and Capricorn Groups. Reports of the Great Barrier Reef Committee 3: 58–63.
Majer, J. D. 1976. The influence of ants and ant manipulation on the cocoa farm fauna. J. Applied Ecol. 13: 157–175.
Malone, C. R. 1965. Killdeer (*Charadrius vociferus* Linnaeus) as a means of dispersal for aquatic gastropods. Ecology 46: 551–552.
Mason, R. 1961. Dispersal of tropical seeds by ocean currents. Nature 191: 408–409.
Mattson, W. J. and N. D. Addy. 1975. Phytophagous insects as regulators of forest primary production. Science 190: 515–522.
Maxwell, W. G. H. 1968. Atlas of the Great Barrier Reef. Elsevier Publ. Co., Amsterdam, 258 pp.
May, R. M. 1975. Island biogeography and the design of wildlife preserves. Nature 254: 177–178.
Mayr, E. 1942. Systematics and the Origin of Species. Columbia Univ. Press, New York.
Mayr, E. 1963. Animal Species and Evolution. Harvard University Press, Cambridge, 797 pp.
Mayr, E. 1965. Summary. Chapter (pp. 553–562) in: The Genetics of Colonizing Species (eds. H. G. Baker and G. L. Stebbins), Academic Press, New York.
McAllister, H. J., A. J. Bass and H. J. van Schoor. 1965. Marine turtles on the coast of Tongaland, Natal. The Lammergeyer 3: 11–40.
McNaughton, S. J. 1977. Diversity and stability of ecological communities: a comment on the role of empiricism in ecology. Amer. Nat. 111: 515–525.
Meikle, R. D. 1979. Supplementary notes on *Commicarpus* (Nyctaginaceae). Kew Bull. 34: 341–343.
Menhinick, E. F. 1963. Estimation of insect population density in herbaceous vegetation with emphasis on removal sweeping. Ecology 44: 617–621.
Merrill, E. 1946. Plant Life of the Pacific World. (Chapter 3, Plants of the Seashore), Macmillan Co., New York.

Moorehouse, F. W. 1933. Notes on the Green Turtle (*Chelonia mydas*). Reports of the Great Barrier Reef Committee 4: 1–22.

Moulton, J. M. 1961. Heron Island, Capricorn Group, Australia. III. Some observations on the Heron Island fauna. Atoll Res. Bull No. 82: 15–16.

Muenscher, W. C. 1955. Weeds (second edition), The Macmillan Co., New York, 560 pp.

Mühlenberg, M., D. Leipold, H. J. Mader, and B. Steinhauer. 1977a. Island ecology of arthropods. I. Diversity, niches, and resources on some Seychelles Islands. Oecologia 29: 117–134.

Mühlenberg, M., D. Leipold, H. J. Mader and B. Steinhauer. 1977b. Island ecology of arthropods. II. Niches and relative abundances of Seychelles ants (Formicidae) in different habitats. Oecologia 29: 135–144.

Muir, J. 1937. The seed-drift of South Africa and some influences of ocean currents on the strand vegetation. Union of S. Afr. Dept. Agric. Forestry. Botanical Surv. Mem. No. 16: 1–108 + 22 plates.

Muir, R. J. 1975. Ant respirometric studies. Chapter 8 (pp. 94–100) in: U.S./I.B.P. Desert Biome, Tunisian pre-Saharan Project, Progress Report No. 4, 239 pp.

Musgrave, A. 1925–27a. The biology of North-west Islet. A. Narrative. Australian Zoologist 4: 199–209.

Musgrave, A. 1925–27b. The biology of North-west Islet, Capricorn Group (E.). Insects. Australian Zoologist 4: 241–246.

Nebe, J. 1928. A naturalist's holiday on the Great Barrier Reef. Qld. Nat. 6: 102–108.

Newberry, D. M. and R. A. Spicer. 1979. The terrestrial vegetation of an Indian Ocean coral island: Wilingili, Addu Atoll, Maldive Islands. II. A limited quantitative analysis of the vegetation distribution. Atoll Res. Bull. No. 231: 15–25.

Nielsen, B. O. 1978. Above ground food resources and herbivory in a beech forest ecosystem. Oikos 31: 273–279.

Nilsson, S. G. and I. N. Nilsson. 1978. Species richness and dispersal of vascular plants to islands in Lake Mockein, southern Sweden. Ecology 59: 473–480.

Nordenskiöld, E. 1928. The History of Biology, a Survey. Tudor Publ. Co., New York, 629 pp.

Odum, E. P. 1969. The strategy of ecosystem development. Science 164: 262–270.

Olsson-Seffer, P. 1909. Hydrodynamic factors influencing plant-life on sandy sea shores. The New Phytologist 8: 37–50.

Oosting, H. J. 1954. Ecological processes and vegetation of the maritime strand in the southeastern United States. The Botanical Review 20: 226–262.

Oosting, H. J. 1956. The Study of Plant Communities. Second Edition. W. H. Freeman and Co., San Francisco, 440 pp.

Osborn, T. G. B. 1922. Flora and fauna of Nuyt's Archipelago. No. 3. – A sketch of the ecology of Franklin Islands. Trans. and Proc. Roy. Soc. S. Aust. 46: 194–206 plus 4 plates.

Osborn, T. G. B. 1925. The flora and fauna of Nuyt's Archipelago and the Investigator Group. No. 18. – Notes on the vegetation of Flinders Island. Trans. and Proc. Roy. Soc. S. Aust. 49: 276–289 + 3 plates.

Osman, R. W. 1978. The influence of seasonality and stability on the species equilibrium. Ecology 59: 383–399.

Plowman, K. P. 1979. Litter and soil fauna of two Australian subtropical forests. Aust. J. Ecol. 4: 87–104.

Porter, D. M. 1976. Geography and dispersal of Galapagos Islands vascular plants. Nature 264: 745–746.

Preston, F. W. 1962. The canonical distribution of commoness and rarity: Part II. Ecology 43: 410–432.

Proctor, V. W., C. R. Malone and V. L. de Vlaming. 1967. Dispersal of aquatic organisms: viability of disseminules recovered from the intestinal tract of captive killdeer. Ecology 48: 672–676.

Ranwell, D. S. 1972. Ecology of Salt Marshes and Sand Dunes. Chapman and Hall, London, 258 pp.

Rao, T. A. and A. R. K. Sastry. 1970. New distributional records for coastal plants from Andhra Pradesh. J. Bombay Nat. Hist. Soc. 67: 614–615.
Recher, H. F. 1972. Territorial and agonistic behaviour of the Reef Heron. The Emu 72: 126–130.
Recher, H. F. and J. A. Recher. 1972a. Herons leaving the water to defecate. The Auk 89: 896–897.
Recher, H. F. and J. A. Recher 1972b. The foraging behaviour of the Reef-Heron. The Emu 72: 85–90.
Rehder, H. A. and J. E. Randall. 1975. Ducie Atoll: Its history, physiography and biota. Atoll Res. Bull. No. 183: 1–55.
Revill, D. L., K. W. Stewart and H. E. Schlichting, Jr. 1967. Passive dispersal of viable algae and protozona by certain craneflies and midges. Ecology 48: 1023–1027.
Ridley, H. N. 1896. Dispersal of seeds by birds. Natural Science 8: 186–199.
Ridley, H. N. 1930. The Dispersal of Plants throughout the World. L. Reeve and Co., Ltd., Ashford, Kent, 744 pp.
Riek, E. F., C. D. Michener, W. L. Brown, Jr. and R. W. Taylor. 1970. Hymenoptera (wasps, bees, ants). Chapter 37 (pp. 867–959) in: The Insects of Australia (CSIRO), Melbourne Univ. Press, Melbourne.
Roberts, T. F. 1978. Gazeteer of named reefs of the Great Barrier Reef. Great Barrier Reef Committee, St. Lucia.
Rodman, J. E. 1974. Systematics and evolution of the genus *Cakile* (Cruciferae). Contrib. Gray Herbarium No. 205: 1–146.
Rosen, D. E. 1975. A vicarious model of Caribbean biogeography. Systematic Zoology 24: 431–464.
Rosenzweig, M. L. 1975. On continental steady states of species diversity. Chapter 5 (pp. 121–140) in: Ecology and Evolution of Communities (eds. M. L. Cody and J. M. Diamond), The Belknap Press of Harvard University, Cambridge.
Royal Society of London. 1979. The Terrestrial Ecology of Aldabra. University Press, Cambridge, 263 pp.
Safford, W. E. 1921. Dispersal of plants by ocean currents. Record of Proc. 1st Pan-Pacific Scientific Conference (Species Publication No. 7) 1: 535–536.
Saint Augustine. 413–426. *De Civitate Dei.* English translation (1620), The City of God, by John Healey; 1945 reprint (ed. R. V. G. Tasker), Everyman's Library Nos. 982–983. J. M. Dent and Sons Ltd., London, 2 vols., 371 and 444 pp.
St. John, H. 1951. The distribution of "*Pisonia grandis*" (Nyctaginaceae). Pacific plant studies No. 10. Webbia 8: 225–228.
St. John, H. 1970. Classification and distribution of the *Ipomoea pes-caprae* group (convolvulaceae). Bot. Jb. 89: 563–583.
St. John, H. and A. C. Smith. 1971. The vascular plants of the Horne and Wallis Islands. Pacific Science 25: 313–348.
Sale, P. F., 1978. Coexistence of coral reef fishes – a lottery for living space. Env. Biol. Fish. 3: 85–102.
Salisbury, E. 1952. Downs and Dunes, their Plant Life and its Environment. G. Bell and Sons, Ltd., London, 328 pp.
Sauer, J. D. 1962. Effects of recent tropical cyclones on the coastal vegetation of Mauritius. J. Ecol. 50: 275–290.
Sauer, J. D. 1972. Revision of *Stenotaphrum* (Graminae: Paniceae) with attention to its historical geography. Brittonia 24: 202–222.
Schimper, A. F. W. 1903. Plant-Geography upon a Physiological Basis (translated into English by W. R. Fisher, revised and edited by P. Groom and I. B. Balfour); Clarendon Press, Oxford, 839 pp.
Schoener, A. 1974a. Colonization curves for planar marine islands. Ecology 55: 818–827.
Schoener, A. 1974b. Experimental zoogeography: colonization of marine mini-islands. Amer. Nat. 108: 715–738.

Schoener, A., E. R. Long and J. R. DePalma. 1978. Geographic variation in artificial island colonization curves. Ecology 59: 367–382.
Schoener, T. W. and D. H. Janzen. 1968. Notes on environmental determinants of tropical versus temperate insect size patterns. Amer. Nat. 102: 207–224.
Serventy, D. L., V. Serventy and J. Warham. 1971. The Handbook of Australian Sea-birds. A. H. and A. W. Reed, Sydney.
Shipway, A. K. 1969. The numbers of terns and shearwaters nesting on Heron Island in 1965. The Emu 69: 108–109.
Simberloff, D. S. 1969. Experimental zoogeography of islands: a model for insular colonization. Ecology 50: 296–314.
Simberloff, D. 1976a. Species turnover and equilibrium island biogeography. Science 194: 572–578.
Simberloff, D. S. 1976b. Trophic structure determination and equilibrium in an arthropod community. Ecology 57: 395–398.
Simberloff, D. S. 1978. Using island biogeographic distributions to determine if colonization is stochastic. Amer. Nat. 112: 713–726.
Simberloff, D. S. and E. O. Wilson. 1969. Experimental zoogeography of islands. The colonization of empty islands. Ecology 50: 278–296.
Simberloff, D. S. and E. O. Wilson. 1970. Experimental zoogeography of islands: a two-year record of colonization. Ecology 51: 934–937.
Slater, P. 1970. A Field Guide to Australian Birds, Non-passerines. Rigby Ltd., Adelaide, 428 pp.
Slater, P. 1974. A Field Guide to Australian Birds, Passerines. Rigby Ltd., Adelaide, 309 pp.
Smithers, C. N. 1975. Records of two species of *Ectopsocus* McLachlin (Psocoptera: Extopsocidae) from One Tree Island, Queensland. J. Aust. Ent. Soc. 14: 243–245.
Sobey, D. G. and J. B. Kenworthy. 1979. The relationship between Herring Gulls and the vegetation of their breeding colonies. J. Ecol. 67: 469–496.
Spaull, V. W. 1979. Distribution of soil and litter arthropods on Aldabra Atoll. Phil. Trans. R. Soc. London B. 286: 109–117.
Spicer, R. A. and D. M. Newbery. 1979. The terrestrial vegetation of an Indian Ocean coral island: Wilingili, Addu Atoll, Maldive Islands. I. Transect analysis of the vegetation. Atoll Res. Bull. No. 231: 1–14.
Steers, J. A. 1938. Detailed notes on the islands surveyed and examined by the geographical expedition to the Great Barrier Reef in 1936. Reports of the Great Barrier Reef Committee 4: 51–104.
Stemmerik, J. F. 1964. Nyctaginaceae, pp. 450–468 in: Flora Malesiana, Series I, vol. 6 (ed. C. G. G. J. van Steenis), Wolters-Noordhoff Publ., Groningen.
Strong, D. R. Jr., L. A. Sziska and D. S. Simberloff. 1979. Tests of community-wide character displacement against null hypothesis. Evolution 33: 897–913.
Sykes, W. R. 1970. *Ipomoea pes-caprae* (L.) Ooster in the New Zealand botanical region. N.Z. J. Bot. 8: 249–253.
Sykes, W. R. and E. J. Godley. 1968. Transoceanic dispersal in *Sophora* and other genera. Nature 218: 495–496.
Talbot, F. H. 1967. Expedition to One Tree Island. Aust. Nat. Hist. 8: 343–346.
Taylor, R. W. 1967. Entomological survey of the Cook Islands and Niue. 1. – Hymenoptera – Formicidae. New Zealand J. Science 10: 1092–1095.
Tramer, E. J. 1975. The regulation of plant species diversity on an early successional old-field. Ecology 56: 905–914.
van der Pijl, L. 1969. Principles of Dispersal in Higher Plants. Springer-Verlag, Heidelberg, 154 pp.
van der Valk, A. G. 1974. Environmental factors controlling the distribution of forbs on coastal foredunes in Cape Hatteras National Seashore. Canadian J. Bot. 52: 1057–1073.
van Ooststroom, S. J. 1953. Convolvulaceae. pp 388–512 in: Flora Malesiana, Series I, vol. 4

(ed. C. G. G. J. van Steenis), Noordhoff-Kolff N.V., Djakarta, 631 pp.
van Steenis, C. G. G. J. 1948. Malaysian species of *Sophora* (Legum.). Bull. Botanical Garden Buitenzorg, Series III, 17: 421–428.
van Steenis, C. G. G. J. 1949a. Zygophyllaceae *l. tribulus*. p. 64 in: Flora Malesiana, Series I, vol. 4 (ed. C. G. G. J. van Steenis), Noordhoff-Kolff N.V., Djakarta, 631 pp.
van Steenis, C. G. G. J. 1949b. Plumbaginaceae, pp. 107–112 in: Flora Malesiana, Series I, vol. 4 (ed. C. G. G. J. van Steenis), Noordhoff-Kolff N.V., Djakarta, 631 pp.
Vats, L. K. and J. S. Singh. 1978. Population, biomass and secondary net production of above-ground insects in a tropical grassland. Tropical Ecology 19: 51–64.
Wallace, A. R. 1895. Island Life. Second and revised edition. MacMillan and Co., London, 563 pp.
Walsh, G. E. 1974. Mangroves: a review. A chapter (pp. 51–174) in: Ecology of Halophytes (eds. R. J. Reimold and W. H. Queen), Academic Press, New York, 605 pp.
Walter, H. 1964. Die Vegetation der Erde in öko-physiologischer Betrachtung. Vol. I. Die tropischen und subtropischen Zonen. Veb Gustav Fischer Verlag, Jena.
White, A. D. 1895. A History of the Warfare of Science with Theology in Christendom. 1955 reprint by George Braziller, New York, 474 pp.
White, C. T. and W. MacGillivray. 1925–27. The biology of North-west Islet, Capricorn Group. (H.). Botany. Australian Zoologist 4: 251–252.
Whitehead, D. R. and C. E. Jones. 1969. Small islands and the equilibrium theory of insular biogeography. Evolution 23: 171–179.
Whitton, B. A. and M. Potts. 1977. Observation on redox potential in freshwater pools on Aldabra. Atoll Res. Bull. No. 214: 1–5.
Wickens, G. E. 1979a. The propagales of the terrestrial flora of the Aldabra Archipelago, Western Indian Ocean. Atoll Res. Bull. No. 229: 1–39.
Wickens, G. E. 1979b. Speculations on seed dispersal and the flora of the Aldabra archipelago. Phil. Trans. R. Soc. London B. 286: 85–97.
Wild, H. 1965. The African species of the genus *Melanthera* Rohr. Kirkia 5: 1–17.
Willis, A. J., B. F. Folkes, J. F. Hope-Simpson and E. W. Yemm. 1959. Braunton Burrows: The dune system and its vegetation. Part II. J. Ecol. 47: 249–288.
Wilson, A. T. 1959. Surface of the ocean as a source of air-borne nitrogenous material and other plant nutrients. Nature 184: 99–101.
Wilson, D. E. 1977. Ecological observations on the tropical strand plants *Ipomoea pes-caprae* (L.) R. Br. (Convolvulaceae) *Canavalia maritima* (Aubl.) Thou. (Fabaceae). Brenesia 10/11: 31–42.
Wilson, E. O. 1969. The species equilibrium. pp. 38–47 in: Diversity and Stability in Ecological Systems (eds. G. M. Woodwell and H. H. Smith), Brookhaven Symposia in Biology No. 22, Brookhaven National Laboratory.
Wilson, E. O. and D. S. Simberloff. 1969. Experimental zoogeography of islands. Defaunation and monitoring techniques. Ecology 50: 267–278.
Wilson, E. O. and R. W. Taylor. 1967a. The ants of Polynesia (Hymenoptera: Formicidae). Pac. Insects Monogr. 14: 1–109.
Wilson, E. O. and R. W. Taylor. 1967b. An estimate of the potential evolutionary increase in species density in the Polynesian ant fauna. Evolution 21: 1–10.
Wilson, P. 1932. *Portulaca* (pp. 328–336) in: North American Flora vol. 21, N.Y. Bot. Gardens, New York.
Woodell, S. R. J. 1979. The rôle of unspecialized pollinators in the reproductive success of Aldabran plants. Phil. Trans. Roy. Soc. London B. 286: 99–108.
Woodward, T. E., J. W. Evans and V. F. Eastop. 1970. Hemiptera (bugs, leafhoppers, etc.). Chapter 26 (pp. 387–457) in: The Insects of Australia (CSIRO), Melbourne Univ. Press, Melbourne.
Wright, S. 1921. Systems of mating. Genetics 6: 111–178.
Zimmerman, C. A. 1977. A comparison of breeding systems and seed physiologies in three species of *Portulaca* L. Ecology 58: 860–868.

Index